Lecture Notes in Mathematics

Edited by A. Dold, B. Eckmann and F. Takens

T0225998

1401

J. Steprāns S. Watson (Eds.)

Set Theory and its Applications

Proceedings of a Conference
held at York University, Ontario, Canada,
Aug. 10–21, 1987

Springer-Verlag
Berlin Heidelberg New York London Paris Tokyo Hong Kong

Editors

Juris Steprāns
Stephen Watson
York University,
North York, Ontario
M3J 1P3 Canada

Mathematics Subject Classification (1980): 54-06; 04-06

ISBN 3-540-51730-8 Springer-Verlag Berlin Heidelberg New York
ISBN 0-387-51730-8 Springer-Verlag New York Berlin Heidelberg

Printed in Germany

Printing and binding: Druckhaus Beltz, Hemsbach/Bergstr.
2146/3140-543210 – Printed on acid-free paper

Preface

The Set Theory and its Applications Conference at York University was held in Toronto, Canada during the two weeks of August 10–21, 1987. It was attended by 80 mathematicians from 12 different countries.
Financial support for this conference was provided by the Natural Sciences and Engineering Research Council under the auspices of the Canadian Mathematical Society and by York University through both the President's Office and the Office of the Dean of the Faculty of Arts. The organizers would like to express their thanks for this financial support without which the conference would not have been possible.
The conference featured both contributed talks and a series of invited lectures on topics central to the study of set theory and its applications, particularly to general topology. The organizers would like to thank the invited speakers James Baumgartner, Arnold Miller, Andreas Blass, Neil Hindman, Stevo Todorcevic, Ronald Jensen, Hugh Woodin, Jan van Mill, Boban Velickovic, Menachem Magidor, A. V. Arhangel'skii and Mary Ellen Rudin for a series of highly informative and stimulating lectures. We would also like to thank the many speakers who contributed lectures to the conference, in many cases announcing new and important results for the first time.
The organizers would like to thank the staff of the Mathematics Department of York University for their help in the organization of the conference. We would like, in particular, to thank Joan Young for her knowledgeable assistance.
All of the articles in this volume have been refereed. We would like to thank the referees for their many helpful comments which have greatly improved the quality of the articles which make up this volume. We would also like to thank Eadie Henry for her tireless work in the preparation of one manuscript from handwritten original draft through several series of revisions to a well-written and clearly presented final copy.
This volume of Springer Lecture Notes in Mathematics constitutes the proceedings of this conference. We would like to thank the editors for accepting the proceedings.

Alan Dow
Donald Pelletier
Juris Steprāns
Stephen Watson

Dedication

Two weeks before the conference began, we received the sad news that Eric van Douwen had died. We had been musing about the conference for more than two years and even from the beginning we had decided that Eric would be invited and that he would give a series of lectures. We were curious about what topics he would choose because we knew that when Eric lectured, people paid attention to his ideas and mathematics would advance a little bit in his direction. We knew we would see a solution to some well-known problem, a completely original proof for some well-known result, and we knew that we would all aim a little bit higher. We also simply looked forward to seeing Eric again.
Eric wrote over seventy papers in sixteen years but we always felt that the greater part of what he knew had never been written down. We wanted to talk to him and listen to him until we tired from sheer stimulation. General topologists form a community and Eric Karel van Douwen was an elder who possessed the knowledge of its oral traditions, but was denied the opportunity to pass it down. The loss to us was devastating, for we lost not only a mathematician but a friend.

Table of Contents

SPACES OF IDEALS OF PARTIAL FUNCTIONS

Murray G. Bell
Department of Mathematics, University of Manitoba
Winnipeg, Canada R3T 2N2

1. Introduction

We present three spaces with interesting properties. They are all constructed in a similar fashion using ideals of partial functions. Examples 1 and 2 use only ZFC but Example 3 requires the continuum hypothesis CH.

Example 1 is a first countable, countably compact, ccc, non-separable space. In [1] and [2], the author had constructed first countable, ccc, non-separable spaces that were sigma-compact and pseudocompact respectively. We cannot get local compactness in ZFC alone because Martin's Axiom plus not CH implies, cf. Juhasz [6], that every first countable, locally compact, ccc space is separable. On the other hand, V = L implies, cf. Jech [5], that there is a compact, hereditarily Lindelof, ordered space which does not have property K.

Example 2 is a bit of an oddity. It is a compact, nowhere ccc space which admits a countable to one continuous map onto the Cantor space 2^{ω}. It follows that it is first countable. Moreover, it is also an Eberlein compact.

Example 3, assuming CH, is a first countable, compact, ccc space which does not have property K. It is the presence of both first countable and compact which makes this a new result from CH. There are first countable ccc spaces which do not have property K but none are known, under CH, which have first countable compactifications. We mention that Kunen [7] has an example, under CH, of a compact, hereditarily Lindelof, non-separable space which was the support of a measure algebra, hence has property K. Our example is a Corson compact as well.

2. Notations and Definitions

If X is a set, then $\mathcal{P}(X)$ denotes the set of all subsets of X. The first infinite ordinal is ω, the first uncountable ordinal is ω_1 and c is the first ordinal equipotent with $\mathcal{P}(\omega)$. Two collections of sets S and T are said to be combinatorially similar, written $S \equiv T$, if there exists a bijection $h : S \to T$ such that for all $n < \omega$ and for all $s_1, \ldots, s_n$ in S, we have $\cap\{s_i : i \le n\} \ne \phi$ iff $\cap\{h(s_i) : i \le n\} \ne \phi$. Let A and B be subsets of ω. We write $A \subset_* B$ if $A - B$ is finite and $B - A$ is infinite and we write $A =^* B$ if $A - B$ is finite and $B - A$ is finite.

A space is ω-bounded if every countable subspace has compact closure. A space is ccc if every disjoint collection of open sets is countable. A space has property K if every uncountable collection of open sets contains an uncountable subcollection whose every two members have a non-empty intersection. A space is nowhere ccc if no nonempty open set is ccc. Whenever we refer to a topology on 2^X we mean the product topology where 2 is the discrete space $\{0,1\}$.

3. Partial Functions and Ideals

Define $P = \{p : p$ is a partial function from ω to $2\}$. We think of a $p \in P$ as a set of ordered pairs and if $p \in 2^A$ then we write dom $p = A$. Define a partial order on P by $p \leq q$ iff $p \subseteq q$. Then P with $\leq$ is a conditionally complete poset and if $A \subseteq P$ has an upper bound then its least upper bound is $\cup A$. Let Q be a subset of P such that Q contains all the finite functions and $Q = \{\cup F : F$ is a finite subset of Q and $\cup F \in P\}$. A subset I of Q is called a total Q-ideal if $\cup I$ is a total function from ω to 2 and $I = \{q \in Q :$ there exists a finite subset F of I with $q \leq \cup F\}$.

Put $J(Q) = \{I \subseteq Q : I$ is a total Q-ideal$\}$. If we identify $\mathcal{P}(Q)$ with 2^Q via characteristic functions, then we can endow $J(Q)$ with the subspace topology from 2^Q. $J(Q)$ is seen to be closed in 2^Q and thus $J(Q)$ is a compact, Hausdorff, zero-dimensional space. For each $q \in Q$, put $q^+ = \{I \in J(Q) : q \in I\}$ and $q^- = J(Q) - q^+$. Internally, $J(Q)$ has the topology which has $\cup\{\{q^+,q^-\} : q \in Q\}$ as a closed or open subbase. Define $\lambda : J(Q) \to 2^\omega$ by the union map $\lambda(I) = \cup I$ and put $\mathcal{P} = \{q^+ : q \in Q\}$. It is easy to see that λ is a continuous and onto mapping.

3.1. Lemma $\mathcal{P}$ is a T_0-separating binary π-base for $J(Q)$ which consists of clopen sets.

Proof: In [3], where we first looked at these spaces $J(Q)$, we showed that $\mathcal{P}$ was a π-base for $J(Q)$, i.e., for each non-empty open set U in $J(Q)$, there exists $q \in Q$ with $q^+ \subseteq U$. If I and J are distinct total Q-ideals then choose $q \in (I - J) \cup (J - I)$. Then q^+ contains at most one of I and J, so $\mathcal{P}$ is T_0-separating. If $\{q^+ : q \in R\}$ is a linked collection, then $\cup R \in P$. Let f be any total function with $\cup R \leq f$. Put $I = \{q \in Q : q \leq f\}$. $I \in J(Q)$ and $I \in \cap\{q^+ : q \in R\}$.

The following $J(Q)$ was used in [3] for a completely different reason. Our first example is a dense subspace of it.

Example 1. A first countable, countably compact, ccc, non-separable space X.

Let $\{A_\alpha : \alpha < \omega_1\}$ be a collection of subsets of ω such that $A_0 = \phi$ and $\alpha < \beta$ implies $A_\alpha \subset^* A_\beta$. Put $Q = \{p \in P :$ there exists $\alpha < \omega_1$ with dom $p =^* A_\alpha$ and $p^{-1}(1) =^* A_\alpha\}$. Define $\delta : \mathcal{P}(Q) \to \omega_1 + 1$ by $\delta(R) = \sup\{\alpha < \omega_1 :$ there exists $r \in R$ with dom $r =^* A_\alpha\}$.

For each $A \subseteq \omega$, let f_A be the characteristic function of A and let $I_A = \{q \in Q : q \leq f_A\}$. Using $\mathcal{P}$ as a π-base, we see that $D = \{I_A : A =^* \omega\}$ is a countable dense subspace of $J(Q)$, hence $J(Q)$ is separable. Put $X = \{I \in J(Q) : \delta(I) < \omega_1\}$. Note that $X \cap D = \phi$.

Since X is also dense in $J(Q)$, X is ccc, even sigma-centered. For each

$\alpha < \omega_1$, put $X_\alpha = \{I \in J(Q) : \delta(I) \le \alpha\}$. If $q \in Q$ and $\delta(\{q\}) > \alpha$ then $X_\alpha \cap q^+ = \phi$; thus X_α, as a subspace of $J(Q)$, has a countable base. Furthermore, $X = \cap\{q^- : \delta(\{q\}) = \alpha + 1\}$, so each X_α is a compact G-delta subspace of $J(Q)$. Since $X = \cup\{X_\alpha : \alpha < \omega_1\}$, which is a strictly increasing union, we deduce that X is first countable, ω-bounded and non-separable.

Our next two spaces have a common structure which we now set up.

Let $\{(A_\alpha,B_\alpha) : \alpha < \omega_1\}$ be a collection of pairs of subsets of ω such that Q1 : $A_0 = \phi = B_0$, Q2 : $A_\alpha \cap B_\alpha = \phi$ for each $\alpha < \omega_1$, and Q3 : for each uncountable $A \subseteq \omega_1$, there exists α,β in A with $A_\alpha \cap B_\beta \ne \phi$.

Put $Q = \{p \in P :$ there exists $\alpha < \omega_1$ with $A_\alpha \cup B_\alpha \subseteq \text{dom } p$, $\text{dom } p =^* A_\alpha \cup B_\alpha$ and $p^{-1}(1) =^* A_\alpha\}$. Similar to Example 1, define $\delta : P(Q) \to \omega_1 + 1$ by $\delta(R) = \sup\{\alpha < \omega_1 :$ there exists $r \in R$ with $\text{dom } r =^* A_\alpha \cup B_\alpha\}$. We would like to thank Juris Stephrans for communicating the following fact to us: Q2 and Q3 imply that for every uncountable $A \subseteq \omega_1$, there does not exist a subset B of ω such that for each $\alpha \in A$, both $A_\alpha - B$ and $B_\alpha \cap B$ are finite; hence we get that for every $I \in J(Q)$, $\delta(I) < \omega_1$. For each $\alpha < \omega_1$, put $M_\alpha = \{I \in J(Q) : \delta(I) \le \alpha\}$. Similar to Example 1, we get that $J(Q) = \cup\{M_\alpha : \alpha < \omega_1\}$, which is a strictly increasing union, and each M_α is a compact, second countable, G-delta subspace of $J(Q)$. In particular, $J(Q)$ is first countable. As before, let $\lambda : J(Q) \to 2^\omega$ be the union map and let $\mathcal{P} = \{q^+ : q \in Q\}$. But now put $\mathcal{P}_0 = \{q^+ :$ there exists $\alpha < \omega_1$ with $\text{dom } q = A_\alpha \cup B_\alpha$ and $q^{-1}(1) = A_\alpha\}$.

3.2 Lemma. $\mathcal{P}$ is a point-countable collection and $\mathcal{P} = \cup\{\mathcal{P}_n : n < \omega\}$, where for each $n < \omega$, $\mathcal{P}_n \equiv \mathcal{P}_0$. Consequently, $J(Q)$ does not have property K and $J(Q)$ is Corson compact.

Proof. If $I \in \cap\{q^+ : q^+ \in \mathcal{P}'\}$, where $\mathcal{P}'$ is uncountable, then $\delta(I) = \omega_1$; a contradiction. Hence $\mathcal{P}$ is point-countable. Since $\mathcal{P}$ is a binary collection, $\mathcal{P}$ does not have an uncountable linked subcollection. So, $J(Q)$ does not have property K.

For every sequence F, G, H of pairwise disjoint finite subsets of ω and for each $s \in 2^H$ put $\mathcal{P}(F, G, H, s) = \{q^+ \in \mathcal{P} :$ there exists $\alpha < \omega_1$, with $F \subseteq A_\alpha$, $G \subseteq B_\alpha$, $H \cap (A_\alpha \cup B_\alpha) = \phi$, $\text{dom } q = A_\alpha \cup B_\alpha \cup H$ and $q^{-1}(1) = (A_\alpha - F) \cup G \cup s^{-1}(1)\}$. Clearly, $\mathcal{P}$ is the union over all of these $\mathcal{P}(F, G, H, s)$'s, of which there are only countably many. It is tedious, but straightforward, to prove that each $\mathcal{P}(F, G, H, s)$ is combinatorially similar to $\mathcal{P}_0$ by the natural bijection. So, $\mathcal{P} = \cup\{\mathcal{P}_n : n < \omega\}$, where each $\mathcal{P}_n \equiv \mathcal{P}_0$. Finally, a well-known characterization of Corson compacts is that they have a T_0-separating, point-countable collection of open F-sigma sets; $\mathcal{P}$ is such a collection.

Example 2. A compact nowhere ccc space which admits a countable to one continuous map onto 2^ω.

Replace Q3 by the stronger Q3' : for each $\alpha < \beta$; $A_\beta \cap B_\alpha \ne \phi$; and add Q4: for each $\alpha < \beta$, $A_\alpha \subset^* A_\beta$ and $B_\alpha \subset^* B_\beta$. In other words, we have a canonical Hausdorff Gap. Because of the chain condition Q4, the total Q-ideals can be identified as follows: choose an $f \in 2^\omega$ and choose α with $0 < \alpha < \omega_1$; then you get the total Q-ideal $I(f,\alpha) = \{q \in Q : q \leq f \text{ and } \delta(\{q\}) < \alpha\}$. For each $f \in 2^\omega$, there exists $\beta < \omega_1$, such that if $\delta(\{q\}) \geq \beta$ then $q \nleq f$, otherwise $f^{-1}(1)$ would fill the gap. Hence, if $\alpha \geq \beta$ then $I(f,\alpha) = I(f,\beta)$. We conclude, that for each $f \in 2^\omega$, $\lambda^{-1}(f)$ is a countable set; hence λ is a countable to one map.

Condition Q3' implies that $\mathcal{P}_0$ is a disjoint family; hence by lemma 3.2, $\mathcal{P}$ is a sigma-disjoint π-base. Together with Lemma 3.1, we see that $J(Q)$ is a special kind of Corson compact, an Eberlein compact. Since for each $q \in Q$ and for each $\alpha > \delta(\{q\})$ there exists $r \in Q$ with $\delta(\{r\}) = \alpha$ and $q^+ \cap r^+ \neq \phi$, we see that each q^+ has uncountable cellularity. Hence, $J(Q)$ is nowhere ccc.

Example 3. (CH) A first countable compact ccc space without property K.

Assume $c = \omega_1$. To Q1 thru Q3 add Q5: for each uncountable $A \subseteq \omega_1$, there exists $\alpha \neq \beta$ in A such that $A_\alpha \cap B_\beta = \phi$ and $B_\alpha \cap A_\beta = \phi$. This can be achieved in exactly the same way that van Douwen and Kunen proved that CH implies † in [4]: Let $C\ell$ denote closure in 2^ω. Identify $\mathcal{P}(\omega)$ and 2^ω so that $A_\alpha = a_\alpha^{-1}(1)$ and $B_\alpha = b_\alpha^{-1}(1)$. Enumerate all countable subsets of ω_1 as $\{I_\alpha : \alpha < \omega_1\}$ such that for each $\alpha < \omega_1$, $I_\alpha \subseteq \alpha$. Inductively construct $\{(a_\alpha, b_\alpha) : \alpha < \omega_1\}$ such that $A_\alpha \cap B_\alpha = \phi$ and such that if $a_\alpha \in C\ell\{a_\beta : \beta \in I_\gamma\}$ and $b_\alpha \in C\ell\{b_\beta : \beta \in I_\gamma\}$ for some $\gamma < \alpha$, then there exists $\beta \in I_\gamma$ with $A_\alpha \cap B_\beta \neq \phi$ and also there exists $\pi \in I_\gamma$ with $A_\alpha \cap B_\pi = \phi$ and $B_\alpha \cap A_\pi = \phi$. We were unable to complete the inductive step while insisting that we also satisfy condition Q4 but if the reader only carries along the weaker Q4': for each $\alpha < \beta$, $A_\alpha \cap B_\beta =^* \phi$ and $B_\alpha \cap A_\beta =^* \phi$; he will have no difficulty completing the inductive step. Upon completion, to prove Q3 and Q5, use the hereditary separability of 2^ω.

Condition Q5 implies that $\mathcal{P}_0$ does not have an uncountable disjoint subcollection and so it follows from Lemma 3.2 that $\mathcal{P}$ has no uncountable disjoint subcollection. Hence, $J(Q)$ is a first countable, Corson compact space that is ccc but does not have property K.

4. References

1. M. Bell, A normal first countable ccc nonseparable space, Proc. Amer. Math. Soc. 74(1), 1979, 151-155.
2. M. Bell, First countable pseudocompactifications, Topology Appl. 21, 1985, 159-166.
3. M. Bell, G_κ subspaces of Hyadic spaces, submitted manuscript.
4. E. van Douwen and K. Kunen, L-spaces and S-spaces in $\mathcal{P}(\omega)$, Topology Appl. 14, 1982, 143-149.
5. T. Jech, Nonprovability of Souslin's hypothesis, Comment. Math. Univ. Carolinae 8, 1967, 293-296.
6. I. Juhasz, Cardinal Functions in Topology, Mathematical Centre Tract 34, Amsterdam, 1971.
7. K. Kunen, A compact L-space under CH, Topology Appl. 12, 1981, 283-287.

Remarks on partition ordinals

by

James E. Baumgartner[1]

Abstract. After a brief survey of the theory of partition ordinals, i.e., ordinals α such that $\alpha \to (\alpha, n)^2$ for all $n < \omega$, it is shown that $\mathrm{MA}(\aleph_1)$ implies that $\omega_1\omega$ and $\omega_1\omega^2$ are partition ordinals. This contrasts with an old result of Erdős and Hajnal that $\alpha \not\to (\alpha, 3)^2$ holds for both these ordinals under the Continuum Hypothesis.

1. Partition ordinals.

The theory of ordinary partition relations for cardinal numbers is fairly well understood at present (see [EHMR], for example), but the corresponding theory for non-cardinal ordinal numbers seems still to be in its infancy. This paper begins with a survey of results and problems concerning ordinals α satisfying the partition relation $\alpha \to (\alpha, n)^2$ for all $n < \omega$ and ends with the proof from Martin's Axiom + ¬CH that this relation holds when α is $\omega_1\omega$ or $\omega_1\omega^2$.

As usual, if X is a set and n is a cardinal then $[X]^n$ denotes the collection of all n-element subsets of X. If α, β, and γ are ordinals, then the partition relation $\alpha \to (\beta, \gamma)^2$ means that for all $f : [\alpha]^2 \to 2$ there is $X \subseteq \alpha$ such that either X has order type β and f is constantly 0 on $[X]^2$ or else X has order type γ and f is constantly 1 on $[X]^2$.

A convenient alternative definition says that $\alpha \to (\beta, \gamma)^2$ iff every graph on α either has an independent set of type β or a complete subgraph of type γ. Here, of course, the set E of edges of the graph is identified with $f^{-1}\{1\}$ in the other definition.

---

[1] Preparation of this paper was partially supported by National Science Foundation grant number DMS-8704586.

For cardinal numbers there is a strong result due to Erdős, Dushnik and Miller (see [Wi] for a treatment of it), namely if κ is a cardinal then $\kappa \to (\kappa, \omega)^2$. The machinery of ordinary partition relations for cardinals yields strengthenings of this result for various κ. For example, if κ is regular and $\kappa > \omega$ then $\kappa \to (\kappa, \omega+1)^2$; if in addition $\lambda^{\aleph_0} < \kappa$ whenever $\lambda < \kappa$ then $\kappa \to (\kappa, \omega_1 + 1)^2$; and so on.

For non-cardinals the situation is quite different. The following result is well-known:

Proposition 1.1. *If* $\alpha \neq |\alpha|$ *then* $\alpha \not\to (|\alpha|+1, \omega)^2$.

Proof. Let $<$ be the usual ordering on α and let $<_1$ be a well-ordering of α in order type $|\alpha|$. Let $E = \{\{\xi, \eta\} : <$ and $<_1$ disagree on $\{\xi, \eta\}\}$. Then E determines a graph on α. An independent set must have the same order type under both $<$ and $<_1$, hence must have order type $\leq |\alpha|$, while an independent set must be both well-ordered by $<$ and conversely well-ordered by $<_1$, hence must be finite.

Thus the strongest relation we can have for non-cardinals is $\alpha \to (\alpha, n)^2$ for all finite n. Let us refer to ordinals satisfying $\alpha \to (\alpha, n)^2$ for all $n < \omega$ as *partition ordinals.* We will be interested in the project of identifying all partition ordinals.

More generally, we are interested in identifying all pairs α, n satisfying $\alpha \to (\alpha, n)^2$. The primary open problem in this area, due to Erdős, is whether these projects are different. So far, in every case where $\alpha \to (\alpha, 3)^2$ has been found to hold (and this includes consistency results as well) α has turned out to be a partition ordinal. Is this a fact provable in ZFC? Is it a proposition consistent with ZFC?

A couple of observations will help narrow the search for partition ordinals. If $\alpha = \beta + \gamma$ where $\beta, \gamma < \alpha$ then α cannot be a partition ordinal, for the graph on α obtained by drawing an edge between each element of the β-piece and each element of the γ-piece has no independent set of type α and no (complete) triangles. Thus $\alpha \not\to (\alpha, 3)^2$. If α cannot be expressed in the form $\beta + \gamma$ with $\beta, \gamma < \alpha$ then α is said to be *indecomposable.* It is well known that the indecomposable ordinals are exactly the (ordinal) powers of ω.

The first nontrivial candidate to be a partition ordinal is thus $\alpha = \omega^2$, and it is an old result of Specker [Sp] that ω^2 is in fact a partition ordinal. Let us outline Specker's argument very briefly since it is relevant to the Martin's Axiom results obtained later in the paper. The ordinal ω^2 may be identified with the set of all strictly increasing pairs in $\omega \times \omega$ under the lexicographic ordering, i.e., with $W = \{(m, n) \in \omega \times \omega : m < n\}$. Now, given $f : [W]^2 \to 2$, partition $[\omega]^4$ into 16 parts, where the part in which $\{m, n, p, q\}$ is placed depends exactly on

$$f\{(m,n),(p,q)\},\ f\{(m,p),(n,q)\},\ f\{(m,q),(n,p)\},\ \text{and}\ f\{(m,n),(m,p)\},$$

and we assume $m < n < p < q$. Applying Ramsey's Theorem to the latter partition, we obtain an infinite set $X \subseteq \omega$ such that all elements of $[X]^4$ lie in the same piece of the partition. It is now a simple matter either to find an element x of $[(X \times X) \cap W]^n$ such that f is 1 on $[x]^2$ or else to build a subset Y of $(X \times X) \cap W$ of order type ω^2 such that f is 0 on $[Y]^2$.

A similar approach to ω^3 using Ramsey's Theorem applied to a partition of $[\omega]^6$ will very nearly succeed, but it will end by finding the following counterexample proving $\omega^3 \not\to (\omega^3, 3)^2$. Let $W = \{(m,n,p) \in \omega \times \omega \times \omega : m < n < p\}$, ordered lexicographically, and let

$$E = \{\{(m_1,n_1,p_1),(m_2,n_2,p_2)\} : m_1 < n_1 < m_2 < p_1 < n_2 < p_2\}.$$

It is easily checked that this graph on W has no triangles and no independent set of type ω^3.

One can argue similarly that $\omega^k \not\to (\omega^k, 3)^2$ for $3 < k < \omega$, but there is an easier way to draw this conclusion. An ordinal α can be ***pinned to*** β, written $\alpha \twoheadrightarrow \beta$, iff there is $f : \alpha \to \beta$ such that every subset of α of order type α is carried by f to a subset of β of order type β. We refer to f as a ***pinning map.*** Then we have

Proposition 1.2. *If $\alpha \twoheadrightarrow \beta$ and $\alpha \to (\alpha, n)^2$ then $\beta \to (\beta, n)^2$.*

Proof. Let $f : \alpha \to \beta$ be a pinning map and let $E \subseteq [\beta]^2$. Define $E' \subseteq [\alpha]^2$ by $\{\xi,\eta\} \in E'$ iff $f(\xi) \neq f(\eta)$ and $\{f(\xi), f(\eta)\} \in E$. If $X \subseteq \alpha$ is a complete subgraph then $f``X$ is a complete subgraph of $\beta$ of the same cardinality. If $X \subseteq \alpha$ is $E'$-independent of type $\alpha$ then $f``X$ is E-independent of type β.

Proposition 1.2 is almost always used in its contrapositive form to lift counterexamples, i.e., if $\alpha \twoheadrightarrow \beta$ and $\beta \not\to (\beta, n)^2$ then $\alpha \not\to (\alpha, n)^2$. Since we know $\omega^3 \not\to (\omega^3, 3)^2$ it is interesting to ask for which α do we have $\alpha \twoheadrightarrow \omega^3$. For countable α, we have the result of Galvin and Larson [GL] that if $\omega^3 \leq \alpha < \omega_1$ and α is indecomposable then $\alpha \not\twoheadrightarrow \omega^3$ iff $\alpha = \omega^{\omega^\beta}$ for some β.

Thus the next possible partition ordinal after ω^2 is ω^ω, and it is a result of Chang (for $n = 3$) and Milner (for arbitrary n) that $\omega^\omega \to (\omega^\omega, n)^2$ for all n. A much shorter proof was found by Larson; one version of her argument may be found in [Wi].

This leaves us with an open problem frequently mentioned by Erdős: is it true that ω^{ω^β} is a partition ordinal whenever $1 < \beta < \omega_1$? Note that if α is countable then the assertion that α is a partition ordinal is a Π^1_2 statement. Thus Shoenfield's Absoluteness Theorem applies and one cannot expect to get independence results, at least not in the usual way via forcing.

Let us look at uncountable ordinals. It is known that if $2^\kappa = \kappa^+$ then $\kappa^+\kappa \not\rightarrow (\kappa^+\kappa, 3)^2$ (Erdős-Hajnal [EH]) and $(\kappa^+)^2 \not\rightarrow ((\kappa^+)^2, 3)^2$ (Hajnal [EH] for regular κ; the author [Ba] for singular κ). For the case $\kappa = \omega$ this says that under CH we have $\omega_1\omega \not\rightarrow (\omega_1\omega, 3)^2$ and $\omega_1^2 \not\rightarrow (\omega_1^2, 3)^2$. It is easy to see that for every indecomposable α, if $\omega_1 < \alpha < \omega_2$ then either $\alpha \rightarrow \omega_1\omega$ (if cf $\alpha = \omega$) or else $\alpha \rightarrow \omega_1^2$ (if cf $\alpha = \omega_1$). It follows that $\alpha \not\rightarrow (\alpha, 3)^2$ for all α with $\omega_1 < \alpha < \omega_2$.

Under GCH, therefore, the next interesting problem is whether $\omega_2\omega$ is a partition ordinal, and Shelah and Stanley [SS] have recently proved that it is. They also showed that it is consistent (relative to the existence of a weakly compact cardinal) that GCH holds and $\omega_3\omega_1$ is a partition ordinal, and also that it is consistent relative to ZF alone that GCH holds and $\omega_3\omega_1 \not\rightarrow (\omega_3\omega_1, 3)^2$. The latter argument may be regarded as adjoining a special kind of $(\omega_2, 1)$-morass and then deducing $\omega_3\omega_1 \not\rightarrow (\omega_3\omega_1, 3)^2$. See [Mi].

Specker's argument yields the result that if κ is weakly compact then κ^2 is a partition ordinal. More generally, if κ is weakly compact, $\alpha < \kappa$ and $\alpha \rightarrow (\alpha, n)^2$ then $\kappa\alpha \rightarrow (\kappa\alpha, n)^2$ and $\kappa^2\alpha \rightarrow (\kappa^2\alpha, n)^2$ (See [L1]). Larson [L1] has shown that if κ is Ramsey then both κ^ω and $\kappa^\omega\omega$ are partition ordinals (but note that $\kappa^\omega\omega^2$ is not since $\kappa^\omega\omega^2 \rightarrow \omega^3$).

The author [Ba] has shown that if κ is regular and there is a κ-Souslin tree then $\kappa^2 \not\rightarrow (\kappa^2, 3)^2$. The counterexample is very easy to describe. Let $(\kappa, <_T)$ be a κ-Souslin tree such that $\alpha <_T \beta$ implies $\alpha < \beta$ and every $\alpha \in \kappa$ has κ successors. Then $W = \{(\alpha, \beta) : \alpha <_T \beta\}$ has order type κ^2 under the lexicographic ordering. Define $E \subseteq [W]^2$ so that $\{(\alpha, \beta), (\gamma, \delta)\} \in E$ provided that $\alpha <_T \gamma <_T \beta$ and the immediate $<_T$-successor of γ lying below δ is not comparable with β. Then there are no triangles and an easy argument shows that there are no independent sets of type κ^2 either. Another easy argument combined with Jensen's well-known result (see [DJ]) that in the constructible universe L, κ is weakly compact iff there are no κ-Souslin trees allows one to conclude that in L, $\kappa^2 \rightarrow (\kappa^2, 3)^2$ iff cf κ is weakly compact (iff κ^2 is a partition ordinal).

The counterexample for ω^3 lifts to show $\kappa^3 \not\rightarrow (\kappa^3, 3)^2$ for any κ, and a rather more elaborate counterexample constructed by Larson [L2] shows that $\kappa^{\omega+1} \not\rightarrow (\kappa^{\omega+1}, 3)^2$. (Note that while $\omega^{\omega+1} \to \omega^3$ we do not in general have $\kappa^{\omega+1} \to \kappa^3$.)

The principal result of this paper is the fact that MA($\aleph_1$) (Martin's Axiom for collections of $\aleph_1$ dense sets) implies that both $\omega_1\omega$ and $\omega_1\omega^2$ are partition ordinals. The proof occupies Sections 2 and 3. In Section 2 the notion of a Ramsey near-ordering is introduced and a weak partition result is proved in ZFC. In Section 3 MA($\aleph_1$) is applied to complete the proof.

In addition to the problems mentioned earlier in this section there are many interesting open questions.

What are the partition ordinals under MA($\aleph_1$)? In particular, what happens for $\omega_1\omega^\omega$, ω_1^2 and ω_1^ω? Conceivably if $\alpha < \omega_1$ and $\alpha \to (\alpha, n)^2$ then $\omega_1\alpha \to (\omega_1\alpha, n)^2$ and even $\omega_1^2\alpha \to (\omega_1^2\alpha, n)^2$ (in analogy with the situation for weakly compact cardinals).

Is it consistent that $\omega_2\omega_1$ is a partition ordinal?

Is it consistent that $\omega_2\omega$ is a partition ordinal and CH fails? (It is easily consistent with ¬CH that $\omega_2\omega \not\to (\omega_2\omega, 3)^2$. See [SS].)

Under GCH, is $\omega_2\omega^2$ a partition ordinal? How about ω_2^ω?

2. Ramsey near-orderings and a weak partition relation in ZFC.

A *near-ordering* is a binary relation $<$ on a set X that is irreflexive, antisymmetric and *near-transitive*, that is to say, if $x_1 < x_2 < \ldots < x_n$ and $x_1 < x_n$ then $\{x_1, \ldots, x_n\}$ is a *chain*, i.e., a set linearly ordered by $<$. If $Y \subseteq X$ then let $[Y]^n_<$ denote the set of n-element chains of elements of Y. (Thus $[X]^n = [X]^n_<$, where $<$ is any linear ordering of X.)

Let α be a countable ordinal and suppose $<_R$ is a near-ordering on α. We call $<_R$ a *Ramsey* near-ordering if the following conditions are satisfied:

(1) $\forall \beta < \alpha\ \{\gamma < \alpha : \gamma <_R \beta\}$ is finite and $<_R$ is well-founded

(2) $\forall X \subseteq \alpha$ if X has order type α (in the usual ordering) then $\forall n < \omega\ [X]^n_{<_R}$ is nonempty

(3) $<_R$ has the *Ramsey property*, i.e., $\forall X \subseteq \alpha$ if X has type α and $[X]^n_{<_R} = P_0 \cup \ldots \cup P_{k-1}$ then $\exists Y \subseteq X$ Y has type α and $[Y]^n_{<_R} \subseteq P_i$ for some $i < k$

(4) $<_R$ has the *doubling property*, i.e., $\forall X \subseteq \alpha$ if X has type α then $\exists Y \subseteq X$ Y has type α and $\exists h : Y \to X - Y$ $\forall y \in Y$ $y <_R h(y)$ and if $z \in Y$ and $y <_R z$ then $h(y) <_R z$ and $\{y, h(y), z, h(z)\}$ is a chain (which by near-transitivity is only to say that $y <_R h(z)$).

Of course the simplest Ramsey near-ordering is the usual ordering on $\alpha = \omega$, and this is the best example to keep in mind while reading the rest of this section. We will also have to treat the case $\alpha = \omega^2$, when four different Ramsey near-orderings are involved.

Note that $\alpha \times \omega_1$, with the lexicographic ordering, has order type $\omega_1 \cdot \alpha$. If $\beta < \alpha$ let $C_\beta = \{\beta\} \times \omega_1$, the βth *column* of $\alpha \times \omega_1$. Suppose E is a graph on $\alpha \times \omega_1$, i.e., $E \subseteq [\alpha \times \omega_1]^2$. If $K \subseteq [\alpha]^2$ let us say H is *independent mod* K if the set $Z = \{\{x,y\} \in H : \exists \beta, \gamma\ \{\beta,\gamma\} \in K,\ x \in C_\beta,\ y \in C_\gamma\}$ has the property that $Z \cap E = 0$. In the case of a near-ordering $<_R$ we say H is *independent mod* $<_R$ if it is independent mod K when K is the comparability graph of $<_R$, i.e., $K = \{\{x,y\} : x <_R y\}$. Let us write

$$\omega_1 \alpha \rightarrow (\omega_1 \alpha, n)^2 \bmod <_R$$

to mean that $\forall D \subseteq \alpha \times \omega_1$ if D has order type $\omega_1 \alpha$ then $\forall E \subseteq [D]^2\ \exists H \subseteq D$ either $|H| = n$ and $[H]^2 \subseteq E$ or H has order type $\omega_1 \alpha$ and H is independent mod $<_R$.

If A and B are uncountable subsets of $\alpha \times \omega_1$ we say (A, B) is *E-good* iff for any uncountable $A' \subseteq A$ and $B' \subseteq B$ there are uncountable $A'' \subseteq A'$ and $B'' \subseteq B'$ such that $\{\{x,y\} : x \in A'',\ y \in B''\}$ is disjoint from E. A set $H \subseteq \alpha \times \omega_1$ is *weakly independent mod* $K \subseteq [\alpha]^2$ iff $\forall \{\beta,\gamma\} \in K\ (H \cap C_\beta, H \cap C_\gamma)$ is E-good. We replace K by $<_R$ as above, and we write

$$\omega_1 \alpha \rightarrow_w (\omega_1 \alpha, n)^2 \bmod <_R$$

to mean that $\forall D \subseteq \alpha \times \omega_1$ if D has order type $\omega_1 \alpha$ then $\forall E \subseteq [D]^2\ \exists H$ either $|H| = n$ and $[H]^2 \subseteq E$ or H has order type $\omega_1 \alpha$ and H is weakly independent mod $<_R$.

Theorem 2.1. *Suppose $<_R$ is a Ramsey near-ordering on α. Then*

$$\forall n < \omega\ \ \omega_1 \alpha \rightarrow_w (\omega_1 \alpha, n)^2 \bmod <_R.$$

Proof. Fix $n < \omega$, D and $E \subseteq [D]^2$. Let

$$E_1 = \{\{(\beta,\xi),(\gamma,\eta)\} \in E : \beta <_R \gamma \text{ and } \xi < \eta\}$$

and

$$E_2 = \{\{(\beta,\xi),(\gamma,\eta)\} \in E : \beta <_R \gamma \text{ and } \xi > \eta\}.$$

We say the edges of E_1 *go up* and the edges of E_2 *go down*. It will suffice to prove the theorem for E_1 and E_2 separately since then it may be applied to E_1 and E_2 successively to yield the result for E.

For the rest of the proof we use E to refer either to E_1 or E_2. This makes a difference in only a few places, and there the difference is slight.

Suppose A and B are uncountable subsets of $\alpha \times \omega_1$. We say (A, B) is *E-bad* iff for any uncountable sets $A' \subseteq A$ and $B' \subseteq B$ there are $x \in A$ and $y \in B$ with $\{x,y\} \in E$.

Lemma 2.2. *Let A and B be uncountable subsets of $\alpha\times\omega_1$. If (A,B) is not E-good then there are $A'\subseteq A$ and $B'\subseteq B$ such that (A',B') is E-bad.*

Proof. Trivial.

Lemma 2.3. *Suppose $\beta <_R \gamma$, $A\subseteq C_\beta$, $B\subseteq C_\gamma$ and (A,B) is E-bad.*

(a) If $E=E_1$ then $\{x\in A:\{y\in B:\{x,y\}\in E\}$ is uncountable$\}$ is uncountable.

(b) If $E=E_2$ then $\{y\in B:\{x\in A:\{x,y\}\in E\}$ is uncountable$\}$ is uncountable.

Proof. (a) Since the edges of E go up, every element of B is connected to only countably many elements of A. If the lemma is false then almost all the elements of A are connected to only countably many elements of B, and now it is easy to thin out A and B to uncountable sets $A'\subseteq A$, $B'\subseteq B$ such that $\forall x\in A'\ \forall y\in B'\ \{x,y\}\notin E$, contrary to the assumption that (A,B) is E-bad.

The proof of (b) is exactly similar.

Note also that E-goodness and E-badness are hereditary downward, i.e., if (A,B) is E-good or E-bad then so is (A',B'), where $A'\subseteq A$, $B'\subseteq B$ are uncountable.

Finally, before we get into the meat of the proof, let us observe that the doubling property can be improved.

Lemma 2.4. *Let $k<\omega$. For any $X\subseteq\alpha$, if X has type α then $\exists Y\subseteq X$ Y has type α and there are $f_1,\ldots,f_k:Y\to X-Y$ such that $\forall y\in Y$ $y<_R f_1(y)<_R f_2(y)<_R\cdots<_R f_k(y)$ and $\{y,f_1(y),\ldots,f_k(y)\}$ is a chain, and if $z\in Y$ is such that $y<_R z$ then $f_k(y)<_R z$ and $\{y,f_1(y),\ldots,f_k(y)\}\cup\{z,f_1(z),\ldots,f_k(z)\}$ is a chain.*

Proof. For $k=1$ this is the doubling property. Suppose $k>1$ and Y satisfies the lemma for $k-1$. Find $Y'\subseteq Y$ of type α and $f_k:Y'\to Y-Y'$ satisfying the doubling property. Let $y\in Y'$. Then $y<_R f_k(y)$ so by the inductive hypothesis and near-transitivity $\{y,f_1(y),\ldots,f_k(y)\}$ is a chain. Let $z\in Y'$, $y<_R z$. Then $f_k(y)<_R z$ and $y<_R f_k(z)$ by the doubling property for Y' so, again by near-transitivity, $\{y,f_1(y),\ldots,f_k(y)\}\cup\{z,f_1(z),\ldots,f_k(z)\}$ is a chain, as desired.

Now we begin the main construction. Recall that we were given $D\subseteq\alpha\times\omega_1$ of order type $\omega_1\alpha$ with $E\subseteq[D]^2$. Let $X=\{\beta<\alpha: C_\beta\cap D$ is uncountable$\}$ and for $\beta\in X$ let $C'_\beta=C_\beta\cap D$. Then X has type α and without loss of generality we may assume $D=\bigcup\{C'_\beta:\beta\in X\}$.

For each k with $1 \le k \le n$ and each $a \in [X]^k_{<_R}$ we will choose an uncountable set $B_a \subseteq C'_{\min(a)}$, where $\min(a)$ refers to the minimum element of a under $<_R$. A similar convention applies to the expression $\max(a)$. The construction proceeds by $<_R$-recursion on $\max(a)$. First suppose $|a| = 1$, say $a = \{\gamma\}$. Let $K(\gamma)$ be the set of all chains b in X such that $1 \le |b| \le n$, $\max(b) <_R \gamma$ and $b \cup \{\gamma\}$ is a chain. By condition (1) in the definition of a Ramsey near-ordering, $K(\gamma)$ is finite. Hence we may find $B_a \subseteq C'_\gamma$ such that for all $b \in K(\gamma)$ either (B_b, B_a) is E-good or else there is $B \subseteq B_b$ such that (B, B_a) is E-bad. This can be accomplished by thinning out C'_γ successively, once for each $b \in K(\gamma)$, and using the hereditary property of E-goodness and E-badness.

If now $|a| > 1$ then we may write $a = b \cup \{\gamma\}$ where $b \in K(\gamma)$. If $(B_b, B_{\{\gamma\}})$ is E-good, then let $B_a = B_b$; otherwise choose $B_a \subseteq B_b$ so that $(B_a, B_{\{\gamma\}})$ is E-bad. Thus $(B_a, B_{\{\gamma\}})$ will be E-good or E-bad. This completes the construction.

For the rest of the proof, let us assume that there is no weakly independent set $H \subseteq D$ mod $<_R$ with order type $\omega_1\alpha$.

Lemma 2.5. *There is $Y \subseteq X$ such that Y has order type α and whenever $1 \le k \le n$, $a, b \in [Y]^k_{<_R}$, $\max(a) < \min(b)$ and $a \cup b$ is a chain, then (B_a, B_b) is not E-good.*

Proof. It will suffice to prove this for each k since then Y may be obtained in n steps. Suppose $a \cup b \in [X]^{2k}_{<_R}$, $|a| = |b| = k$, $\max a < \min b$. Put $a \cup b \in P_0$ if (B_a, B_b) is E-good and $a \cup b \in P_1$ otherwise. By the Ramsey property for $<_R$ we find $Y \subseteq X$ with $[Y]^{2k}_{<_R} \subseteq P_i$ for $i = 0$ or 1. If $i = 1$ we are done, so suppose $i = 0$.

Let us apply Lemma 2.4 (the strong doubling property) with k replaced by $k-1$ to find $Y' \subseteq Y$ satisfying that lemma. For each $\gamma \in Y'$ let $a(\gamma) = \{\gamma, f_1(\gamma), \ldots, f_{k-1}(\gamma)\}$, and let $A_\gamma = B_{a(\gamma)}$. We claim $H = \bigcup\{A_\gamma : \gamma \in Y'\}$ is weakly independent mod $<_R$, and it clearly has order type $\omega_1\alpha$. Let $\beta, \gamma \in Y'$, $\beta <_R \gamma$. Then $a(\beta) \cup a(\gamma)$ is a chain, hence belongs to $[Y]^{2k}_{<_R}$ so $a(\beta) \cup a(\gamma) \in P_0$ and (A_β, A_γ) is E-good. Thus H is weakly independent mod $<_R$ as claimed, and Lemma 2.5 is proved.

Fix $a \in [Y]^{2n}_{<_R}$, and suppose the elements of a are $\gamma_1 <_R \gamma_2 <_R \cdots <_R \gamma_{2n}$. For $1 \le i \le n$ let $a_i = \{\gamma_i, \gamma_{i+1}, \ldots, \gamma_n\}$ and let $A_i = B_{a_i}$.

Lemma 2.6. *If $1 \le i < j \le n$ then (A_i, A_j) is E-bad.*

Proof. Let $b = \{\gamma_i, \gamma_{i+1}, \ldots, \gamma_{j-1}\}$ and let $c = \{\gamma_j, \gamma_{j+1}, \ldots, \gamma_{2j-i-1}\}$. Then $|b| = |c| \le n$, $\max(b) < \min(c)$ and $b \cup c$ is a chain. Hence, since $b \cup c \subseteq a \subseteq Y$, Lemma 2.5 says that (B_b, B_c) is not E-good. Since $B_c \subseteq B_{\{\gamma_j\}}$ it follows that $(B_b, B_{\{\gamma_j\}})$ is not E-good, and hence by the construction of the B's, $(B_{b'}, B_{\{\gamma_j\}})$ is E-bad, where $b' = b \cup \{\gamma_j\}$. But we also have $A_i \subseteq B_{b'}$ and $A_j \subseteq B_{\{\gamma_j\}}$ so by the hereditary property of E-badness, (A_i, A_j) is E-bad.

The following lemma will complete the proof.

Lemma 2.7. *Suppose $\{\beta_1, \ldots, \beta_k\}$ is a chain, $F_i \subseteq C'_{\beta_i}$ and (F_i, F_j) is E-bad whenever $i < j$. Then we may find $x_i \in F_i$ such that $[\{x_1, \ldots, x_k\}]^2 \subseteq E$.*

Proof. The proof is by induction on k, and it depends on whether we are dealing with E_1 or E_2. We treat the case $E = E_1$; the case $E = E_2$ is exactly similar. If $k = 1$ the lemma is trivial so suppose $k > 1$. By the hereditary property of E-badness and Lemma 2.3(a) we may find uncountable $F \subseteq F_1$ such that $\forall x \in F\ \forall j > 1\ \{y \in F_j : \{x, y\} \in E_1\}$ is uncountable. Fix $x_1 \in F$ and for $j > 1$ let $F'_j = \{y \in F_j : \{x_1, y\} \in E_1\}$. By inductive hypothesis we may find $x_j \in F'_j$ with $[\{x_2, \ldots, x_k\}]^2 \subseteq E_1$, and now it is clear that $[\{x_1, \ldots, x_k\}]^2 \subseteq E_1$ as well. To treat $E = E_2$, use Lemma 2.3(b) and choose $F \subseteq F_k$.

3. Applying Martin's Axiom.

Let us emphasize that Theorem 2.1 was proved in ZFC. Of course, since one application of the theorem is to the case $\alpha = \omega$ and $<_R = <$, we know that we cannot pass from weakly independent sets to independent sets in ZFC. However, Martin's Axiom will suffice.

Theorem 3.1. *Assume $MA(\aleph_1)$. Suppose $E \subseteq [\alpha \times \omega_1]^2$, $K \subseteq [\alpha]^2$ and $H \subseteq \alpha \times \omega_1$ is weakly independent mod K. Then there is $H' \subseteq H$ that is independent mod K, and H' may be chosen so that whenever $H \cap C_\beta$ is uncountable then so is $H' \cap C_\beta$.*

Proof. Let P consist of all finite subsets of H that are independent mod K, ordered by reverse inclusion, and for $\xi < \omega_1$ let $P_\xi = \{p \in P : p \subseteq \alpha \times (\omega_1 - \xi)\}$. We will apply Martin's Axiom to one of the P_ξ.

Lemma 3.2. *P (hence each of the P_ξ) has the countable chain condition.*

Proof. Suppose not. Let $\langle p_\nu : \nu < \omega_1 \rangle$ be an antichain. By standard Δ-system techniques we may assume that the p_ν's are pairwise disjoint (removing the common part does not affect incompatibility) and that they all "look the same", more precisely, for each ν there is a bijection $\pi_\nu : p_0 \to p_\nu$ that preserves columns in the sense that $x \in C_\beta$ iff $\pi_\nu(x) \in C_\beta$.

Let $\langle (x_i, y_i) : i < k \rangle$ enumerate all pairs (x, y) such that $x, y \in p_0$ and for some $\beta, \gamma < \alpha$ we have $x \in C_\beta$, $y \in C_\gamma$ and $\{\beta, \gamma\} \in K$. We construct inductively uncountable subsets $A_0 \supseteq A_1 \supseteq \ldots \supseteq A_k$ and $B_0 \supseteq B_1 \supseteq \ldots \supseteq B_k$ of ω_1 as follows. Let $A_0 = B_0 = \omega_1$. Suppose A_i and B_i are given. Let $X_i = \{ \pi_\nu(x_i) : \nu \in A_i \}$ and $Y_i = \{ \pi_\nu(y_i) : \nu \in B_i \}$. If $x_i \in C_\beta$, $y_i \in C_\gamma$ then $X_i \subseteq C_\beta \cap H$, $Y_i \subseteq C_\gamma \cap H$ and since H is weakly independent mod K, (X_i, Y_i) is E-good. Choose uncountable $X_i' \subseteq X_i$, $Y_i' \subseteq Y_i$ such that $\forall x \in X_i'\ \forall y \in Y_i'$ $\{x, y\} \notin E$, and let $A_{i+1} = \{ \nu \in A_i : \pi_\nu(x_i) \in X_i' \}$, $B_{i+1} = \{ \nu \in B_i : \pi_\nu(y_i) \in Y_i' \}$. Finally, suppose $\nu \in A_k$, $\rho \in B_k$. We claim p_ν and p_ρ are compatible. If not then there are $x \in p_\nu$, $y \in p_\rho$, $\beta, \gamma < \alpha$ such that $\{\beta, \gamma\} \in K$ and $\{x, y\} \in E$. Clearly for some $i < k$ $\pi_\nu(x_i) = x$, $\pi_\rho(y_i) = y$. But then since $\nu \in A_k \subseteq A_{i+1}$, $\rho \in B_k \subseteq B_{i+1}$ we have $\pi_\nu(x_i) = x \in X_i'$, $\pi_\rho(y_i) = y \in Y_i'$ so $\{x, y\} \notin E$, a contradiction.

Now let $X = \{ \beta < \alpha : H \cap C_\beta \text{ is uncountable} \}$, and for each $\beta \in X$ and $\eta < \omega_1$ let $D_{\beta\eta} = \{ p \in P : \exists \zeta > \eta\ (\beta, \zeta) \in p \}$. If each $D_{\beta\eta}$ is dense in P then we are done easily by Martin's Axiom. Unfortunately it may happen that not all the $D_{\beta\eta}$ are dense in P, so we must work a little harder.

Lemma 3.3. *$\exists \xi < \omega_1\ \forall \beta \in X\ \forall \eta \in \omega_1\ D_{\beta\eta} \cap P_\xi$ is dense in P_ξ.*

Proof. The proof is similar to Lemma 3.2. Suppose the lemma is false. For each $\xi < \omega_1$ choose $p_\xi \in P_\xi$ and β_ξ, η_ξ such that no extension of p_ξ lies in $D_{\beta_\xi \eta_\xi}$. Let us make the same "Δ-system" assumptions about $\langle p_\xi : \xi < \omega_1 \rangle$ as before. Also, since there are only countably many possibilities for β_ξ we may assume that they are all the same, say β. Now let $\langle x_i : i < k \rangle$ enumerate all $x \in p_0$ such that for some γ, $x \in C_\gamma$ and $\{\beta, \gamma\} \in K$. We construct uncountable A_i and B_i much as before, except that now $A_i \subseteq \omega_1$ and $B_i \subseteq H \cap C_\beta$. Let $A_0 = \omega_1$, $B_0 = H \cap C_\beta$. Given A_i and B_i, let $X_i = \{ \pi_\xi(x_i) : \xi \in A_i \}$. Then (X_i, B_i) is E-good, so we may find $X_i' \subseteq X_i$ $B_{i+1} \subseteq B_i$ so that $\forall x \in X_i'\ \forall y \in B_{i+1}$ $\{x, y\} \notin E$. Let $A_{i+1} = \{ \xi \in A_i : \pi_\xi(x_i) \in X_i' \}$. If $\xi \in A_k$ then p_ξ is compatible with $\{y\}$ for all $y \in B_k$. But since B_k is uncountable there is $y \in B_k$ so that $p_\xi \cup \{y\} \in D_{\beta\eta_\xi}$, a contradiction.

Of course, Lemma 3.3 completes the proof.

Corollary 3.4. *Assume $MA(\aleph_1)$. If $<_R$ is a Ramsey near-ordering on $\alpha < \omega_1$, then*

$$\forall n < \omega \quad \omega_1\alpha \rightarrow (\omega_1\alpha, n)^2 \text{ mod } <_R.$$

Theorem 3.5. *Assume $MA(\aleph_1)$. Then*

(a) $\forall n < \omega \quad \omega_1\omega \to (\omega_1\omega, n)^2$.

(b) $\forall n < \omega \quad \omega_1\omega^2 \to (\omega_1\omega^2, n)^2$.

Proof. (a) is almost immediate from Corollary 3.4, using the fact that $<$ is a Ramsey near-ordering of ω. The only other point to check is that if H is independent mod $<$ then we can thin out the columns of H so that they, too, are independent. But this is easy using the well known partition relation $\omega_1 \to (\omega_1, \omega)^2$.

For (b) we can treat columns the same way, so it is just a matter of finding some near-orderings on ω^2. The method we use is an elaboration of the approach of Specker [Sp] in proving $\omega^2 \to (\omega^2, n)^2$ for all $n < \omega$. Since that relation is implied by Theorem 3.5(b) this is not particularly surprising.

The representation of ω^2 that we will use is the lexicographic ordering on $\omega \times \omega$. If $(m_1, n_1), (m_2, n_2) \in \omega \times \omega$ then let

$$(m_1, n_1) <_0 (m_2, n_2) \quad \text{iff} \quad m_1 = m_2 < n_1 < n_2$$

$$(m_1, n_1) <_1 (m_2, n_2) \quad \text{iff} \quad m_1 < m_2 < n_1 < n_2$$

$$(m_1, n_1) <_2 (m_2, n_2) \quad \text{iff} \quad m_1 < n_1 < m_2 < n_2$$

$$(m_1, n_1) <_3 (m_2, n_2) \quad \text{iff} \quad m_2 < m_1 < n_1 < n_2.$$

Note that $n_1 < n_2$ in every case so all these relations are well-founded. It is easy to see that $<_0$, $<_2$ and $<_3$ are partial orderings; $<_1$ is not transitive but it is near-transitive, as one quickly checks. It follows that all the $<_i$ are near-orderings. We claim they are all Ramsey. Condition (1) is easy, and (2) is straightforward. Before we look at the Ramsey and duplication properties it will help to have a couple of lemmas.

It is not the case that every pair of distinct elements of $\omega \times \omega$ is comparable with respect to some $<_i$, but we can nearly make it so as follows.

Lemma 3.6. *There is $W \subseteq \omega \times \omega$ such that W has order type ω^2, for all $(m, n) \in W$ we have $m < n$, and whenever $(m_1, n_1), (m_2, n_2)$ are distinct elements of W, if $m_1 \neq m_2$ then m_1, n_1, m_2, n_2 are all distinct. It follows that every pair of elements of W is comparable with respect to one and only one $<_i$.*

The proof is easy. Of course, a similar result holds when instead of $\omega \times \omega$ we start with $A \times A$ where A is any infinite subset of ω.

There is also a kind of converse to Lemma 3.6.

Lemma 3.74. *Suppose $X \subseteq \omega \times \omega$ has order type ω^2 and $\pi : \omega \times \omega \to X$ is the order-isomorphism. Then there is infinite $A \subseteq \omega$ such that $\forall x, y \in A \times A\ \forall i < 4$ if $x <_i y$ then $\pi(x) <_i \pi(y)$.*

Proof. We will define a mapping f of $[\omega]^4$ into ${}^4 2$, the set of functions from 4 into 2, as follows. Suppose $\{a, b, c, d\} \in [\omega]^4$ with $a < b < c < d$. Then $f\{a, b, c, d\} = \sigma$, where $\sigma : 4 \to 2$ is defined by

$$\begin{aligned}
\sigma(0) = 1 &\quad \text{iff} \quad \pi(a,b) <_0 \pi(a,c) \\
\sigma(1) = 1 &\quad \text{iff} \quad \pi(a,c) <_1 \pi(b,d) \\
\sigma(2) = 1 &\quad \text{iff} \quad \pi(a,b) <_2 \pi(c,d) \\
\sigma(3) = 1 &\quad \text{iff} \quad \pi(b,c) <_3 \pi(a,d).
\end{aligned}$$

By Ramsey's Theorem there is infinite $A \subseteq \omega$ such that f is constant on $[A]^4$. Suppose the constant value is σ. It will suffice to show that $\sigma(i) = 1$ for all $i < 4$. Let $i = 1$, for example. It is easy to find $a, b, c, d \in A$ with $a < b < c < d$ and $\pi(a,c) <_1 \pi(b,d)$. Hence if $f\{a, b, c, d\} = \tau$ we must have $\tau(1) = 1$. But $\tau = \sigma$ by the choice of A, so $\sigma(1) = 1$. The other cases are all similar.

Now let us check the Ramsey property for the $<_i$. For concreteness let us assume again that $i = 1$. Suppose $X \subseteq \omega \times \omega$ has type ω^2 and $[X]^n_{<_1} = P_0 \cup \ldots \cup P_{k-1}$. Find A and π as in Lemma 3.7. Suppose $s \in [A]^{2n}$, $s = \{k_0, k_1, \ldots, k_{2n-1}\}$. Then $a_s = \{ (k_j, k_{n+j}) : j < n \}$ is an element of $[A \times A]^n_{<_1}$ and every such element arises in this way. Define $g : [A]^{2n} \to k$ by $g(s) = m$ iff π"$a_s \in P_m$. Let $B \subseteq A$ be infinite and homogeneous for g. Let $W \subseteq B \times B$ satisfy Lemma 3.6, and let $Y = \pi$"W. Note that if $x, y \in W$ and $x <_1 y$ then $\pi(x) <_1 \pi(y)$ by Lemma 3.7, while if $\pi(x) <_1 \pi(y)$ then $x <_1 y$ since x and y must be comparable with respect to some $<_i$ and $i \neq 1$ would imply $\pi(x) <_i \pi(y)$ also, which is a contradiction. Thus, on W π preserves $<_1$ in both directions. Hence if $a \in [Y]^n_{<_1}$, $\pi^{-1}a \in [W]^n_{<_1} \subseteq [B \times B]^n_{<_1}$ and by the homogeneity of B all such a must lie in the same P_m. The treatment of $<_0$, $<_2$ and $<_3$ is entirely similar, except that for $<_0$ we take $s \in [A]^{n+1}$ and let $a_s = \{ (k_0, k_{j+1}) : j < n \}$.

Finally, we check the doubling property. Let $X \subseteq \omega \times \omega$ have type ω^2, and let π and A be as in Lemma 3.7. Again we treat the case $<_1$. Let $\langle a_n : n < \omega \rangle$ enumerate A in increasing order. Let $B = \{ a_{2n} : n < \omega \}$. If $a_m, a_n \in B$ and $m < n$ then let $f^*(a_m, a_n) = (a_{m+1}, a_{n+1})$. Let $W \subseteq B \times B$ satisfy Lemma 3.6 and let $Y = \pi$"W and $f = \pi f^* \pi^{-1}$. If

$x, y \in Y$ and $x <_1 y$ then $\pi^{-1}x <_1 \pi^{-1}y$ and clearly $a = \{\pi^{-1}x, f^*\pi^{-1}x, \pi^{-1}y, f^*\pi^{-1}y\}$ is a $<_1$-chain in the order given. But now $\pi``a = \{x, f(x), y, f(y)\}$ is also a $<_1$-chain.

For $<_0$ let $f^*(a_m, a_n) = (a_m, a_{n+1})$. For $<_2$ let $B = \{a_{3n} : n < \omega\}$ and let $f^*(a_m, a_n) = (a_{n+1}, a_{n+2})$. For $<_3$ let $B = \{a_{2n+1} : n < \omega\}$ and let $f^*(a_m, a_n) = (a_{m-1}, a_{n+1})$.

This now completes the proof of Theorem 3.5.

References

[Ba] J. Baumgartner, *Partition relations for uncountable cardinals*, **Israel J. Math. 21** (1975), 296–307.

[DJ] K. Devlin and H. Johnsbråten, *The Souslin Problem*, **Lecture Notes in Mathematics 405**, Springer-Verlag, 1974.

[EH] P. Erdős and A. Hajnal, *Ordinary partition relations for ordinal numbers*, **Periodica Math. Hung. 1** (1971), 171–185.

[EHMR] P. Erdős, A. Hajnal, A. Máté and R. Rado, **Combinatorial Set Theory: Partition Relations for Cardinals**, North-Holland, 1984.

[GL] F. Galvin and J. Larson, *Pinning countable ordinals*, **Fund. Math. 82** (1975), 357–361.

[L1] J. Larson, *Partition theorems for certain ordinal products*, in *Infinite and Finite sets*, **Coll. Math. Soc. J. Bolyai 10**, Keszthely, Hungary, 1973, 1017–1024.

[L2] J. Larson, *A counter-example in the partition calculus for an uncountable ordinal*, **Israel J. Math. 36** (1980), 287–299.

[Mi] T. Miyamoto, Ph.D. dissertation, Dartmouth College, 1987.

[SS] S. Shelah and L. Stanley, *A theorem and some consistency results in partition calculus*, to appear.

[Sp] E. Specker, *Teilmengen von Mengen mit Relationen*, **Comment. Math. Helv. 31** (1957), 302–314.

[Wi] N. H. Williams, **Combinatorial Set Theory**, North-Holland, 1977.

Dartmouth College

Hanover, NH 03755 USA

APPLICATIONS OF SUPERPERFECT FORCING AND ITS RELATIVES

Andreas Blass
Mathematics Department
University of Michigan
Ann Arbor, MI 48109

1. INTRODUCTION.

This paper is a survey of some recent developments concerning the combinatorial principle of near coherence of filters (NCF) [5,9,10], some extensions and generalizations of it [7], and some applications [6,7]. Most of the material here has appeared or will appear in the papers just cited, but there are some new results in Sections 2 and 5.

Since NCF and the extensions to be discussed here were originally motivated by applications outside set theory, we devote the rest of this introduction to a (somewhat) historical overview of these applications and their connection to NCF. The later sections of the paper are arranged in a more logical order, proceeding from general principles to particular consequences, the reverse of the historical order.

Consider the following three questions.

1. [12] Is the ideal of compact operators on Hilbert space the sum of two properly smaller ideals?

2. [3] Does the Stone-Čech remainder of a closed half-line, $\beta[0,\infty) - [0,\infty)$, have more than one composant?

3. [15] Are there more than 4 slenderness classes of abelian groups?

In Question 1, "Hilbert space" means infinite-dimensional, separable, complex Hilbert space, and "ideal" means two-sided ideal in the ring of all bounded linear operators. This question was answered affirmatively in [11] under the assumption of the continuum hypothesis (CH), or Martin's axiom (MA), or even weaker assumptions to be discussed below. The answer remains affirmative (under the same assumptions) if the ideal of compact operators is replaced by any ideal containing an operator of infinite rank.

In Question 2, "composant" means an equivalence class with respect to the equivalence relation "lies in a proper subcontinuum with"; that this relation is an equivalence follows from Bellamy's theorem [3] that $\beta[0,\infty) - [0,\infty)$ is indecomposable, i.e., not the union of two proper subcontinua. Question 2 was answered affirmatively, assuming CH, by Rudin

[21]; her proof, which also works under MA, actually shows that there are 2^c composants. (We use c for $2^{\aleph_0}$, the cardinality of the continuum.)

Question 3 is based on the following concepts introduced by Specker [25]. A subgroup U of the additive group $\mathbb{Z}^\omega$ (direct product) is called monotone if it is obtainable from some family T of non-decreasing functions $\omega \longrightarrow \omega - \{0\}$ by taking all the functions $\omega \longrightarrow \mathbb{Z}$ whose absolute values are eventually majorized by some function in T. A group G is U-slender if every homomorphism $U \longrightarrow G$ annihilates all but finitely many of the "unit vectors" $(0,\cdots,0,1,0,\cdots) \in U$. To each monotone U, one associates the slenderness class consisting of all U-slender groups; different U's may give rise to the same slenderness class. Göbel and Wald showed [15] that the number of distinct slenderness classes is at least four and that it is 2^c if MA holds.

The weakest assumption from which an affirmative answer to Question 1 was deduced in [11] is that there are two non-principal ultralfilters $\mathcal{U}$ and $\mathcal{V}$ on ω such that $f(\mathcal{U}) \neq f(\mathcal{V})$ for every finite-to-one $f:\omega \longrightarrow \omega$. This assumption follows easily from CH or MA, since any two non-isomorphic Ramsey ultrafilters can serve as $\mathcal{U}$ and $\mathcal{V}$. My attempts to prove the assumption in ZFC alone failed. When I raised the question at the 1977 meeting in honor of Rothberger, van Douwen pointed out that the same assumption suffices for Rudin's proof of the affirmative answer to Question 2. It is shown in [6] that Question 3 also has an affirmative answer under the same hypothesis.

It will be convenient to adopt, for the remainder of this paper, the convention that "filter" means proper filter on ω containing all of the cofinite sets and "ultrafilter" means non-trivial ultrafilter on ω.

The principle of near coherence of filters (NCF) is the negation of the assumption discussed two paragraphs ago. That discussion implies that NCF follows from negative answers to any of the three questions and that NCF contradicts MA (and therefore contradicts CH). It is shown in [5] that NCF is equivalent to each of the following statements.

(a) For every two filters $\mathcal{F}$ and $\mathcal{G}$, there are finite-to-one functions $f,g:\omega \longrightarrow \omega$ such that $f(\mathcal{F}) \cup g(\mathcal{G})$ generates a (proper) filter (i.e., $f(\mathcal{F})$ and $g(\mathcal{G})$ cohere; hence the terminology "near coherence").

(b) For every two ultrafilters $\mathcal{U}$ and $\mathcal{V}$, there are finite-to-one functions $f,g:\omega \longrightarrow \omega$ such that $f(\mathcal{U}) = g(\mathcal{V})$

(c) The same as (a) or (b) with the additional requirements that $f = g$ and f is monotone.

(d) For every ultrafilter $\mathcal{U}$, there is a finite-to-one function $f:\omega \longrightarrow \omega$ such that the ultrafilter $f(\mathcal{U})$ is generated by fewer than $\mathfrak{d}$ sets. (Here $\mathfrak{d}$ is the dominating number, discussed in more detail below.)

It turns out that NCF is equivalent to a negative answer to Question 1; the implication not in [11] is established in [6]. Furthermore, NCF is equivalent to a negative answer to Question 2; the direction not implicit in [21] is due to Mioduszewski [19,20] (and was rediscovered in [6]). The corresponding equivalence for Question 3 remains an open problem. It is known [7] that a negative answer to Question 3 implies the following "filter dichotomy" principle, which in turn implies NCF, but both converses are open problems.

Filter Dichotomy (FD): For every filter $\mathcal{F}$, there is a finite-to-one function $f:\omega \longrightarrow \omega$ such that $f(\mathcal{F})$ is the cofinite filter or an ultrafilter.

To deduce NCF from this, apply it with $\mathcal{F}$ being the intersection of two arbitrary ultrafilters, $\mathcal{U}$ and $\mathcal{V}$. For any finite- to-one $f:\omega \longrightarrow \omega$, the filter $f(\mathcal{F}) = f(\mathcal{U}) \cap f(\mathcal{V})$ is not the cofinite filter; indeed, if we partition ω into three infinite pieces, one must be in $f(\mathcal{U})$ and one in $f(\mathcal{V})$ as these are ultrafilters, so the union of these two (possibly equal) pieces is a coinfinite set in $f(\mathcal{F})$. So FD requires that, for some finite-to-one f, $f(\mathcal{F}) = f(\mathcal{U}) \cap f(\mathcal{V})$ is an ultrafilter $\mathcal{W}$. But then $f(\mathcal{U}) = \mathcal{W} = f(\mathcal{V})$, so we have established NCF.

The consistency of NCF relative to ZFC was proved by Shelah in the fall of 1984. The original proof is in [9], and a simplified version is in [10] (except for one section unaffected by the simplification). The model used in the simplified proof is obtained from a model of ZFC + GCH by iterating Miller's forcing with superperfect (= rational perfect) trees [18] $\aleph_2$ times with countable supports. This model will be discussed in more detail below.

In [7], Shelah's proof is extended to show that both of his models of NCF (from [9] and [10]) also satisfy FD and in fact have only four slenderness classes of abelian groups. The proofs of these facts involve changing only one lemma in Shelah's argument. It therefore seems worthwhile to encapsulate the rest of Shelah's argument in a combinatorial principle, true in these models, which would imply NCF, FD, and the negative answer to Question 3. Such a principle was presented in [7]; it is the inequality $\mathfrak{u} < \mathfrak{g}$, where $\mathfrak{u}$ is the minimum possible cardinality of an ultrafilter base and where $\mathfrak{g}$ is a new cardinal invariant of the continuum.

The invariant $\mathfrak{g}$ and the inequality $\mathfrak{u} < \mathfrak{g}$ are the primary topic of this paper. In Section 2, we discuss $\mathfrak{g}$ along with some other (old) cardinal invariants of the continuum. We also show that $\mathfrak{u} < \mathfrak{g}$ implies FD and therefore NCF. In Section 3, we sketch the proof [7] that the superperfect forcing model satisfies $\mathfrak{u} < \mathfrak{g}$; most of this proof is just like the proof in [10] that the same model satisfies NCF. In Section 4, we deduce from $\mathfrak{u} < \mathfrak{g}$ that there are only four slenderness classes of abelian groups. Finally, in Section 5, we describe a different model for $\mathfrak{u} < \mathfrak{g}$, one motivated directly by the definition of $\mathfrak{g}$.

2. $\mathfrak{g}$ AND SOME OTHER CARDINALS.

We shall need to consider four of the standard "cardinal invariants of the continuum" [13] (not counting $\mathfrak{c} = 2^{\aleph_0}$ and $\aleph_1$) and a new invariant $\mathfrak{g}$. Their definitions are as follows.

$\mathfrak{b}$, the <u>bounding number</u>, is the smallest cardinality of any unbounded family $\mathcal{B} \subseteq {}^{\omega}\omega$; here "unbounded" means that for every $f \in {}^{\omega}\omega$ there is $g \in \mathcal{B}$ such that $f(n) \le g(n)$ for infinitely many n.

$\mathfrak{d}$, the <u>dominating number</u>, is the smallest cardinality of any dominating family $\mathcal{D} \subseteq {}^{\omega}\omega$; here "dominating" means that for every $f \in {}^{\omega}\omega$ there is $g \in \mathcal{D}$ such that $f(n) \le g(n)$ for all but finitely many n.

$\mathfrak{u}$ is the smallest cardinality of any ultrafilter base; an ultrafilter base is a family $\mathcal{B} \subseteq \mathcal{P}(\omega)$ such that the members of $\mathcal{B}$ and their supersets in $\mathcal{P}(\omega)$ constitute an ultrafilter.

$\mathfrak{h}$, the <u>distributivity number</u> is the smallest cardinal κ such that some κ dense families in $[\omega]^{\omega}$ have no common member. Here $[\omega]^{\omega}$ is the collection of all infinite subsets of ω, and $\mathcal{D} \subseteq [\omega]^{\omega}$ is <u>dense</u> if

(a) for all $X \in \mathcal{D}$ and $Y \in [\omega]^{\omega}$, if $Y - X$ is finite, then $Y \in \mathcal{D}$ (i.e., $\mathcal{D}$ is closed under subsets and finite changes), and

(b) for every $X \in [\omega]^{\omega}$, there is $Y \in \mathcal{D}$ with $Y \subseteq X$.

(If we view $[\omega]^{\omega}$ as a notion of forcing, then $\mathfrak{h}$ is the smallest ordinal that acquires, in the generic extension, a new function into the ground model, i.e., the smallest cardinal such that the associated complete Boolean algebra is not $(\mathfrak{h},\infty)$-distributive. In fact, $\mathfrak{h}$ acquires a new subset, so even $(\mathfrak{h},2)$-distributivity fails. $\mathfrak{h}$ is the height of the base matrix trees of [1].)

$\mathfrak{g}$ is the smallest cardinal κ such that some κ groupwise dense families in $[\omega]^\omega$ have no common number. Here a family $\mathcal{G} \subseteq [\omega]^\omega$ is <u>groupwise dense</u> if

(a) for all $X \in \mathcal{G}$ and $Y \in [\omega]^\omega$, if $Y - X$ is finite, then $Y \in \mathcal{G}$, and

(b′) for every family Π of infinitely many pairwise disjoint finite subsets of ω, the union of some (necessarily infinite) subfamily of Π is in $\mathcal{G}$.

In the definition of "groupwise dense", (a) is the same as in the definition of "dense", while (b′) is stronger than (b) since, given X as in (b), we can apply (b′) to $\Pi = \{\{x\} \mid x \in X\}$ and obtain the desired Y. Thus, every groupwise dense family is dense, and therefore $\mathfrak{g} \geq \mathfrak{h}$. Not every dense family is groupwise dense; easy counterexamples are $\{X \in [\omega]^\omega \mid X$ contains only finitely many pairs of consecutive integers$\}$ and $\{X \in [\omega]^\omega \mid$ the function that enumerates X in increasing order eventually majorizes $n^2\}$. We shall see later that $\mathfrak{h} < \mathfrak{g}$ is consistent with ZFC.

In clause (b′) of the definition of groupwise dense, we can assume without loss of generality that Π is a partition of ω into intervals. Indeed, given an arbitrary infinite family Π of disjoint finite subsets of ω, we can trivially find a partition Π' of ω into finite intervals such that each interval from Π' includes at least one set from Π. If the conclusion of (b′) holds for Π', then it also holds for Π because of clause (a).

Observe, for future reference, that if Π is a counterexample to (b′) for a particular $\mathcal{G}$, then so is any Π' whose members are finite unions of members of Π. We shall sometimes say that such a Π' is obtained from Π by merging some of its members.

The ZFC-provable order relationships between these cardinal invariants are summarized by the following Hasse diagram, in which κ is below λ if and only if it is provable in ZFC that $\kappa \leq \lambda$.

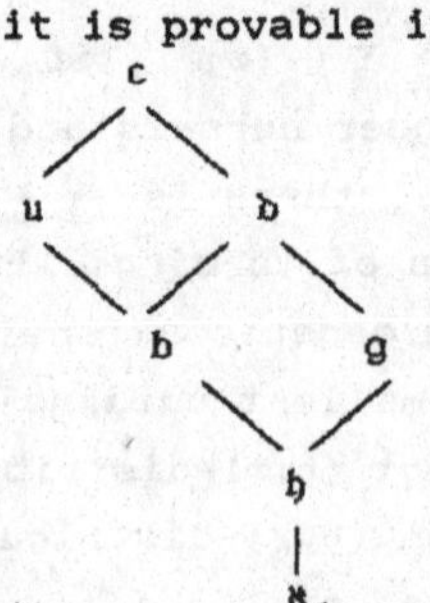

(A larger diagram, involving more of the standard cardinal invariants [13], is given in the Appendix.)

By way of partially justifying this Hasse diagram and also providing an easy example of how the definition of $\mathfrak{g}$ is used in practice, we shall prove that $\mathfrak{g} \le \mathfrak{d}$. The only other inequality involving $\mathfrak{g}$, namely $\mathfrak{g} \ge \mathfrak{h}$, was established above. The inequalities not involving $\mathfrak{g}$ are all well known; the only non-trivial ones are $\mathfrak{b} \le \mathfrak{u}$, which is due to Solomon [24], and $\mathfrak{h} \le \mathfrak{b}$ which can be proved quite similarly to $\mathfrak{g} \le \mathfrak{d}$.

<u>Theorem 1</u>. $\mathfrak{g} \le \mathfrak{d}$.

<u>Proof</u>. Let $\mathcal{D}$ be a dominating family of cardinality $\mathfrak{d}$. For each $f \in \mathcal{D}$, let

$$\mathcal{G}_f = \{X \in [\omega]^\omega \mid \text{For infinitely many } n,\ f(n) < \text{next}(X,n)\},$$

where next(X,n) is the smallest number $\ge n$ in X. We verify that each $\mathcal{G}_f$ is groupwise dense. For (a), notice that, if $Y - X$ is finite, then $\text{next}(Y,n) \ge \text{next}(X,n)$ for all sufficiently large n. For (b′), given a partition Π of ω into intervals $[a_i, a_{i+1})$, choose an increasing sequence of integers $i(k)$ inductively so that $a_{i(k+1)} > f(a_{i(k)+1})$, and let X be the union of the intervals $[a_{i(k)}, a_{i(k)+1})$ of Π. Then $X \in \mathcal{G}_f$ because

$$\text{next}(X, a_{i(k)+1}) = a_{i(k+1)} > f(a_{i(k)+1})$$

for all k.

If the $\mathfrak{d}$ groupwise dense families $\mathcal{G}_f$ had a common member X, then $\mathcal{D}$ would not dominate the function $\text{next}(X,-)$, contrary to our choice of $\mathcal{D}$. So $\mathfrak{g} \le \mathfrak{d}$. □

As mentioned above, a similar argument establishes that $\mathfrak{h} \le \mathfrak{b}$; the essential change is to replace "For infinitely many n, $f(n) < \text{next}(X,n)$" with "For all but finitely many n, $f(n) < \text{next}(X, \text{next}(X,n) + 1)$".

To complete the justification of the Hasse diagram, we must show that no additional inequalities are provable. For this, it suffices to exhibit four models of ZFC, one for each of the following strict inequalities:

$\aleph_1 < \mathfrak{h}$,

$\mathfrak{b} < \mathfrak{u}$,

$\mathfrak{u} < \mathfrak{g}$, and

$\mathfrak{g} < \mathfrak{b}$.

The two not involving $\mathfrak{g}$ are well known. MA implies $\mathfrak{h} = \mathfrak{c}$, so the first inequality holds in models of MA + $\neg$ CH. The second inequality holds in the model obtained from a model of CH by adjoining $\aleph_2$ random reals. The third inequality $\mathfrak{u} < \mathfrak{g}$ is the main subject of this paper; a consistency proof is sketched in Section 3. The last inequality, $\mathfrak{g} < \mathfrak{b}$, holds in the model obtained from a model of MA + $\mathfrak{c} = \aleph_2$ by adjoining $\aleph_1$ random reals. It is well known that $\mathfrak{b} = \aleph_2$ in this model, since every $f \in {}^{\omega}\omega$ is dominated by a function in the ground model of MA and this ground model has $\mathfrak{b} = \aleph_2$. To show that $\mathfrak{g} = \aleph_1$ in the random extension, and thus to complete the justification of the Hasse diagram, we use a partition theorem, which sheds more light on the significance of $\mathfrak{g}$ and seems to be of some independent interest.

Call a family $\mathfrak{X}$ of subsets of ω _Turing cofinal_ if every subset of ω is Turing reducible to (i.e., recursive in, $\leq_T$) at least one member of $\mathfrak{X}$; in other words, $\mathfrak{X}$ is cofinal in the usual ordering of Turing degrees. Call $\mathfrak{X}$ _almost Turing cofinal_ if there is a single set $Q \subseteq \omega$ such that $\{X \vee Q \mid X \in \mathfrak{X}\}$ is Turing cofinal (where $\vee$ means recursive join); in other words, $\mathfrak{X}$ is cofinal in the degrees with respect to reducibility relative to Q.

Theorem 2. _If an almost Turing cofinal set is partitioned into fewer than_ $\mathfrak{g}$ _pieces, then at least one of these pieces is almost Turing cofinal_.

Proof. Suppose $\varkappa < \mathfrak{g}$ and $\bigcup_{\alpha<\varkappa} \mathfrak{X}_\alpha$ is almost Turing cofinal, with witness Q. For each $\alpha < \varkappa$, set

$$\mathfrak{G}_\alpha = \{Z \in [\omega] \mid \text{No infinite subset of } Z \text{ is } \leq_T X \vee Q, \text{ for any } X \in \mathfrak{X}_\alpha\}.$$

Since, by our choice of Q, every Z is $\leq_T X \vee Q$ for some X in _some_ $\mathfrak{X}_\alpha$, the intersection of the $\varkappa$ families $\mathfrak{G}_\alpha$ is empty. As $\varkappa < \mathfrak{g}$, at least one $\mathfrak{G}_\alpha$ fails to be groupwise dense; fix such an α. Condition (a) in the definition of "groupwise dense" clearly holds for $\mathfrak{G}_\alpha$, so condition (b′) must fail. Fix a partition Π of ω into intervals I_n, no union of which is in $\mathfrak{G}_\alpha$; let $P \subseteq \omega$ encode the partition, e.g., by having P consist of the left endpoints of the I_n's. We shall complete the proof by showing that $\mathfrak{X}_\alpha$ is almost Turing cofinal, with $P \vee Q$ as witness.

Given any $Y \subseteq \omega$, find $Y' \in [\omega]^{\omega}$ such that Y is recursive in every infinite subset of Y'; for example, let Y' consist of codes for all the finite initial segments of the characteristic function of Y. By choice of Π, we know that the set $\bigcup_{n \in Y'} I_n$ is not in $\mathcal{G}_{\alpha}$, so it has an infinite subset $W \leq_T X \vee Q$ for some $X \in \mathfrak{X}_{\alpha}$. Set $Z = \{n \mid W \text{ meets } I_n\}$. Then Z is an infinite subset of Y', so $Y \leq_T Z$. Also clearly, $Z \leq_T W \vee P$. Combining all these Turing reductions, we obtain $Y \leq_T X \vee P \vee Q$. As Y was arbitrary and $X \in \mathfrak{X}_{\alpha}$, this completes the proof. $\square$

<u>Remark</u>. For this paragraph only, we suspend the convention that filters are on ω, as we shall discuss certain filters on $\mathcal{P}(\omega)$. For each $Q \subseteq \omega$, we can define the Turing cone filter relative to Q as the filter $\mathcal{F}(Q)$ on $\mathcal{P}(\omega)$ generated by the cones

$$\mathcal{C}_X(Q) = \{Y \in \mathcal{P}(\omega) \mid X \leq_T Y \vee Q\}.$$

Let $\mathcal{F}$ be the intersection of all the filters $\mathcal{F}(Q)$. Then a set $\mathfrak{X}$ is almost Turing cofinal if and only if its complement is not in $\mathcal{F}$. Theorem 2 therefore asserts that $\mathcal{F}$ is g-complete, i.e., closed under intersections of fewer than g sets. By contrast, each $\mathcal{F}(Q)$, in particular the ordinary Turing cone filter $\mathcal{F}(\emptyset)$, is only $\aleph_1$-complete. In fact, the intersection of any $\aleph_1$ distinct Turing cones is empty.

We are now in a position to complete the proof that $g = \aleph_1$ in the model M obtained by adding $\aleph_1$ random reals to a ground model that satisfies $MA + c = \aleph_2$. In fact, the nature of the ground model is irrelevant; only the $\aleph_1$ random reals matter. Let these random reals be r_{ξ} for $\xi < \aleph_1$, and let $\mathfrak{X}_{\alpha}$ be the set of reals in the model obtained by adding $(r_{\xi} : \xi < \alpha)$ to the ground model. A well known consequence of the countable antichain condition is that $\bigcup_{\alpha<\aleph_1} \mathfrak{X}_{\alpha}$ contains all the reals of M, hence is certainly almost Turing cofinal in M. If g were larger than $\aleph_1$, then Theorem 2 would require some single $\mathfrak{X}_{\alpha}$ to be almost Turing cofinal, say with witness Q. By the countable antichain condition, Q is in some $\mathfrak{X}_{\beta}$, and we may assume $\beta \geq \alpha$ by increasing β if necessary. Since $\mathfrak{X}_{\beta}$ is clearly closed under recursive join and under Turing reduction, it would include $\{X \vee Q \mid X \in \mathfrak{X}_{\alpha}\}$ and therefore all reals. This is absurd, as $r_{\beta} \notin \mathfrak{X}_{\beta}$. This contradiction shows that g cannot be $> \aleph_1$, so $g = \aleph_1$ in M.

<u>Corollary</u>. <u>In any model obtained by adjoining uncountably many Cohen or random reals to any model of</u> ZFC, $g = \aleph_1$.

Proof. The preceding discussion covers the case of $\aleph_1$ random reals. The case of more than $\aleph_1$ random reals can be viewed as first adjoining all but $\aleph_1$ of the random reals and then, over this new ground model, adjoining $\aleph_1$ random reals, so this case reduces to the previous one. Finally, the proof for Cohen reals is exactly the same as for random reals. □

We now begin our study of the combinatorial principle $\mathfrak{u} < \mathfrak{g}$. The consistency proof will occupy the next section, but we close this section with two remarks and a theorem about this inequality.

The first remark is that the "next stronger" inequality, $\mathfrak{u} < \mathfrak{h}$, is inconsistent, by the Hasse diagram discussed above. In fact, there is an easy argument that shows that every filter generated by fewer than $\mathfrak{h}$ sets is feeble, i.e., is sent to the cofinite filter by some finite-to-one function $f:\omega \longrightarrow \omega$. (This also follows from $\mathfrak{h} \le \mathfrak{b}$ together with Solomon's proof in [24].) Suppose that $\mathfrak{B}$ is a filter base of cardinality $< \mathfrak{h}$. For each $B \in \mathfrak{B}$, the family

$$\mathfrak{D}_B = \{X \in [\omega]^\omega \mid \text{Between every two sufficiently large elements of } X \text{ there is an element of } B\}$$

is clearly dense (but not groupwise dense). As there are fewer than $\mathfrak{h}$ of these families, there is an X common to them all. Then the function $f(n) = |X \cap n|$, constant on intervals between successive members of X, takes all but finitely many values on every $B \in \mathfrak{B}$. So this function sends the filter generated by $\mathfrak{B}$ to the cofinite filter.

The second remark is that the inequality $\mathfrak{u} < \mathfrak{b}$, though weaker than $\mathfrak{u} < \mathfrak{g}$, is of considerable interest, because results of Ketonen [16] imply that any ultrafilter generated by fewer than $\mathfrak{b}$ sets is a non-Ramsey P-point. Recent unpublished work of Canjar and Laflamme gives more information about these and related ultrafilters. Formulation (d) of NCF in Section 1 shows that NCF implies $\mathfrak{u} < \mathfrak{b}$. The first model for $\mathfrak{u} < \mathfrak{b}$ was obtained by Shelah shortly before he proved the consistency of NCF. Canjar showed that this model, described in [8], does not satisfy NCF, so $\mathfrak{u} < \mathfrak{b}$ is strictly weaker than NCF.

We show next that $\mathfrak{u} < \mathfrak{g}$ implies the filter dichotomy principle, FD, introduced and shown to imply NCF in Section 1. This result is from [7].

Theorem 3. Assume $\mathfrak{u} < \mathfrak{g}$ and let $\mathcal{F}$ be any filter. Then there is a finite-to-one $f:\omega \longrightarrow \omega$ such that $f(\mathcal{F})$ is either an ultrafilter or the cofinite filter.

Proof. Let $\mathcal{U}$ be an ultrafilter with a base $\mathcal{B}$ of cardinality $< \mathfrak{g}$. We shall find, for every filter $\mathcal{F}$, a finite-to-one f such that either $f(\mathcal{F})$ is the cofinite filter or $f(\mathcal{F}) = f(\mathcal{U})$, an ultrafilter. For each $B \in \mathcal{B}$, let

$$\mathcal{G}_B = \{X \in [\omega]^\omega \mid \text{There is } A \in \mathcal{F} \text{ such that, whenever } x < y \text{ are in } X \text{ and } [x,y) \text{ meets } A, \text{ then } [x,y) \text{ meets } B\}.$$

Case 1. For some $B \in \mathcal{B}$, $\mathcal{G}_B$ is not groupwise dense.

Clearly, $\mathcal{G}_B$ is closed under subsets. It is also closed under finite modifications because $\mathcal{F}$ contains all cofinite sets. So, as $\mathcal{G}_B$ is not groupwise dense, there must be a partition Π of ω into intervals, no union of which is in $\mathcal{G}_B$. By merging intervals, we can assume that each interval in Π meets B. Let $f:\omega \longrightarrow \omega$ be the function that sends the nth interval of Π to n; this is finite-to-one, and we shall show that $f(\mathcal{F})$ is the cofinite filter. Suppose the contrary, and let Y be a coinfinite set in $f(\mathcal{F})$. Then $f^{-1}(Y)$ is a set $A \in \mathcal{F}$ witnessing that $f^{-1}(\omega - Y) \in \mathcal{G}_B$; indeed, if $x < y$ are in $f^{-1}(\omega - Y)$ and $[x,y)$ meets $f^{-1}(Y)$, say at n, then $f^{-1}\{f(n)\}$ is included in $[x,y)$ and, being an interval of Π, meets B. But $f^{-1}(\omega - Y)$ is an infinite union of intervals of Π, so it cannot be in $\mathcal{G}_B$. This contradiction shows that $f(\mathcal{F})$ is the cofinite filter.

Case 2. For every $B \in \mathcal{B}$, $\mathcal{G}_B$ is groupwise dense.

Since $|\mathcal{B}| < \mathfrak{g}$, there is an X common to all the $\mathcal{G}_B$'s. Let $f(n) = |X \cap [0,n]|$, so f is constant on the intervals $[x,y)$ into which ω is divided by the elements x,y of X. For each $B \in \mathcal{B}$, there exists (because $X \in \mathcal{G}_B$) an $A \in \mathcal{F}$ with $f(A) \subseteq f(B)$. Since $\mathcal{B}$ is a basis for $\mathcal{U}$, it follows that $f(\mathcal{U}) \subseteq f(\mathcal{F})$. Since $f(\mathcal{U})$ is an ultrafilter, $f(\mathcal{U}) = f(\mathcal{F})$. □

3. SUPERPERFECT FORCING YIELDS $\mathfrak{u} < \mathfrak{g}$.

In this section, we define superfect forcing, a notion of forcing introduced by Miller [18], and we show that iterating it $\aleph_2$ times with countable supports over a model of GCH yields a model of $\mathfrak{u} < \mathfrak{g}$. This result is from [7], but most of the proof is as in [10] where the same model was shown to satisfy NCF.

By a <u>tree</u>, we mean a subtree of the tree of finite subsets of ω, ordered by end extension. In such a tree, a node with infinitely many immediate successors is called <u>infinitely branching</u>; a node with at least two immediate successors is called <u>branching</u>. A tree is <u>superperfect</u> if every node has an infinitely branching successor. We consider the notion of forcing in which the conditions are the superperfect trees and extensions are subtrees. If G is a generic set (of superperfect trees), then the intersection of all the trees in G is a single branch, and the union of this branch is an infinite subset W of ω. W determines G as the set of all superperfect trees that contain, as nodes, all the initial segments of W. Thus, this notion of forcing can be viewed as adjoining a generic subset W of ω.

A straightforward fusion argument shows that this forcing is proper [23] and in fact satisfies Baumgartner's Axiom A [2].

It will be useful at times to confine attention to superperfect trees of a particularly simple sort. A superperfect tree has <u>interval structure</u> if ω can be partitioned into intervals $I_0 = [0,a_0)$, $I_1 = [a_0,a_1)$, etc., so that the first branching node (also called the trunk) of the tree is a subset of I_0, for each branching node a, if $\max(a) \in I_k$, then $\{n > \max(a) \mid a \cup \{n\}$ is a node$\}$ consists of one member from each I_j, $j > k$ (so a is infinitely branching), and, if $a \cup \{n\}$ is a successor of a with $n \in I_j$; then $a \cup \{n\}$ has an (infinitely) branching successor b with $\max(b) \in I_j$. It is not hard to prune an arbitrary superperfect tree down to one with interval structure, so the conditions with interval structure are dense in the notion of forcing.

The concept of a tree with interval structure $I_0, I_1, \ldots$ may be clarified by the following anthropomorphic description of what is involved in choosing a path through such a tree ("climbing the tree"). The tree amounts to a specification, for each $j > 0$, of 2^{j-1} subsets of I_j, indexed by the j-tuples of 0's and 1's that start with 1. At each stage of the climbing process, we decide whether to take or leave a certain set. At stage 0, we must take the trunk. At stage $j > 0$, the j decisions from previous stages are coded as a j-tuple of 0's and 1's (1 for "take", 0 for "leave"), and the subset of I_j indexed by this j-tuple is offered to us, to take or leave. Any sequence of choices that contains "take" infinitely often represents an infinite path through the tree, a path whose union is the union of all the taken sets.

If a superperfect tree has interval structure $I_0, I_1, \cdots$ and if we are given another partition of ω into intervals $J_0, J_1, \cdots$, each of which is the union of one or more of the I's, then that tree has a superperfect subtree with interval structure $J_0, J_1, \cdots$. Indeed, it suffices to take the subtree of all nodes that meet each J_n in at most its first subinterval I_k. We refer to such pruning as merging intervals.

We now embark on a sequence of lemmas leading up to the consistency of $\mathfrak{u} < \mathfrak{g}$.

Lemma 1. If $\mathscr{G}$ is, in the ground model, a groupwise dense subset of $[\omega]^\omega$, then the generic W is a subset of some $X \in \mathscr{G}$.

Proof. Let any condition p be given. Extend it to a condition p' with interval structure $I_0, I_1, \cdots$. As $\mathscr{G}$ is groupwise dense, it contains the union X of some (infinitely many) of these intervals I_n. We may assume $I_0 \subseteq X$, since $\mathscr{G}$ is closed under finite modifications. Let J_n denote the union of the I_k's from the nth one that is $\subseteq X$ up to but not including the $(n+1)^{st}$ such. Thus, the I's that are $\subseteq X$ are just those that are initial segments of the J's that contain them. Let p'' be obtained from p' by merging intervals to achieve interval structure $J_0, J_1, \cdots$, as described above. Then clearly p'' forces W to be a subset of X, since every node of p'' is a subset of X. We have shown that every condition p has an extension p'' forcing the desired conclusion, so, by genericity, this conclusion holds. □

Lemma 2. If the infinitely branching nodes of a superperfect tree are 2-colored, then there is a superperfect subtree all of whose branching nodes have the same color.

Proof. Try to build, inductively, a superperfect subtree all of whose branching nodes have the first color. The only way the construction can fail is if, for some node a, all infinitely branching successors of a have the second color. But then the subtree of nodes comparable with a is superperfect and homogeneous for the second color. □

<u>Lemma 3. [18] If $\mathcal{U}$ is a P-point in the ground model, then $\mathcal{U}$ generates an ultrafilter, in fact a P-point, in the superperfect forcing extension.</u>

<u>Proof</u>. "In fact a P-point" follows from the rest of the lemma because the forcing is proper; any countable subset of $\mathcal{U}$ in the extension is included in a countable subset of $\mathcal{U}$ in the ground model, to which we can apply the assumption that $\mathcal{U}$ is a P-point in the ground model. So we concentrate on proving that $\mathcal{U}$ generates an ultrafilter in the extension. So suppose p forces "$A \subseteq \omega$". We seek an extension p' of p and a set $B \in \mathcal{U}$ (in the ground model) such that p' forces either "$B \subseteq A$" or "$B \subseteq \omega - A$".

Successively extend p (i.e., prune the tree) to arrange that:

(1) If a is an immediate successor of an infinitely branching node and if $j \leq \max(a)$, then the subtree of nodes comparable with a decides whether $j \in A$. Call this decision a's <u>opinion</u> about j.

(2) If a is an infinitely branching node, then, for each $j \in \omega$, all but finitely many immediate successors of a have the same opinion about j. Set

$$A_a = \{j \in \omega \mid \text{"} j \in A \text{" is the opinion of almost all immediate successors of } a\}.$$

(3) All or none of the sets A_a are in $\mathcal{U}$; this uses Lemma 2. Replacing A by $\omega - A$ if necessary, we may assume that all A_a are in $\mathcal{U}$.

(4) p has interval structure.

As $\mathcal{U}$ is a P-point, we can fix a set $B \in \mathcal{U}$ that is almost included in A_a for every branching node a of p. Merge intervals, i.e., further extend p, to arrange that the interval structure, say $[0,i_0),[i_0,i_1),\cdots,$ satisfies:

(5) If $a \subseteq i_k$ is a branching node, then $B - A_a \subseteq i_{k+1}$. (Remember that $B - A_a$ is finite, so we just need to take i_{k+1} large enough relative to i_k.)

(6) If $a \subseteq i_k$ is a branching node and $a \cup \{n\}$ is an immediate successor of it with $n > i_{k+1}$, and if $j \in A_a \cap i_k$, then $a \cup \{n\}$ has the opinion that $j \in A$. (Again, just take i_{k+1} large enough relative to i_k.)

(7) The first branching node is $\subseteq i_0$.

Since $\mathcal{U}$ is an ultrafilter, we can shrink B to a set, still in $\mathcal{U}$ and still to be called B, that meets at most every fourth interval (of the interval structure) and has no members $< i_2$. Since B meets at most every fourth interval, we can extend p to a condition p' with the same trunk as p, and whose nodes minus the trunk are disjoint from the intervals meeting B and from the adjacent intervals. Replace B by B minus the trunk of p'; of course the new B is still in $\mathcal{U}$.

We complete the proof by showing that p' forces $B \subseteq A$. If not, then there would exist $j \in B$ and an extension q of p' forcing "$j \notin A$". Let $[i_k, i_{k+1})$ be the interval containing j. As $j \in B$, all nodes of p' are disjoint from $[i_{k-1}, i_{k+2})$. Extending q if necessary, we may assume that its trunk b has $\max(b) \geq i_{k+2}$. Let a be the last branching predecessor of b in p' with $\max(a) < i_{k+2}$. (This exists by (7).) As a is disjoint from $[i_{k-1}, i_{k+2})$, we have $\max(a) < i_{k-1}$, and b is a successor of $a \cup \{n\}$ for some $n \geq i_{k+2}$ (because of the "last" in the choice of a, and the definition of interval structure). By (5), $B - A_a \subseteq i_k$. But $j \in B$ and $j \geq i_k$, so $j \in A_a$. Thus, $j \in A_a \cap i_{k+1}$, and (6) tells us that $a \cup \{n\}$ has (in p) the opinion that $j \in A$. That is, the subtree of p consisting of the nodes comparable with $a \cup \{n\}$ forces $j \in A$. But q is an extension of this condition and forces $j \notin A$. This contradiction completes the proof of the lemma. □

We now turn to the iteration of superperfect forcing. Over a model of CH, we iterate superperfect forcing $\aleph_2$ times with countable support. Because superperfect forcing is proper, Shelah's work implies

<u>Lemma 4</u> [23, III.4.1, see also 9, p. 235]. <u>The iteration satisfies the $\aleph_2$-antichain condition.</u> □

<u>Lemma 5</u>. <u>Cardinals and cofinalities are preserved by the iteration.</u> □

<u>Lemma 6</u> [23, V.4.4]. <u>Every real in the iteration occurs first at a stage of cofinality $\leq \omega$.</u> □

<u>Lemma 7</u>. <u>Every P-point in the ground model generates an ultrafilter, in fact a P-point, in the iterated forcing extension.</u>

<u>Proof</u>. Use Lemma 3 and [9, Section 4]. □

We are now in a position to prove the consistency of $u < g$.

Theorem 4 [7]. <u>If superperfect forcing is iterated</u> $\aleph_2$ <u>times over a model of</u> CH, <u>then the resulting model satisfies</u> $u = \aleph_1$ <u>and</u> $g = \aleph_2$.

<u>Proof</u>. By Lemma 7 and the theorem [22] that CH implies the existence of P-points, we immediately have $u = \aleph_1$.

For the rest of the proof, let us write M_α for the model obtained after the first α stages of the iteration. It follows from properness [23,III.3.2 and III.4.1; see also 9, pp. 236-237] that every set $\mathfrak{X}$ of reals in the final model $M_{\aleph_2}$ has $\mathfrak{X} \cap M_\alpha \in M_\alpha$ for an $\aleph_1$-closed unbounded set of $\alpha < \aleph_2$. If $\mathcal{G}$ is a groupwise dense family in $M_{\aleph_2}$, then, for an $\aleph_1$-closed unbounded set of α's, M_α not only has $\mathcal{G} \cap M_\alpha$ as a member but satisfies "$\mathcal{G} \cap M_\alpha$ is groupwise dense". If we are given $\aleph_1$ groupwise dense families in $M_{\aleph_2}$, then, as any $\aleph_1$ $\aleph_1$-closed unbounded sets in $\aleph_2$ intersect, there is an $\alpha < \aleph_2$ such that each of the given families restricts to a groupwise dense family in M_α. But then, by Lemma 1, the next generic real W_α (adjoined to M_α to form $M_{\alpha+1}$) has supersets in each of these restrictions, hence belongs to all of the given families. Therefore, $g > \aleph_1$. Since $g \le c = \aleph_2$, we have $g = \aleph_2$. □

4. GROWTH TYPES

Let $\mathbb{N}{\nearrow}\mathbb{N}$ be the set of non-decreasing functions from $\mathbb{N} = \omega-\{0\}$ to itself. By an <u>ideal</u>, we shall mean a subset of $\mathbb{N}{\nearrow}\mathbb{N}$ that is closed downward and closed under binary (pointwise) maxima. A growth type is a nonempty ideal closed under doubling (hence also closed under addition). Ideals, and therefore growth types, are pre-ordered by the relation $I \le J$ defined by

$$(\exists\, r \in \mathbb{N}{\nearrow}\mathbb{N})(\forall\, f \in I)(\exists\, g \in J)(\forall \text{ sufficiently large } n) f(n) \le g(r(n)).$$

The associated equivalence relation, $I \le J \le I$, is written $I \equiv J$. A partial description of the preordered set of ideals is given by the following diagram, in which underlining is used to indicate ideals equivalent to growth types.

$\mathbb{N}\nearrow\mathbb{N}$

? ? ? ?

$\underline{L} = \{f \in \mathbb{N}\nearrow\mathbb{N} \mid (\forall n)\ f(n) \le n\}$

|

$\underline{B} = \bigcup_{k\in\omega} B_k$

|
⋮
|

$B_k = \{f \in \mathbb{N}\nearrow\mathbb{N} \mid (\forall n)\ f(n) \le k\}$

|
⋮
|

$B_1 = \{1\}$

|

$B_0 = \emptyset$

The only ideals that are not $\ge L$ are those indicated below L in the diagram. L is equivalent to $\{f \in \mathbb{N}\nearrow\mathbb{N} \mid (\forall n)\ f(n) \le g(n)\}$ for any fixed unbounded $g \in \mathbb{N}\nearrow\mathbb{N}$. More generally, L is equivalent to any ideal that contains an unbounded function but is not an unbounded family. Every Borel ideal is equivalent to one of those indicated in the diagram.

The preordering of growth types was introduced by Göbel and Wald [14] in their study of slenderness classes of abelian groups. The connection with group theory is as follows. If I is a growth type, then the functions $f:\omega \longrightarrow \mathbb{Z}$ such that I contains the function

$$n \mapsto \max_{k\le n} |f(k)|$$

constitute a subgroup $[I]$ of the additive group $\mathbb{Z}^\omega$. Such subgroups are called <u>monotone</u>; they clearly contain the standard unit vectors $e_i \in \mathbb{Z}^\omega$, where e_i has all components equal to 0 except $e_i(i) = 1$. A group G is $[I]$-<u>slender</u> if every homomorphism $[I] \longrightarrow G$ sends all but finitely many e_i to 0. Göbel and Wald proved [14] that every $[I]$-slender group is $[J]$-slender if and only if $I \le J$. They also proved [15] that there are at least four inequivalent growth types and therefore at least four slenderness classes. Specifically, they showed that, if $\mathcal{U}$ is an ultrafilter, then the ideal

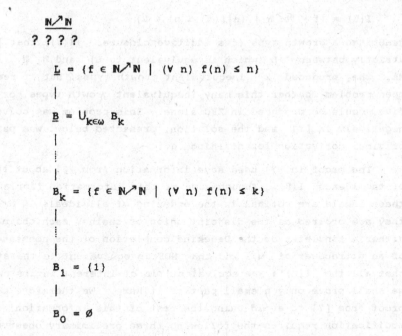

$$I(\mathcal{U}) = \{f \in \mathbb{N}\nearrow\mathbb{N} \mid \{n|f(n) \le n\} \in \mathcal{U}\}$$

generates a growth type (its additive closure, $+I(\mathcal{U})$) that lies strictly between $+L$ (which is equivalent to L) and $\mathbb{N}\nearrow\mathbb{N}$. Assuming MA, they produced 2^c inequivalent growth types, but it remained an open problem whether this many inequivalent growth types, or even just five, could be produced in ZFC alone. This problem was solved negatively in [7], and the solution, presented below, was part of the original motivation for defining g.

The proof in [7] used some information from [6] about the ordering of the ideals $I(\mathcal{U})$. Though some of this information (for example that these ideals are cofinal in the ordering of all ideals $< \mathbb{N}\nearrow\mathbb{N}$, that they are ordered as the disjoint union of chains, each chain being either a singleton or the Dedekind completion of the nonstandard part of an ultrapower of ω, and that NCF is equivalent to the statement that all the $I(\mathcal{U})$'s are equivalent) is of interest in its own right, we shall prove only a small part of it here. We therefore modify the proof from [7] to avoid using the rest of this information; the modification requires the following three preliminary observations.

First, the functions $\text{next}(X,-)$ for $X \in \mathcal{U}$ (or X in any base for $\mathcal{U}$) generate $I(\mathcal{U})$. Indeed, $\text{next}(X,n) \le n$ for all $n \in X$ so $\text{next}(X,-) \in I(\mathcal{U})$, and if $f \in \mathbb{N}\nearrow\mathbb{N}$ and $f(n) \le n$ for all $n \in X$ then $f \le \text{next}(X,-)$.

Second, if $f \in \mathbb{N}\nearrow\mathbb{N}$ and $f(n) \ge n$ for all n, then $I(\mathcal{U}) \le I(f(\mathcal{U}))$ for any ultrafiltr $\mathcal{U}$. Indeed, the function r required by the definition of $\le$ can be taken to be $r(n) = f(n) + 1$ because, if $X \in \mathcal{U}$, then

$$\text{next}(X,n) \le f(\text{next}(X,n)) \le \text{next}(f(X),f(n) + 1)$$

so $\text{next}(X,-)$ is majorized by $\text{next}(f(X),-) \circ r$.

Third, every ideal $J < \mathbb{N}\nearrow\mathbb{N}$ is $\le I(\mathcal{U})$ for some ultrafilter $\mathcal{U}$. Indeed, if $f \in (\mathbb{N}\nearrow\mathbb{N}) - J$ then the sets $\{n|f(n) > g(n)\}$ for $g \in J$ have the strong finite intersection property (i.e., intersections of finitely many are infinite), so there is an ultrafilter $\mathcal{V}$ containing them all. Then $J \le I(f(\mathcal{V}))$, as witnessed by $r(n) = f(n) + 1$, because, if $g \in J$, then $X = \{n \mid f(n) > g(n)\} \in \mathcal{V}$, so $f(X) \in f(\mathcal{V})$ and

$$g(n) \le g(\text{next}(X,n)) \le f(\text{next}(X,n)) \le \text{next}(f(X), f(n) + 1),$$

so g is majorized by $\text{next}(f(X),-) \circ r$.

<u>Theorem 5</u> [7]. <u>If</u> $\mathfrak{u} < \mathfrak{g}$, <u>then all ideals strictly between</u> L <u>and</u> $\mathbb{N}\nearrow\mathbb{N}$ <u>are equivalent.</u>

<u>Proof</u>. Fix an ultrafilter $\mathcal{V}$ with a base $\mathcal{B}$ of cardinality $< \mathfrak{g}$. We show first that, if J is any ideal, then $J \le L$ or $J \ge I(\mathcal{V})$. For each $B \in \mathcal{B}$, let

$$\mathcal{G}_B = \{X \in [\omega]^\omega \mid (\exists f \in J)(\forall \text{ sufficiently large } z) \quad \text{next}(B,z) \le f(\text{next}(X,z))\}.$$

If all the families $\mathcal{G}_B$ are groupwise dense, then, since there are fewer than $\mathfrak{g}$ of them, they have a common member X. This means that, for each $B \in \mathcal{B}$, there is $f \in J$ such that, for all sufficiently large z, $\text{next}(B,z) \le f(\text{next}(X,z))$; in other words, $\text{next}(B,-)$ is eventually majorized by $f \circ r$, where $r = \text{next}(X,-)$. Thus, $I(\mathcal{V}) \le J$, as desired.

Suppose, therefore, that $\mathcal{G}_B$ fails to be groupwise dense for a certain $B \in \mathcal{B}$. Since $\mathcal{G}_B$ is clearly closed under subsets and finite modifications, there must be a partition of ω into intervals $[a_i, a_{i+1})$, no infinite union of which is in $\mathcal{G}_B$. Merging intervals if necessary, we assume that each $[a_i, a_{i+1})$ meets B. We shall show that J is dominated by the function which, on each $[a_i, a_{i+1})$, is constant with value a_{i+3}; this means that $J \le L$, as desired. Suppose, for a contradiction, that some $f \in J$ were not so dominated. Then

$$X = \bigcup \{[a_{i+1}, a_{i+2}) \mid \text{For some } n \in [a_i, a_{i+1}),\ f(n) > a_{i+3}\},$$

an infinite union of our intervals, is nevertheless in $\mathcal{G}_B$, with witness f, contrary to our choice of the intervals. Indeed, for any z, say in $[a_j, a_{j+1})$, let $x = \text{next}(X,z)$. So $x \in [a_{i+1}, a_{i+2})$ for some i with $j \le i+1$, and there is (by definition of X) $n \in [a_i, a_{i+1})$ with $f(n) > a_{i+3}$. Then, since B meets each of our intervals,

$$\text{next}(B,z) < a_{j+2} \le a_{i+3} < f(n) \le f(x) = f(\text{next}(X,z)),$$

so $X \in \mathcal{G}_B$, as claimed.

This completes the proof that $J \le L$ or $J \ge I(\mathcal{V})$, for all ideals J and for any ultrafilter $\mathcal{V}$ generated by fewer than $\mathfrak{g}$ sets.

Now suppose $J \le L$ and $J < \mathbb{N}\nearrow\mathbb{N}$. Let $\mathcal{V}$ still be an ultrafilter generated by fewer than $\mathfrak{g}$ sets, and, by the third preliminary remark, let $\mathcal{U}$ be an ultrafilter with $J \le I(\mathcal{U})$. Since

$u < g$, we have FD by Theorem 3 and therefore NCF, as pointed out in Section 1. So let $f:\omega \longrightarrow \omega$ be a finite-to-one non-decreasing function with $f(\mathcal{U}) = f(\mathcal{V})$. Replacing f with $g \circ f$ for a suitable one-to-one g, we may assume that $f(n) \geq n$ for all n. Then we have

$$
\begin{aligned}
J &\leq I(\mathcal{U}) && \text{by our choice of } \mathcal{U} \\
&\leq I(f(\mathcal{U})) && \text{by the second preliminary remark} \\
&= I(f(\mathcal{V})) && \text{as } f(\mathcal{U}) = f(\mathcal{V}) \\
&\leq I(\mathcal{V}) && \text{as } f(\mathcal{V}) \text{ is generated by fewer than } g \text{ sets} \\
&\leq J && \text{as } \mathcal{V} \text{ is generated by fewer than } g \text{ sets.}
\end{aligned}
$$

(At the last two steps, we have invoked the first part of the proof and the fact that neither $I(\mathcal{V})$ nor J is $\leq L$.) Thus, $J \equiv I(\mathcal{V})$. Since $\mathcal{V}$ didn't depend on J, this shows that all such ideals J are equivalent. □

It is shown in [7], by considering ideals $I(\mathcal{F})$ (defined like $I(\mathcal{U})$ but for filters $\mathcal{F}$ that are not ultrafilters) or their additive closures, that the conclusion of Theorem 5, or the immediate consequence that there are only four inequivalent growth types, implies the filter dichotomy principle. (Thus, Theorem 5 subsumes Theorem 3, but Theorem 3 was used in the proof of Theorem 5.) For all the implications in the chain

$u < g \Longrightarrow$ All ideals strictly between L and $\mathbb{N}^{\nearrow}\mathbb{N}$ are equivalent

$\Longrightarrow$ Only four inequivalent growth types

$\Longrightarrow$ FD

$\Longrightarrow$ NCF,

the converses are open problems.

5. ANOTHER NOTION OF FORCING

The following notion of forcing is designed to adjoin, as straightforwardly as possible, a subset W of ω that has supersets in all groupwise dense families in the ground model. A condition is a pair (a,X) where a is a finite subset of ω and X is an infinite family of pairwise disjoint finite subsets of ω disjoint from a. The "intended meaning" of (a,X) is that $a \subseteq W$ and $W - a$ is a union of members of X. Accordingly, an extension of (a,X) is a condition (b,Y) such that $a \subseteq b$ and such that $b - a$ and all members of Y are unions of members of X. The set W corresponding to a generic set of conditions is the union of the first components of the conditions in the generic set; it is easy to check that (a,X) forces "$a \subseteq W$ and $W - a$ is a union of members of X", as intended.

It is easy to see that every condition can be extended to one of the special form $(a, \{x_k | k \in \omega\})$, where $\max(a) < \min(x_0)$ and $\max(x_k) < \min(x_{k+1})$ for all k. So we may, at our convenience, confine attention to such <u>normalized</u> conditions.

This forcing can be decomposed as a two-step iteration. The first step is to force with infinite families of pairwise disjoint finite subsets of ω, where Y is an extension of X if every member of Y is a union of members of X. It was shown in [4] that this forcing adjoints a special sort of ultrafilter $\mathcal{U}$ (there called a stable ordered-union ultrafilter) on the set of finite nonempty subsets of ω. If G is a generic set, then the closures, under finite unions, of the members of G constitute a base for $\mathcal{U}$.

The second step is to force with conditions (a,X) as above, except that the closure of X under finite unions is required to belong to $\mathcal{U}$.

Since both steps of the iteration were, as far as I know, first considered by Matet [17], I shall call their composite, the forcing now under consideration, <u>Matet forcing</u>.

The first of the two forcing notions (or rather its separative quotient) is countably closed. The second satisfies the countable antichain condition; in fact, it is σ-centered since conditions with the same first component are compatible. It follows that Matet forcing is proper.

If $\mathcal{G}$ is a groupwise dense family in the ground model and (a,X) is any condition, then X has an infinite subset Y whose union is in $\mathcal{G}$. Then (a,Y) is an extension of (a,X) forcing "$W - a \subseteq \bigcup Y \in \mathcal{G}$" and therefore, since $\mathcal{G}$ is closed under finite modifications, "W has a superset in $\mathcal{G}$". It follows, just as for superperfect forcing, that an $\aleph_2$-step countable-support iteration of Matet forcing over a model of GCH produces a model where $\mathfrak{g} = \aleph_2$. Laflamme and I have (independently) verified that a P-point in the ground model generates an ultrafilter after (one-step) Matet forcing and therefore also after iterated Matet forcing. We thus have another model for $\mathfrak{u} < \mathfrak{g}$.

The similarity between Matet forcing and Miller's superperfect forcing can be clarified by considering the anthropomorphic description of superperfect trees with interval structure in Section 3. If that description is modified so that the jth interval has only one specified subset (independent of the previous decisions) rather than 2^{j-1}, the result is equivalent to Matet forcing. A normalized Matet condition can be viewed as a superperfect tree in which the branches from any one infinitely branching node are essentially the same as those from any other. The tree corresponding to (a,X) has as nodes the initial segments of sets of the form $a \cup y$, where y is a finite union of members of X; the sets $a \cup y$ themselves are the branching nodes. The situation can be roughly but succinctly summarized by the proportion

$$\frac{\text{Matet forcing}}{\text{Miller forcing}} = \frac{\text{Mathias forcing}}{\text{Laver forcing}}.$$

Notice that the two-step decomposition of Matet forcing described above is analogous to a well-known decomposition of Mathias forcing, namely first generically adjoin a (selective) ultrafilter on ω, and then shoot a real through it.

APPENDIX: OTHER CARDINAL INVARIANTS

We define here the standard cardinal invariants of the continuum, other than those already defined in Section 2, and we extend the Hasse diagram of provable inequalities in Section 2 to include these additional invariants.

$\mathfrak{a}$ is the smallest cardinality of any infinite maximal almost disjoint family of infinite subsets of ω; here "almost disjoint" means that every two sets in the family have a finite intersection.

$\mathfrak{i}$ is the smallest cardinality of any maximal independent family of subsets of ω; here "independent" means that any finitely many sets from the family and the complements of any finitely many other sets from the family have an infinite intersection.

$\mathfrak{s}$, the <u>splitting number</u>, is the smallest cardinality of any splitting family $\mathcal{S} \subseteq \mathcal{P}(\omega)$; here "splitting" means that, for every infinite $A \subseteq \omega$, there is $S \in \mathcal{S}$ such that both $A \cap S$ and $A - S$ are infinite.

$\mathfrak{r}$ is the smallest cardinality of any unsplittable family $\mathcal{R} \subseteq [\omega]^{\omega}$; here "unsplittable" means that, for every $A \subseteq \omega$, there is $R \in \mathcal{R}$ such that $R \cap A$ and $R - A$ are not both infinite.

$\mathfrak{p}$ is the smallest cardinality of any family $\mathfrak{B} \subseteq \mathcal{P}(\omega)$ such that every finite subfamily has an infinite intersection but there is no set A such that $A - B$ is finite for all $B \in \mathfrak{B}$.

$\mathfrak{t}$ is defined like $\mathfrak{p}$ except that $\mathfrak{B}$ must be linearly ordered by inclusion modulo finite sets.

These cardinals and the ones defined in Section 2 satisfy (provably in ZFC) the inequalities indicated in the following Hasse diagram. I thank Peter Nyikos for pointing out the inequality $\mathfrak{r} \leq \mathfrak{i}$, which was missing from the diagram at STACY.

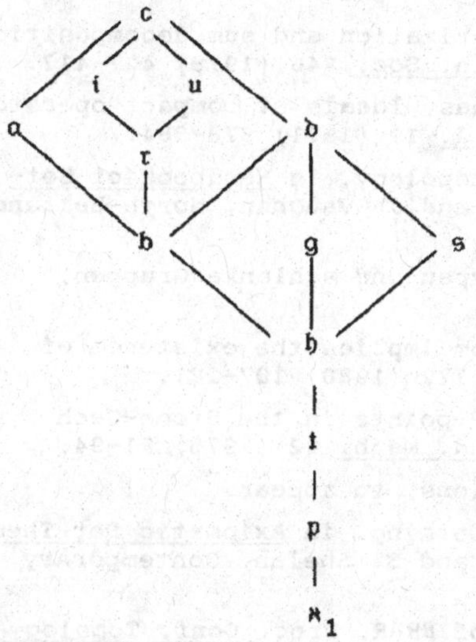

REFERENCES

1. B. Balcar, J. Pelant, and P. Simon, The space of ultrafilters on N covered by nowhere dense sets, Fund. Math. 110 (1980) 11-24.
2. J. Baumgartner, Iterated forcing, in Surveys in Set Theory, ed. A.R.D. Mathias, London Math. Soc. Lecture Notes 87, 1983, pp. 1-59.
3. D. Bellamy, A non-metric indecomposable continuum, Duke Math. J. 38 (1971) 15-20.
4. A. Blass, Ultrafilters related to Hindman's finite unions theorem and its extensions, in Logic and Combinatorics, ed. S. Simpson, Contemporary Mathematics 65 (1987) 89-124.
5. ________, Near coherence of filters, I: Cofinal equivalence of models of arithmetic, Notre Dame J. Formal Logic 27 (1986) 579-591.

6. ________, Near coherence of filters, II: Applications to operator ideals, the Stone-Cech remainder of a half-line, order ideals of sequences, and slenderness of groups, Trans. Amer. Math. Soc. 300 (1987) 557-581.

7. ________ and C. Laflamme, Consistency results about filters and the number of inequivalent growth types, to appear in J. Symbolic Logic.

8. ________ and S. Shelah, Ultrafilters with small generating sets, to appear.

9. ________, ________, There may be simple $P_{\aleph_1}$ and $P_{\aleph_2}$ points and the Rudin-Keisler order may be downward directed, Ann. Pure Appl. Logic 83 (1987) 213-243.

10. ________, ________, Near coherence of filters, III: A simplified consistency proof, to appear.

11. ________ and G. Weiss, A characterization and sum decomposition of operator ideals, Trans. Amer. Math. Soc. 246 (1978) 407-417.

12. A. Brown, C. Pearcy, and N. Salinas, Ideals of compact operators on Hilbert space, Michigan Math. J. 19 (1971) 373-384.

13. E. van Douwen, The integers and topology, in Handbook of Set-Theoretic Topology, ed. K. Kunen and J. Vaughan, North-Holland, 1984, pp. 111-167.

14. R. Göbel and B. Wald, Wachstumstypen und schlanke Gruppen, Symposia Math. 23 (1979) 201-239.

15. ________, ________, Martin's axiom implies the existence of certain slender groups, Math. Z. 172 (1980) 107-121.

16. J. Ketonen, On the existence of P-points in the Stone-Čech compactification of integers, Fund. Math. 92 (1976) 91-94.

17. P. Matet, Some filters of partitions, to appear.

18. A. Miller, Rational perfect set forcing, in Axiomatic Set Theory, ed. J. Baumgartner, D.A. Martin, and S. Shelah, Contemporary Mathematics 31 (1964) 143-159.

19. J. Midoduszewski, On composants of βR-R, Proc. Conf. Topology and Measure, I (Zinnowitz), ed. J. Flachsmeyer, Z. Frolík, and F. Terpe, Ernst-Moritz-Arndt-Universität zu Greifwald, 1978, 257-283.

20. ______________, An approach to $\beta R\setminus R$, in Topology, ed. Á. Császár, Colloq. Math. Soc. János Bolyai 23 (1980) 853-854.

21. M.E. Rudin, Composants and βN, Proc. Washington State Univ. Conf. General Toplogy 1970, pp. 117-119.

22. W. Rudin, Homogeneity problems in the theory of Cech compactifications, Duke Math. J. 23 (1956) 409-419.

23. S. Shelah, Proper Forcing, Lecture Notes in Mathematics 940, Springer-Verlag 1982.

24. R.C. Solomon, Families of sets and functions, Czechoslovak Math. J. 27 (1977) 556-559.

25. E. Specker, Additive Gruppen von Folgen ganzer Zahlen, Portugal. Math. 9 (1950) 131-140.

TOWARDS A STRUCTURE THEORY FOR IDEALS ON $P_\kappa\lambda$

DONNA M. CARR AND DONALD H. PELLETIER[1]

Introduction

In their now classic paper, Baumgartner, Taylor and Wagon [BTW] extended much of the theory of κ-complete *ultra*filters over an uncountable cardinal κ to the context of κ-complete filters over κ.

Jech [Je1] generalized many combinatorial features of the structure $(\kappa,<)$ to the structure $(P_\kappa\lambda,\subset)$ where for any uncountable cardinals κ and λ with κ regular and $\kappa \le \lambda$, $P_\kappa\lambda$ denotes the set $\{x \subseteq \lambda : |x| < \kappa\}$. In particular, he extended the notions of closed, unbounded and stationary sets to this context, and proved (among other things) that if κ is λ-supercompact, then every λ-supercompact *ultra*filter over $P_\kappa\lambda$ extends the cub filter.

There is now a fully developed theory of *ultra*filters over $P_\kappa\lambda$. As yet however, very few parts of this theory have been extended to the context of κ-complete filters over $P_\kappa\lambda$. Several people have worked on this.

One of the first successes in this program was a result of Carr [C1]. She proved that every normal *filter* on $P_\kappa\lambda$ extends the cub filter thereby generalizing familiar results of Neumer [N] and Fodor [F] to the context of $P_\kappa\lambda$.

Because $(\kappa,<)$ is a total ordering but $(P_\kappa\lambda,\subset)$ is not, certain combinatorial features of the former do not seem to generalize nicely to the latter. These include some partition-theoretic results, and various selectivity properties of ideals.

One approach to this, the one adopted by Zwicker [Z1],[Z2] and Shelah [S1],[S2], is to relativize everything to certain "nice" substructures $(A,\subset)$ of $(P_\kappa\lambda,\subset)$, e.g. where A is a stationary coding set.

We adopt a different approach here. We study the structure $(P_\kappa\lambda,<)$ where the binary relation $<$ is defined in $P_\kappa\lambda$ by $x < y$ iff

---

[1] We wish to thank the Natural Sciences and Engineering Research Council (NSERC) of Canada for partial support, and the referee for some useful suggestions.

$$x \subset y \wedge |x| < |y \cap \kappa|$$

and show that in some respects, $(P_\kappa\lambda,<)$ is a better analogue of $(\kappa,<)$ than is $(P_\kappa\lambda,\subset)$.

In particular, we prove that if κ is λ-supercompact, then *every* λ-supercompact ultrafilter over $P_\kappa\lambda$ is strongly normal (see 1.3, 1.7), and has a certain partition property (1.8, 1.10) thereby generalizing a familiar result of Rowbottom [R] to the context of $(P_\kappa\lambda,<)$. As well, we generalize some results of Weglorz [W] concerning selective, Ramsey and quasinormal ideals to this context (2.3 - 2.8). Finally, in section 3, we shall prove that for strongly normal ideals, the partition property introduced in 1.8 is equivalent to a certain distributivity property.

The partial ordering $<$ defined above has appeared sporadically in the literature - e.g. see [Ma], [Me], [SRK], [DM]. But it had not been examined systematically until Carr introduced it independently in [C2] and obtained some partition-theoretic properties of $(P_\kappa\lambda,<)$ under various large cardinal assumptions on κ.

We conclude this section by detailing the basic concepts and notation we use in the sequel.

The basic combinatorial notions are defined here for $(P_\kappa\lambda,\subset)$ as in Jech [Je1]. For each $x \in P_\kappa\lambda$, $\hat{x}$ denotes the set $\{y \in P_\kappa\lambda : x \subset y\}$. $X \subseteq P_\kappa\lambda$ is said to be *unbounded* iff $(\forall x \in P_\kappa\lambda)(X \cap \hat{x} \neq 0)$, and $I_{\kappa\lambda}$ denotes the *ideal of not unbounded subsets of* $P_\kappa\lambda$. In the sequel, an *ideal on* $P_\kappa\lambda$ is always a proper, non-principal, κ-complete ideal on $P_\kappa\lambda$ extending $I_{\kappa\lambda}$ unless we specify otherwise. Further, for any ideal I on $P_\kappa\lambda$, I^+ denotes the set $\{X \subseteq P_\kappa\lambda : X \notin I\}$, and I^* the filter dual to I; $FSF_{\kappa\lambda}$ denotes $I^*_{\kappa\lambda}$.

$C \subseteq P_\kappa\lambda$ is said to be *closed* iff it is closed under unions of $\subseteq$-chains of length $< \kappa$, and is called a *cub* iff it is both closed and unbounded. Further, $S \subseteq P_\kappa\lambda$ is said to be *stationary in* $P_\kappa\lambda$ iff $S \cap C \neq 0$ for every cub $C \subseteq P_\kappa\lambda$. Finally, $NS_{\kappa\lambda}$ denotes the non-stationary ideal on $P_\kappa\lambda$, and $CF_{\kappa\lambda}$ its dual.

An ideal I on $P_\kappa\lambda$ is said to be *normal* iff every function $f : P_\kappa\lambda \longrightarrow \lambda$ that is regressive on a set in I^+ (i.e. has the property that $\{x \in P_\kappa\lambda : f(x) \in x\} \in I^+$) is constant on a set in I^+. A filter F on $P_\kappa\lambda$ is said to be normal iff its dual ideal $\{X \subseteq P_\kappa\lambda : P_\kappa\lambda - X \in F\}$

is normal. Jech [Je1] proved that $NS_{\kappa\lambda}$ is normal, and Carr proved in [C1] that it is the smallest normal ideal extending $I_{\kappa\lambda}$.

1. Strongly normal ideals on $P_\kappa\lambda$

Let κ be an uncountable regular cardinal, and λ a cardinal $\geq \kappa$. Assume that κ is weakly inaccessible and hence that $\{x \in P_\kappa\lambda : x \cap \kappa \text{ is a cardinal}\}$ is cub in $P_\kappa\lambda$.

1.1 Definition. For every $x, y \in P_\kappa\lambda$, write $x < y$ iff $0 \neq x \subset y \wedge |x| < |y \cap \kappa|$, and for each $x \in P_\kappa\lambda$ let $\tilde{x}$ denote the set $\{y \in P_\kappa\lambda : x < y\}$ and κ_x the cardinal $|x \cap \kappa|$.

1.2 Remark. It is easy to see that $\tilde{x}$ is a cub for every $x \in P_\kappa\lambda$ and hence that $\{\tilde{x} : x \in P_\kappa\lambda\}$ generates a κ-complete filter over $P_\kappa\lambda$. Moreover, this filter is just $FSF_{\kappa\lambda}$.

1.3 Definition For every $X \subseteq P_\kappa\lambda$, a function $f : X \longrightarrow P_\kappa\lambda$ is said to be *regressive w.r.t.* $<$ or *$<$-regressive* iff $(\forall x \in X)(f(x) < x)$.

Further, an ideal I on $P_\kappa\lambda$ is said to be *normal w.r.t.* $<$ or *strongly normal* iff every $f : P_\kappa\lambda \longrightarrow P_\kappa\lambda$ that is $<$-regressive on a set in I^+ is constant on a set in I^+.

Note that the term *strongly normal* has been used elsewhere for different notions - e.g. see [Je2], p 412.

In [C2], Carr proved that *if $\lambda^{<\kappa} = \lambda$ and κ is λ-ineffable or almost λ-ineffable or has the λ-Shelah property, then the associated ideal is strongly normal*, and used these facts to obtain some partition-theoretic results in [C2], and a reflection property in [C3].

1.4 Remark. It is easy to see that *every strongly normal ideal on $P_\kappa\lambda$ is normal.* An observation of Zwicker shows that this does not reverse. In fact, *$NS_{\kappa\lambda}$ is not strongly normal*: Notice that $A = \{x \in P_\kappa\lambda : x \cap \kappa \text{ is a singular cardinal and } cf(x \cap \kappa) = \omega\}$ is stationary in $P_\kappa\lambda$. Furthermore, the function $f : A \longrightarrow P_\kappa\lambda$ defined by $f(x)$ = a cofinal subset of $x \cap \kappa$ is $<$-regressive on A. Notice that for every $y \in P_\kappa\lambda$ and every $x \in A \cap f^{-1}(\{y\})$, $x \cap \kappa = \bigcup y$. Thus $A \cap f^{-1}(\{y\}) \in I_{\kappa\lambda} \subset NS_{\kappa\lambda}$, so f is constant on *no* stationary subset of $P_\kappa\lambda$.

The above argument can also be used to show that *the ideal dual to DiPrisco's weak cub filter is not strongly normal.* Recall that DiPrisco [D] (also see [DM]) defined $X \subseteq P_\kappa\lambda$ to be *weakly closed* iff for every $\delta < \kappa$ and every $x_0 \subset \ldots \subset x_\xi \subset \ldots$ $(\xi < \delta)$ from X such that $|x_0| < \ldots < |x_\xi| < \ldots$ $(\xi < \delta)$, $\bigcup\{x_\xi : \xi < \kappa\} \in X$, and proved that the set of all unbounded and weakly closed subsets of $P_\kappa\lambda$ generates a κ-complete normal filter over $P_\kappa\lambda$ properly extending $CF_{\kappa\lambda}$. This filter is called the *weak cub filter*.

The characterization given in 1.6 below of our notion of strong normality will be useful in the sequel. We need the following definition.

1.5 Definition. The *< - diagonal union* $\nabla_<\{X_a : a \in P_\kappa\lambda\}$ and *< - diagonal intersection* $\Delta_<\{X_a : a \in P_\kappa\lambda\}$ of a $P_\kappa\lambda$-indexed family $\{X_a : a \in P_\kappa\lambda\}$ of subsets of $P_\kappa\lambda$ are defined by

$\nabla_<\{X_a : a \in P_\kappa\lambda\} = \{x \in P_\kappa\lambda : (\exists a < x)(x \in X_a)\}$, and

$\Delta_<\{X_a : a \in P_\kappa\lambda\} = \{x \in P_\kappa\lambda : (\forall a < x)(x \in X_a)\}$.

Routine arguments yield the following useful fact.

1.6. Proposition. *An ideal I on $P_\kappa\lambda$ is strongly normal iff $\nabla_<\{X_a : a \in P_\kappa\lambda\} \in I$ for every $P_\kappa\lambda$-indexed family $\{X_a : a \in P_\kappa\lambda\}$ of sets from I.* □

The next theorem that we wish to present here is that for λ-supercompact ultrafilters over $P_\kappa\lambda$, normality and strong normality are equivalent. This result has appeared previously in various guises - e.g. see [Ma], [Me], [SRK], [DM].

1.7 Theorem. *If κ is λ-supercompact, then every λ-supercompact ultrafilter over $P_\kappa\lambda$ is strongly normal.*

Proof. Assume that κ is λ-supercompact, and let U be a λ-supercompact ultrafilter over $P_\kappa\lambda$. Further, let M be the unique transitive isomorph of $Ult_U(V)$ and $j : V \longrightarrow M$ the canonical elementary embedding.

Pick $X \in U$ and a $<$ - regressive function $f : X \longrightarrow P_\kappa\lambda$. Then $(\forall x \in X)(f(x) < x)$, so $(\forall x \in X)(f(x) \in P_{\kappa_x} x)$. Thus $\{x \in P_\kappa\lambda : f(x) \in P_{\kappa_x} x\} \in U$, so $M \models [f]_U \in P_\kappa[id]_U$, so $M \models [f]_U \in P_\kappa j[\lambda]$. Let $z \in P_\kappa\lambda$ be such that $M \models [f]_U = j[z]$. Since $|z| < \kappa$, $j[z] = j(z)$. Thus $\{x \in P_\kappa\lambda : f(x) = z\} \in U$, so f is constant on a set in U. □

We are especially interested in this result because we can use it to obtain a $P_\kappa\lambda$ generalization of a familiar result of Rowbottom [R], namely that every normal measure ultrafilter over a measurable cardinal has the partition property. This is done in 1.10 below after we provide the notation we need, and some background.

1.8 Definitions. For any $X \subseteq P_\kappa\lambda$, $[X]^2_<$ denotes the set $\{(x,y) \in X \times X : x < y\}$.

For any $X \subseteq P_\kappa\lambda$ and any ideal I over $P_\kappa\lambda$, the symbol $X \underset{<}{\rightarrow} (I^+)^2$ means that for any partition $f : [X]^2_< \rightarrow 2$, there are an $H \subseteq X$ and an $i \in 2$ such that $H \in I^+$ and $f[[H]^2_<] = \{i\}$. Finally, the symbol $I^+ \underset{<}{\rightarrow} (I^+)^2$ means that $X \underset{<}{\rightarrow} (I^+)^2$ holds for every $X \in I^+$.

1.9 Remark. Recall that a λ-supercompact ultrafilter U over $P_\kappa\lambda$ is said to have the *partition property* iff for every $f : [P_\kappa\lambda]^2_\subset \rightarrow 2$ there is an $H \in U$ such that $|f[[H]^2_\subset]| = 1$.

Solovay (see [Me]) proved that if κ is supercompact, then for certain small cardinals $\lambda \geq \kappa$ (e.g. $\lambda = \kappa^+$ and $\lambda = 2^\kappa$) every λ-supercompact ultrafilter over $P_\kappa\lambda$ has the partition property, *but* that for certain large $\lambda > \kappa$ (e.g. for λ measurable), there is a λ-supercompact ultrafilter over $P_\kappa\lambda$ that does *not* have the partition property. Thus this property does not yield a nice $P_\kappa\lambda$ generalization of Rowbottom's result. However, our notion does:

1.10 Theorem. *If κ is λ-supercompact, then for every λ-supercompact ultrafilter U over $P_\kappa\lambda$ and every $X \in U$, $X \underset{<}{\rightarrow} (U)^2$.*

Proof. Assume that κ is λ-supercompact, and let U be a λ-supercompact ultrafilter over $P_\kappa\lambda$. Further, pick $X \in U$ and $f : [X]^2_< \rightarrow 2$. For each $x \in X$ and each $i \in 2$, set $A_{xi} = \{y \in X : f(x,y) = i\}$. Notice that for $x \in X$, either $A_{x0} \in U$ or else $A_{x1} \in U$. Moreover, either $\{x \in X : A_{x0} \in U\} \in U$ or else $\{x \in X : A_{x1} \in U\} \in U$. Assume w.l.o.g. that $B = \{x \in X : A_{x0} \in U\} \in U$, and then define $(X_a : a \in P_\kappa\lambda)$ by

$$X_a = \begin{cases} A_{a0} & \text{if } a \in B \\ P_\kappa\lambda & \text{otherwise.} \end{cases}$$

Now set $H = B \cap \Delta_< \{X_a : a \in P_\kappa\lambda\}$. It is easy to see that $f[[H]^2_<] = \{0\}$. □

In section 3 below, we shall see that the fact established in theorem

1.10 is a special case of a stronger result, namely that *for any strongly normal ideal* I *over* $P_\kappa\lambda$, $I^+ \to (I^+)^2_<$ *holds iff* I *is* $(\lambda^{<\kappa},2)$*-distributive* (theorem 3.2).

We conclude this section by showing that the assumption "I is a strongly normal ideal on $P_\kappa\lambda$" has certain consequences for κ, λ and I.

1.11 Theorem. *Let* I *be a strongly normal ideal over* $P_\kappa\lambda$. *Then*

(i) $\{x \in P_\kappa\lambda : x \cap \kappa$ *is a regular cardinal*$\} \in I^*$,

(ii) if κ *is inaccessible, then*

$$\{x \in P_\kappa\lambda : x \cap \kappa \text{ is an inaccessible cardinal}\} \in I^*,$$

(iii) if κ *is inaccessible, then* κ *is Mahlo, and*

(iv) if κ *is inaccessible and* $\lambda^{<\kappa} = \lambda$, *then for every bijection* $\varphi : P_\kappa\lambda \twoheadrightarrow \lambda$, $\{x \in P_\kappa\lambda : \varphi[P_{\kappa_x} x] = x\} \in I^*$; *hence* $\{x \in P_\kappa\lambda : |x|^{<|x \cap \kappa|} = |x|\} \in I^*$.

Proof. We shall just prove (iv). The proofs of (i)-(iii) are left to the reader.

Assume that κ is inaccessible and $\lambda^{<\kappa} = \lambda$, and let $\varphi : P_\kappa\lambda \twoheadrightarrow \lambda$ be a bijection. Routine arguments show that $\{x \in P_\kappa\lambda : (\forall\alpha \in x)(\varphi^{-1}(\alpha) < x)\} \in CF_{\kappa\lambda} \subseteq I^*$, so it remains to obtain $\{x \in P_\kappa\lambda : \varphi[P_{\kappa_x} x] \subseteq x\} \in I^*$. Suppose by way of contradiction that $\{x \in P_\kappa\lambda : \varphi[P_{\kappa_x} x] \subseteq x\} \notin I^*$. Then $X = \{x \in P_\kappa\lambda : \varphi[P_{\kappa_x} x] \subsetneq x\} = \{x \in P_\kappa\lambda : (\exists y < x)(\varphi(y) \notin x)\} \in I^+$. For each $x \in X$, pick $y_x < x$ such that $\varphi(y_x) \notin x$, and then use strong normality to obtain $y \in P_\kappa\lambda$ such that $Y = \{x \in X : y_x = y\} \in I^+$. *But* note that $(\forall x \in Y)(\varphi(y) \notin x)$ thereby contradicting $Y \in I^+ \subseteq I^+_{\kappa\lambda}$. □

Generic ultrapowers of a transitive model M of ZFC modulo an *I-generic* ultrafilter over $P^M_\kappa\lambda$ where I is (in M) a strongly normal, non-principal, κ-complete ideal over $P_\kappa\lambda$ will be studied in a sequel to this paper.

2. Selective, Ramsey and quasinormal ideals on $P_\kappa\lambda$

Recall that an ideal I over an uncountable regular cardinal κ is said to be

(i) a *p-point* iff for every I-small function $f : \kappa \to \kappa$ there is an $A \in I^*$ such that $(\forall\alpha \in \kappa)(|A \cap f^{-1}[\{\alpha\}]| < \kappa)$.

(ii) a *q-point* iff for every $f : \kappa \longrightarrow \kappa$ with the property that $(\forall\alpha \in \kappa)(|f^{-1}[\{\alpha\}]| < \kappa)$, there is an $A \in I^*$ on which f is one-one.
(iii) *selective* iff for every I-small function $f : \kappa \longrightarrow \kappa$ there is an $A \in I^*$ on which f is one-one, i.e. iff I is both a p-point and a q-point.
(iv) *Ramsey* iff for every $A \in (I \times I)^*$, where $I \times I = \{X \subseteq \kappa \times \kappa : \{\alpha \in \kappa : \{\beta \in \kappa : (\alpha,\beta) \in X\} \in I^+\} \in I\}$, there is an $H \in I^*$ such that $(H \times H) \cap \{(\alpha,\beta) \in \kappa \times \kappa : \alpha < \beta\} \subseteq A$.
(v) *quasinormal* iff for every κ-sequence $\langle X_\alpha : \alpha < \kappa\rangle \in {}^\kappa I$ there is an $A \in I^*$ such that

$$\nabla(X_\alpha : \alpha \in A) = \{\beta \in \kappa : (\exists\alpha \in \beta \cap A)(\beta \in X_\alpha)\} \in I.$$

Kunen (see [B]) and Weglorz [W] established the facts summarized in the following theorem.

2.1 Theorem *Let I be an ideal over an uncountable regular cardinal κ.*

(1) I is normal iff for every I-small function $f : \kappa \longrightarrow \kappa$, $\{\inf(f^{-1}[\{\alpha\}]) : \alpha \in \kappa\} \in I^*$.
(2) If I is normal, then I is selective.
(3) I is selective iff I is Ramsey iff I is quasinormal. □

Our first objective in this section is to provide a $P_\kappa\lambda$ analogue of Weglorz's criterion for normality (2.1(1) above). In particular, we shall prove that *an ideal I over $P_\kappa\lambda$ is strongly normal iff for every I-small function $f : P_\kappa\lambda \longrightarrow P_\kappa\lambda$, $\bigcup\{Y_z : z \in P_\kappa\lambda\} \in I^*$ where for each $z \in P_\kappa\lambda$, $Y_z = \{y \in f^{-1}[\{z\}] : (\forall x < y)(f(x) \neq f(y) = z)\}$* (2.3 below). Our proof is a $P_\kappa\lambda$ version of Pelletier's proof in [P] of Weglorz's result. To obtain the reverse implication we need the following lemma which is interesting in its own right.

2.2 Lemma. *For any ideal I over $P_\kappa\lambda$, any $B \in I^+$ and any $<$-regressive function $g : B \longrightarrow P_\kappa\lambda$, there is a $C \subseteq B$ with $C \in I^+$ such that $g\restriction C : C \longrightarrow (P_\kappa\lambda - C)$.*

Proof. Let I, B and g be as described above, and for each $x \in B$ let n_x be the least integer such that $g^n(x) \notin B$; such must exist, for otherwise we'd have an infinite descending sequence $|x| > |g(x)| > |g(g(x))| > \ldots$. For each $n \in \omega$, set $C_n = \{x \in B : n_x = n\}$. Further, set $D = \bigcup\{C_{2n+1} : n \in \omega\}$ and $E = \bigcup\{C_{2n} : n \in \omega\}$. Clearly B is the disjoint union of D and E. Note that $(\forall n \in \omega)(\forall x \in C_n)(g(x) \in C_{n-1})$. Thus $g D : D \longrightarrow (P_\kappa\lambda - D)$ and $g{\restriction}E : E \longrightarrow (P_\kappa\lambda - E)$. Moreover, at least one of D or E must have positive measure. □

Notice that the $(P_\kappa\lambda, \subset)$ version of the above argument does *not* yield the above result for functions $f : B \longrightarrow P_\kappa\lambda$ with the property that $(\forall x \in B)(f(x) \subset x)$.

2.3 Theorem. *For any ideal* I *over* $P_\kappa\lambda$, *the following are equivalent:*

(i) I *is strongly normal,*

(ii) for every $A \in I^+$ *and every* I*-small function* $f : A \longrightarrow P_\kappa\lambda$, $\bigcup\{Y_z : z \in P_\kappa\lambda\} \in (I|A)^*$ *where for each* $z \in P_\kappa\lambda$, $Y_z = \{y \in f^{-1}[\{z\}] : (\forall x < y)(f(x) \neq f(y) = z)\}$ *and* $I|A$ *is the ideal* $\{X \subseteq P_\kappa\lambda : X \cap A \in I\}$, *and*

(iii) for every I*-small function* $f : P_\kappa\lambda \longrightarrow P_\kappa\lambda$, $\bigcup\{Y_z : z \in P_\kappa\lambda\} \in I^*$ *where for every* $z \in P_\kappa\lambda$, Y_z *is defined as in (ii) above.*

Proof. It is clear that (ii) implies (iii). This leaves (i) implies (ii) and (iii) implies (i).

(i) $\longrightarrow$ (ii). Suppose that I is strongly normal, and pick $A \in I^+$ and an I-small function $f : A \longrightarrow P_\kappa\lambda$. Set $Z = \bigcup\{Y_z : z \in P_\kappa\lambda\}$. We shall prove that $(P_\kappa\lambda - Z) \cap A \in I$ and hence conclude that $Z \cup (P_\kappa\lambda - A) \in I^*$ as required. To this end, notice that for each $y \in A - Z$, $(\exists x_y < y)(f(x_y) = f(y))$ and hence that $g : y \mapsto x_y$, $y \in A - Z$ is an I-small $<$-regressive function on $A - Z$. Because $A - Z = \nabla_<\{g^{-1}[\{y\}] : y \in P_\kappa\lambda\}$ and I is strongly normal, it now follows that $A - Z \in I$.

(iii) $\longrightarrow$ (i) Suppose that I is *not* strongly normal. Choose an $A \in I^+$ and an I-small $<$-regressive function $f : A \longrightarrow P_\kappa\lambda$. Let $g = f \cup \mathrm{id}{\restriction}(P_\kappa\lambda - A)$. This g is I-small, so by (iii),

$$B = \bigcup\{\{y \in g^{-1}[\{z\}] : (\forall x < y)(g(x) \neq g(y))\} : z \in P_\kappa\lambda\} \in I^*.$$

Then $A \cap B \in I^+$, so we may apply lemma 2.2 to $g{\restriction}A \cap B$ to get $C \subseteq A \cap B$ such that $C \in I^+$ and $g{\restriction}C : C \longrightarrow (P_\kappa\lambda - C)$. Define $h = g{\restriction}C \cup \mathrm{id}{\restriction}P_\kappa\lambda\text{-}C$. It is clear that h is I-small, so we may use (iii) again to obtain

$$Z' = \bigcup\{\{y \in h^{-1}[\{z\}] : (\forall x < y)(h(y) \neq h(x))\} : z \in P_\kappa\lambda\} \in I^*.$$

But now notice that $Z' \subseteq P_\kappa\lambda - C$: If $y \in Z' \cap C$, then $(y, g(y)) \in h$ and $(g(y), g(y)) \in h$, and hence $h(g(y)) = h(y)$ with $g(y) < y$ which is impossible. Thus $Z' \cap C = \phi$, so $Z' \subseteq P_\kappa\lambda - C$, so $P_\kappa\lambda - C \in I^*$, thus contradicting $C \in I^+$. □

In 2.6, we shall provide $(P_\kappa\lambda, <)$ analogues of the notions defined in 2.1 above. We need some preliminary definitions.

2.4 Definition. For every $A \subseteq P_\kappa\lambda$, a function $f : A \to P_\kappa\lambda$ is said to be *one-one w.r.t. < on A* or *<-one-one on A* iff $(\forall x,y \in A)(x < y \lor y < x \to f(x) \neq f(y))$.

2.5 Definition. For any ideal I on $P_\kappa\lambda$, the ideal $I \times I$ on $P_\kappa\lambda \times P_\kappa\lambda$ is defined by "$X \in I \times I$ iff $X \subseteq P_\kappa\lambda \times P_\kappa\lambda$ and $\{x \in P_\kappa\lambda : \{y \in P_\kappa\lambda : (x,y) \in X\} \in I^+\} \in I$."

Notice that for any $X \subseteq P_\kappa\lambda \times P_\kappa\lambda$, $X \in (I \times I)^*$ iff $\{x \in P_\kappa\lambda : \{y \in P_\kappa\lambda : (x,y) \in X\} \in I^*\} \in I^*$ iff $X \in I^* \times I^*$.

2.6 Definitions. An ideal I over $P_\kappa\lambda$ is said to be

(i) a *p-point* (w.r.t. <) iff for every I-small function $f : P_\kappa\lambda \to P_\kappa\lambda$ there is an $A \in I^*$ on which f is $I_{\kappa\lambda}$-small.

(ii) a *q-point* (w.r.t. <) iff for every $I_{\kappa\lambda}$-small function $f : P_\kappa\lambda \to P_\kappa\lambda$ there is an $A \in I^*$ on which f is < - one-one.

(iii) *selective* (w.r.t. <) iff for every I-small function $f : P_\kappa\lambda \to P_\kappa\lambda$ there is an $A \in I^*$ on which f is one-one w.r.t. <, i.e. iff it is both a p-point and a q-point.

(iv) *Ramsey* (w.r.t. <) iff for every $A \in (I \times I)^*$ there is a $H \in I^*$ such that $(H \times H) \cap \{(x,y) \in P_\kappa\lambda \times P_\kappa\lambda : x < y\} \subseteq A$.

(v) *quasinormal* (w.r.t. <) iff for every $P_\kappa\lambda$-indexed family $\{X_a : a \in P_\kappa\lambda\}$ from I there is an $A \in I^*$ such that $\nabla_<\{X_a : a \in A\} = \{y \in P_\kappa\lambda : (\exists x \in A)(x < y \land y \in X_x)\} \in I$.

The following result is a $P_\kappa\lambda$ analogue of 2.1(2) above:

2.7 Theorem. *For any ideal I over $P_\kappa\lambda$, if I is strongly normal then I is selective.*

Proof. Let $f : P_\kappa\lambda \to P_\kappa\lambda$ be I-small, and define $g : P_\kappa\lambda \to P_\kappa\lambda$ by

$$g(y) = \begin{cases} \text{an } x < y \text{ such that } f(x) = f(y) & \text{if there is such an } x \\ \text{undefined otherwise.} \end{cases}$$

Set $A = \{y \in P_\kappa\lambda : g(y) \text{ is defined}\}$.

It will suffice to prove that $A \in I$, for then f is one-one w.r.t. < on $P_\kappa\lambda - A \in I^*$. If $A \notin I$, then we may use strong normality to obtain an $x \in P_\kappa\lambda$ such that $g^{-1}[\{x\}] \in I^+$. Since $g^{-1}[\{x\}]$

$\subseteq f^{-1}[\{f(x)\}]$, this contradicts the assumption that f is I-small. Thus $A \in I$ as required. □

We conjecture that the $P_\kappa\lambda$ analogue of 2.1(3) holds too. Towards this, we have obtained the results given in theorem 2.8 below. We are still attempting to establish the missing implications, and believe that some of Zwicker's results from [Z3] will assist us in this task.

2.8 Theorem. *For any ideal I over $P_\kappa\lambda$,*
(i) I is quasinormal iff I is Ramsey, and
(ii) if I is Ramsey, then I is selective.

Proof. (i) Suppose first that I is quasinormal, and pick $A \in (I \times I)^*$. Then $X = \{x \in P_\kappa\lambda : \{y \in P_\kappa\lambda : (x,y) \in A\} \in I^*\} \in I^*$. Define $\{X_x : x \in P_\kappa\lambda\}$ by

$$X_x = \begin{cases} \{y \in P_\kappa\lambda : (x,y) \in A\} & \text{if } x \in X \\ P_\kappa\lambda & \text{otherwise.} \end{cases}$$

Now use quasinormality to obtain $B \in I^*$ such that $\Delta_<\{X_x : x \in B\} = \{y \in P_\kappa\lambda : (\forall x \in B)(x < y \longrightarrow y \in X_x)\} \in I^*$. Finally, set $H = X \cap B \cap \Delta_<\{X_x : x \in B\}$, and notice that $(\forall x,y \in H)(x < y \longrightarrow (x,y) \in A)$. Thus I is Ramsey.

Conversely, suppose that I is Ramsey, and pick $\{X_x : x \in P_\kappa\lambda\} \subseteq I^*$. Set $A = \{(x,y) \in P_\kappa\lambda \times P_\kappa\lambda : y \in X_x\}$. Then $\{x \in P_\kappa\lambda : \{y \in P_\kappa\lambda : y \in X_x\} \in I^*\} = P_\kappa\lambda \in I^*$, so $A \in (I \times I)^*$. Now use the Ramsey property to obtain $H \in I^*$ such that $(\forall x,y \in H)(x < y \longrightarrow (x,y) \in A)$. Finally, notice that $H \subseteq \Delta_<\{X_x : x \in H\} = \{y \in P_\kappa\lambda : (\forall x \in H)(x < y \longrightarrow y \in X_x)\}$. Thus I is quasinormal.

(ii) Suppose that I is Ramsey. Pick an I-small function $f : P_\kappa\lambda \longrightarrow P_\kappa\lambda$, and set $A = \{(x,y) \in P_\kappa\lambda \times P_\kappa\lambda : f(x) \neq f(y)\}$. We shall first prove that $A \in (I \times I)^*$, and then use the Ramsey property to obtain an $H \in I^*$ on which f is <-one-one.

For each $x \in P_\kappa\lambda$, set $A_x = \{y \in P_\kappa\lambda : f(y) \neq f(x)\} = P_\kappa\lambda - f^{-1}[\{f(x)\}]$. Because f is I-small, $f^{-1}[\{f(x)\}] \in I$ for every $x \in P_\kappa\lambda$. Thus $A_x \in I^*$ for every $x \in P_\kappa\lambda$, so $\{x \in P_\kappa\lambda : \{y \in P_\kappa\lambda : f(y) \neq f(x)\} \in I^*\} \in I^*$, so $A \in (I \times I)^*$. Now let $H \in I^*$ be such that $(\forall x,y \in H)(x < y \longrightarrow (x,y) \in A)$. It is clear that f is <-one-one on H. □

3. A distributivity property equivalent to $I^+ \to_< (I^+)^2$ for strongly normal ideals.

In 1.7 above, we proved that if κ is λ-supercompact, then every λ-supercompact ultrafilter over $P_\kappa\lambda$ is strongly normal, and then used this fact in 1.10 to prove that for every such ultrafilter U, $U \to_< (U)^2$ holds. We shall obtain a more general result here; we shall prove that strongly normal ideals having a certain distributivity property have the partition property $I^+ \to_< (I^+)^2$ (theorem 3.2).

3.1 Definitions. For any $A \in I^+$, an *I-partition of A* is a maximal collection W of subsets of A with the properties that $(\forall X \in W)(X \in I^+)$ and $(\forall X,Y \in W)(X \neq Y \to X \cap Y \in I)$.

For any cardinals μ and ν, an ideal I is said to be *(μ,ν)-distributive* iff for every $A \in I^+$ and every μ-indexed set $\{W_\alpha : \alpha \in \mu\}$ of I-partitions of A, each of cardinality $\leq \nu$, there are a subset B of A in I^+ and a μ-indexed set $\{X_\alpha : \alpha \in \mu\}$ such that $(\forall\alpha \in \mu)(X_\alpha \in W_\alpha \wedge B - X_\alpha \in I)$.

C.A. Johnson [Jo] proved that for any normal ideal I over $P_\kappa\lambda$, $I^+ \to_< (I^+)^2$ holds iff I is (λ,λ)-distributive and $\{B \in I^+ : (\forall x,y \in B)(x \subset y \to x < y)\}$ is dense in I^+. Our theorem 3.2 below provides a simpler characterization of the relation $I^+ \to_< (I^+)^2$ for strongly normal ideals on $P_\kappa\lambda$.

3.2 Theorem. *For any strongly normal ideal I over $P_\kappa\lambda$, $I^+ \to_< (I^+)^2$ holds iff I is $(\lambda^{<\kappa},2)$-distributive.*

Proof. Let I be a strongly normal ideal over $P_\kappa\lambda$, and suppose first that $I^+ \to_< (I^+)^2$ holds. Fix $A \in I^+$ and a $P_\kappa\lambda$-indexed set $\{W_x : x \in P_\kappa\lambda\}$ of I-partitions of A, each of size at most 2. For each $x \in P_\kappa\lambda$, set $W_x = \{A_{x0},A_{x1}\}$. Notice that for every $x \in P_\kappa\lambda$, $Y_x = [A - (A_{x0} \cup A_{x1})] \cup (A_{x0} \cap A_{x1}) \in I$. By strong normality, $\nabla_<\{Y_x : x \in P_\kappa\lambda\} = \{y \in P_\kappa\lambda : (\exists x < y)(y \in Y_x)\} \in I$, so $\{y \in P_\kappa\lambda : (\forall x < y)(y \notin Y_x)\} \in I^*$. Then because $A \in I^+$, $C = \{y \in A : (\forall x < y)(y \notin Y_x)\} \in (I|A)^* \subseteq I^+$. Notice that for every $x,y \in C$ satisfying $x < y$, y is in precisely one of A_{x0} or A_{x1}.

Fix a well-ordering $\sqsubset$ of $P_\kappa\lambda$, and define $f : [C]^2_< \to 2$ by

$$f(x,y) = \begin{cases} 0 \text{ if } \begin{cases} (\exists u < x)(x \in A_{ui} \wedge y \in A_{u(1-i)}) \text{ and for the} \\ \sqsubset\text{-least such } u,\ x \in A_{u0} \text{ and } y \in A_{u1} \end{cases} \\ 1 \text{ otherwise.} \end{cases}$$

Now let $H \in \mathcal{P}(C) \cap I^+$ be homogeneous for f.

Proceed by $\sqsubset$-induction over $P_\kappa\lambda$ as follows. Pick $y \in P_\kappa\lambda$ and suppose that for every $x \in P_\kappa\lambda$ such that $x \sqsubset y$ we've obtained $X_x \in W_x$ such that $H - X_x \in I$. Define $\{Z_x : x \in P_\kappa\lambda\} \subseteq I$ by

$$Z_x = \begin{cases} H - X_x & \text{if } x \sqsubset y \\ \emptyset & \text{otherwise.} \end{cases}$$

By strong normality, $\nabla_<\{Z_x : x \in P_\kappa\lambda\} = \{z \in P_\kappa\lambda : (\exists x < z)(z \in Z_x)\} \in I$. Thus $\{z \in P_\kappa\lambda : (\forall x < z)(z \notin Z_x)\} \in I^*$, so

$$\{z \in P_\kappa\lambda : (\forall x < z)(x \sqsubset y \rightarrow z \notin Z_x)\} \in I^*, \text{ so}$$

$D = \{z \in H : (\forall x < z)(x \sqsubset y \rightarrow z \in X_x)\} \in (I|H)^*$. Notice that for every $x \in P_\kappa\lambda$ such that $x \sqsubset y$ and every $u, v \in D$ satisfying $x < u, v$, $u, v \in X_x$, so u and v are in the same element of W_x.

We must now produce $X_y \in W_y$ such that $H - X_y \in I$. If $W_y = \{X\}$, simply set $X_y = X$. So assume that $W_y = \{A_{y0}, A_{y1}\}$ with $A_{y0} \neq A_{y1}$. Suppose by way of contradiction that $H - A_{y0} \in I^+$ and $H - A_{y1} \in I^+$. Then since $H - (A_{y0} \cup A_{y1}) \in I$, $H \cap A_{y0} \in I^+$ and $H \cap A_{y1} \in I^+$. Thus $A_{y0}, A_{y1} \in (I|H)^+$, so $D \cap A_{y0} \in (I|H)^+$ and $D \cap A_{y1} \in (I|H)^+$. Now pick $u \in D \cap A_{y1}$ such that $y < u$, then pick $v \in D \cap A_{y0}$ such that $u < v$ and then pick $w \in D \cap A_{y1}$ such that $v < w$. *But* then $f(u,v) = 1$ and $f(v,w) = 0$ contrary to the assumption that H is homogeneous for f.

Conversely, suppose that I is $(\lambda^{<\kappa},2)$-distributive, and fix $A \in I^+$ and $f : [A]^2_< \rightarrow 2$. For every $x \in A$ and every $i \in 2$, set $A_{xi} = \{y \in A : x < y \wedge f(x,y) = i\}$, and notice that for every $x \in A$, at least one of these is in I^+. Define a $P_\kappa\lambda$-indexed set $\{W_x : x \in P_\kappa\lambda\}$ of I-partitions of A by

$$W_x = \begin{cases} \{A_{x0}, A_{x1}\} & \text{if } x \in A \text{ and both are in } I^+ \\ \{A_{x0}\} & \text{if } x \in A \text{ and } A_{x1} \in I \\ \{A_{x1}\} & \text{if } x \in A \text{ and } A_{x0} \in I \\ \{A\} & \text{if } x \notin A. \end{cases}$$

First, use $(\lambda^{<\kappa},2)$-distributivity to obtain $\{X_x : x \in P_\kappa\lambda\}$ and $B \in \mathcal{P}(A) \cap I^+$ such that $(\forall x \in P_\kappa\lambda)(X_x \in W_x \wedge B - X_x \in I)$. Then use strong normality to infer that $\nabla_<\{B - X_x : x \in P_\kappa\lambda\} \in I$, i.e.

$\{y \in P_\kappa\lambda : (\exists x < y)(y \in B - X_x)\} \in I$. Hence $\{y \in P_\kappa\lambda : (\forall x < y)(y \notin B - X_x)\} \in I^*$. Then because $B \in I^+$, $D = \{y \in B : (\forall x < y)(y \in X_x)\} \in I^+$. Now, either $H_0 = \{x \in D : X_x = A_{x0}\} \in I^+$ or else $H_1 = \{x \in D : X_x = A_{x1}\} \in I^+$. Notice that for any $x,y \in H_0$ satisfying $x < y$, $f(x,y) = 0$, and for any such $x,y \in H_1$, $f(x,y) = 1$. □

References

[BTW] Baumgartner, J.E., Taylor, A.D., and Wagon, S., *Structural properties of ideals,* **Dissertationes Mathematicae** CXCVII, 1982.

[B] Blass, A.R., **Orderings of Ultrafilters,** Doctoral Dissertation, Harvard University, 1970.

[C1] Carr, Donna M., *The minimal normal filter on* $P_\kappa\lambda$, **Proc. Amer. Math. Soc.**, 86 (1982), 316-320.

[C2] Carr, Donna M., $P_\kappa\lambda$ *partition relations,* **Fund. Math.**, 128 (1987), 181-195.

[C3] Carr, Donna M., *A note on the λ-Shelah property,* **Fund. Math.**, 128 (1987) 197-198.

[CP] Carr, Donna M., and Pelletier, Donald H., *Towards a structure theory for ideals on* $P_\kappa\lambda$ *II,* in preparation.

[D] DiPrisco, C.A., **Combinatorial properties and supercompact cardinals,** Ph.D Thesis, M.I.T., 1976.

[DM] DiPrisco, C.A., and Marek, W., *Some aspects of the theory of large cardinals,* **Mathematical Logic and Formal Systems** (L.P. de Alcantara, editor), Marcel Dekker, 1985, 87-139.

[F] Fodor, G., *Eine bemerkung zur theorie der regressiven funktionen,* **Acta. Sci. Math.** (Szeged), 17 (1956), 139-142.

[Je1] Jech, Thomas J., *Some combinatorial problems concerning uncountable cardinals,* **Ann. Math. Logic** 5 (1973), 165-198.

[Je2] Jech, Thomas J., **Set Theory,** Academic Press, 1978.

[Jo] Johnson, C.A., *Some partition relations for ideals on* $P_\kappa\lambda$, preprint.

[KP] Kunen, Kenneth, and Pelletier, Donald H., *On a combinatorial property of Menas related to the partition property for measures on supercompact cardinals,* **J. Sym. Logic**, 48 (1983), 475-481.

[Ma] Magidor, M., *There are many normal ultrafilters corresponding to a supercompact cardinal,* **Israel J. Math.** 9 (1971), 186-92.

[Me] Menas, T.K., *A combinatorial property of* $P_\kappa\lambda$, **J. Sym. Logic**, 41 (1976), 225-234.

[N] Neumer, W., *Verallgemeinerung eines Satzes von Alexandroff und Urysohn,* **Math. Zeit.**, 54 (1951), 254-261.

[P] Pelletier, Donald H., *A simple proof and generalization of Weglorz' characterization of normality for ideals,* **Rocky Mountain J. Math.**, 11 (1981), 605-609.

[R] Rowbottom, F., *Some strong axioms of infinity incompatible with the axiom of constructibility,* **Ann. Math. Logic**, 3 (1971), 1-44.

[S1] Shelah, S., *The existence of coding sets,* Lecture Notes in Mathematics, no. 1182, Springer-Verlag, Berlin, 1986, 188-202.

[S2] Shelah, S., *More on stationary coding,* Lecture Notes in Mathematics, no. 1182, Springer-Verlag, Berlin, 1986, 224-246.

[SRK] Solovay, R., Reinhardt, W., and Kanamori, A., *Strong axioms of infinity and elementary embeddings,* **Ann. Math. Logic** 13 (1977), 73-116.

[W] Weglorz, B., *Some properties of filters,* Lecture Notes in Mathematics, no. 619, Springer-Verlag, Berlin, 1977, 311-329.

[Z1] Zwicker, W.S., *$P_\kappa\lambda$ combinatorics I,* **Axiomatic Set Theory**, (Baumgartner, Martin, Shelah, editors) Contemporary Mathematics, vol. 31, American Mathematical Society, 1984, 243-259.

[Z2] Zwicker, W.S., *Lecture notes on the structural properties of ideals on $P_\kappa\lambda$,* **handwritten notes**, July 1984.

[Z3] Zwicker, W.S., Notes for a lecture delivered to the Conference on Set Theory and its Applications at York University in August 1987.

Department of Mathematics
State University of New York at Plattsburgh
Plattsburgh, New York 12901
U.S.A.

Department of Mathematics
York University
North York, Ontario M3J 1P3
Canada

COMPACT SPACES OF COUNTABLE TIGHTNESS IN THE COHEN MODEL

ALAN DOW

Abstract. We investigate compact spaces of countable tightness in the model obtained by adding Cohen reals to a model of GCH . We discover that some of the results which were recently shown to hold assuming PFA will also hold in this model.

1. INTRODUCTION.

The class of compact spaces of countable tighness is a very interesting but perhaps not so well understood class of spaces. There is only a very limited collection of examples of such spaces , primarily consistency results, and the recent and astounding PFA results of Balogh only serve to increase the interest in this class.

We shall study the structure of compact spaces of countable tightness in the Cohen model. It provides an interesting example of the application to topology of forcing and model theoretic reflection. Our main results are that if Cohen reals are added to a model of GCH then in the resulting model compact spaces of countable tightness must have points with character at most ω_1 , every point is the unique complete accumulation point of some set of size at most ω_1 , separable subspaces have cardinality at most 2^{ω_1} and if the space has a small diagonal it is metrizable. We also show in this model that initially ω_1-compact spaces with countable tightness are compact.

The class of compact spaces of countable tightness has been of interest for a long while because it was soon discovered to present many difficulties. Of course there are compact spaces of countable tightness which are not first countable - namely one point compactifications of discrete spaces . But it was a while before it was proven that it was consistent that such spaces needn't be sequential - Ostaszewski's space [1973] . Finally in 1976 , Fedorcuk produced , from diamond, an example which had no converging sequences at all. In 1986 , came the extremely surprising result of Balogh's that PFA implies that such spaces are sequential. In addition he showed that each point is the limit of a converging sequence. We do not have any results about compact spaces of countable tightness which are not known to hold in some other model but it is surprising how much of the PFA results do hold in the Cohen model giving some indication of how difficult it is to construct "large" compact countably tight spaces.

2. BASIC RESULTS.

In order to show the reader the kinds of questions that have interested people in this area we shall begin by listing the basic well-known facts about compact spaces of countable tightness .

<u>Definition.</u> A space X is said to have *countable tightness* if for each $x \in X$ and $A \subset X$, $x \in \mathrm{cl}A$ iff $x \in \mathrm{cl}A'$ for some countable $A' \subset A$. We shall often use $t(X) = \omega$ or even $t = \omega$ to abbreviate that X has countable tightness.

A space X is said to be *sequential* if a set $A \subset X$ is closed iff every A contains the limit of every converging sequence which is a subset of A .

1. FACT: A sequential space has countable tightness.

2. PROPOSITION: [**A2**] For X a compact space
$t=\omega$ iff $\exists \{x_\alpha : \alpha < \omega_1\} \subset X$ such that for each $\alpha < \omega_1$ $\{x_\beta : \beta < \alpha\}$ and $\{x_\beta : \beta > \alpha\}$ have disjoint closures (such a sequence is called a *free-sequence*).

3. FACT: The space 2^{ω_1} does not have countable tightness and if X is a compact space which maps onto a space of uncountable tightness then X does not have countable tightness.

4. FACT: There are compact sequential spaces with arbitrarily large cardinality and character. (The one-point compactification of a discrete space.)

5. FACT: There are separable compact sequential spaces with character $\underline{c}$. The one-point compactification of a Mrowka "ψ-space" in which the relevant maximal almost disjoint family of subsets of ω has cardinality $\underline{c}$.

6. FACT: If X is sequential then the cardinality of X is at most the density of X raised to the ωth power.

7. PROPOSITION: [**A1**] If X is compact sequential then there must be a point x in X such that $2^{\chi(x,X)} \leq \underline{c}$. (Recall that $\chi(x,X)$ is the character of x in X and is defined to the minimum cardinality of a local base at the point x .)

SOME OLD QUESTIONS.

8: Is there a compact space with $t=\omega$ which is not sequential?

9: If X is compact with $t=\omega$ (sequential) must there be a point in X of countable character?

10: Does there exist a separable compact space with $t=\omega$ which has cardinality greater than $\underline{c}$?

SOME RECENT SOLUTIONS.

8: a) Ostaszewski's well-known space (constructed from diamond) is a countably compact , locally compact hereditarily separable space . Therefore its one point compactification is not sequential but does have countable tightness. Many people have observed that such spaces exist in the Cohen model .

b) Balogh has shown that it follows from PFA that compact spaces of countable tightness must be sequential.

9: a) Fedorcuk constructed (from diamond) a compact hereditarily separable , hence $t=\omega$, space in which there are no converging sequences and , of course , every point has character ω_1 . Malyhin has constructed a compact space of countable tightness in the Cohen model in which every point has character ω_1 . Juhasz further proves that there is a model in which there is such a sequential space.

b) PFA implies that there must be a point of character ω in each compact space of countable tightness.

c) $\underline{c} < 2^{\omega_1}$ implies there is a point of character ω in every compact sequential space.

10: a) Fedorcuk's example from 2 a) above is separable with cardinality $2^{\omega_1} > \underline{c}$.

b) PFA implies that if X is compact separable with $t=\omega$ then X has cardinality at most $\underline{c}$.

A RELATED QUESTION. There is a related question asked by the author and independently by van Douwen .

Is every (first countable) initially ω_1-compact
space of countable tightness compact?

A space is said to be *initially ω_1-compact* if every open cover of cardinality at most ω_1 has a finite subcover. A space is initially ω_1-compact iff every subset of cardinality at most ω_1 has a complete accumulation point (i.e. a point whose every neighbourhood hits the set in full size).

Both van Douwen and the author have shown that the answer is yes if CH holds and very recently Fremlin and Nyikos have shown that the answer is yes if PFA is assumed. A "no" answer is not yet known to be consistent.

3. FORCING PRELIMINARIES

Cohen reals are added to a model by forcing with a poset of the form $Fn(I,2) = \{p \subset I \times 2 : p \text{ is a finite function}\}$ ordered by reverse inclusion. An antichain of a poset P is a set of pairwise incompatible elements . A poset is said to be *ccc* if every antichain of P is countable . When the context is clear we shall use elements of V as names for themselves in a forcing sentence. Furthermore if G is P-generic over V and $A \subset X$ is in $V[G]$ while $X \in V$ then we shall assume that a name for A , say $\dot{A}$, is a subset of $X \times P$ and for each $x \in X$ $\{p \in P : (x,p) \in \dot{A}\}$ is an antichain. Recall that if P is *ccc* , $X \in V$, G is P-generic over V and $A \in V[G]$ is a countable subset of X then there is a countable name for A .

REMARK: Suppose that O and X are Ostaszewski's space and Fedorcuk's space respectively. If G is $Fn(I,2)$-generic over V then we may ask what becomes of O and X in $V[G]$. It is not difficult to see that O remains countably compact and locally compact (because it is scattered) and it has been shown that it remains hereditarily separable . Therefore this provides an example in $V[G]$ of a compact non-sequential space with countable tightness in which there is a point to which no sequence converges.

Recall that Fedorcuk's space is a compact subspace of 2^{ω_1} – but we may view it as the Stone space of a certain Boolean algebra. We are then iterested in the Stone space of the same algebra in $V[G]$. S. Todorčević has pointed out to the author that it is straightforward to check that any *property K* forcing (in particular Cohen forcing) will preserve the fact that the Stone space of a Boolean algebra is hereditary separable. This provides an example in $V[G]$ of a separable space of countable tightness of cardinality 2^{ω_1} regardless of the size of the continuum. However the resulting space will have converging sequences. Indeed each "new" point will be the limit of a sequence of "old" points. In addition, Todorčević has noticed that (under very general circumstances) the Stone space will acquire points of first countability.

4. ELEMENTARY SUBMODELS AND FORCING

In the next section we shall frequently be discussing a set X , a $Fn(I,2)$-name $\dot{\tau}$ for a topology on X and occasionally some $Fn(I,2)$-name of another subset of the power set of X , say $\dot{\mathcal{Y}}$. There is some cardinal θ large enough so that any sentence we wish to discuss about these objects will be absolute for $\mathrm{H}(\theta)$ (i.e. they hold in $\mathrm{H}(\theta)$ iff they hold in V) . We shall always assume without mention that θ is this large enough cardinal. Furthermore , when we speak of an elementary submodel we shall mean an elementary submodel of this $H(\theta)$ and that the X , $\dot{\tau}$ (and perhaps $\dot{\mathcal{Y}}$) under discussion are in the submodel. Recall that M is an elementary submodel of $H(\theta)$ iff for each finite sequence $< m_1, \ldots, m_n >$ of elements of M and any formula of set theory $\varphi(v_1, \ldots, v_n)$

$$M \models \varphi(m_1, \ldots, m_n) \qquad \text{iff} \qquad H(\theta) \models \varphi(m_1, \ldots, m_n)$$

(and by our assumption on θ iff $V \models \varphi(m_1, \ldots, m_n)$ providing the m_i's are things we are going to talk about).

When we investigate an elementary submodel M we are interested in the following. From X , $\dot{\tau}$ and $\dot{\mathcal{Y}}$ we define new names which are in a sense the restriction of these names to M . We then use elementarity to deduce what properties the objects named by these new names will have in the extension obtained by forcing with $P \cap M$. Finally we are then interested in the relationship of this object to the space X in the final extension.

If G is P-generic over V we can define, in V[G] , the set M[G] by { val($\dot{A}$,G) : $\dot{A} \in M$} . If P is *ccc* (or if M , G are assumed to have additional properties) it can be shown that M[G] is an elementary submodel of $H(\theta)^{V[G]}$ (for example see [Sh]) . To best exploit this relationship we use the fact that M[G] can often be (essentially) obtained by just forcing with $P \cap M$.

Indeed, suppose we are given X a set , P a forcing poset and $\dot{\mathcal{Y}}$ a P-name such that $1 \Vdash_P \dot{\mathcal{Y}} \subset \mathcal{P}(X)$. Let M be an elementary submodel and define $X_M = X \cap M$ and $P_M = P \cap M$. For each P-name $\dot{Y} \in M$ such that $1 \Vdash_P \dot{Y} \subset X$, we can define a P_M-name $\dot{Y}_M$ so that, for each $p, x \in M$

$$p \Vdash_{P_M} x \in \dot{Y}_M \qquad \text{iff} \qquad p \Vdash_P x \in \dot{Y} .$$

Indeed, for each $x \in X_M$, let A_x be any antichain which is maximal with respect to being a subset of $\{p \in P_M : p \Vdash_P x \in \dot{Y}\}$. Then define $\dot{Y}_M$ to be the name $\cup\{ \{x\} \times A_x : x \in X_M\}$. In fact, since we are assuming that we are only working with "nice" names we can (by elementarity) define Y_M to simply be $Y \cap M$. Note that if P is *ccc* , we have that $1 \Vdash_P \dot{Y} \cap M = \dot{Y}_M$. Now of course we define $\dot{\mathcal{Y}}_M$, a P_M-name , such that $1 \Vdash_{P_M} \dot{\mathcal{Y}}_M = \{\dot{Y}_M : \dot{Y} \in M \text{ and } 1 \Vdash_P \dot{Y} \in \dot{\mathcal{Y}}\}$. In this fashion , we have, for example, that

$$\text{if } 1 \Vdash_P (< X, \dot{\tau} > \text{ is a regular topological space })$$

then

$$1 \Vdash_{P_M} (< X_M, \dot{\tau}_M > \text{ is a base for a regular topological space}) .$$

We will often be careless in distinguishing between a topology and a base for a topology.

If we assume now that P is *ccc* and that G is P-generic over V then it follows that

$$V[G] \models \text{val}(\dot{Y}, G) \cap M = \text{val}(\dot{Y}_M, G \cap M) = \text{val}(\dot{Y}_M, G) .$$

Notice then that we also have that, in V[G] , the topology induced on X_M by val($\dot{\tau}_M$, G$\cap M$) is the same as that induced by val($\dot{\tau}$,G) $\cap M$[G] .

Now suppose we make the additional assumption on M, that $M^\omega \subset M$ (i.e. that M is closed under ω-sequences). Recall that the Lowenheim Skolem theorems give us that for any countable set $A \in H(\theta)$ and any set $B \in H(\theta)$ with $|B| \leq \mathfrak{c}$ there are elementary submodels of $H(\theta)$ M_A and M_B such that $|M_A| = \omega$, $|M_B| = \mathfrak{c}$, $A \subset M_A$, $B \subset M_B$ and M_B is closed under ω-sequences. If M is closed under ω-sequences and $P \in M$ is *ccc* then $P \cap M$ is *completely embedded* in P (i.e. every maximal antichain of $P \cap M$ is maximal in P) hence if G is P-generic over V then G $\cap M$ is P_M-generic over V as well. Therefore, in this case, V[G $\cap M$] is obtained by forcing over V by the poset P_M. Yet another consequence of the fact that P is *ccc* and M is closed under ω-sequences is that we get a kind of ω-absoluteness for M. For example, suppose that $1 \Vdash_P \dot{\mathcal{Y}}$ is a countably complete filter on X, then we get $1 \Vdash_{P_M} \dot{\mathcal{Y}}_M$ is a countably complete filter on X_M. This is almost like saying that, in V[G $\cap \dot{M}$], the model M[G] is closed under ω-sequences. Henceforth when we say "by ω-absoluteness", we shall mean "since $M^\omega \subset M$ and M is an elementary submodel of $H(\theta)$".

As indicated above we shall be interested in $< X_M$, val($\dot{\tau}_M$, G$\cap M$) $>$ in both models, V[G$\cap M$] and V[G] (we are still assuming M is closed under ω-sequences). One thing to be careful of is that, in general, val($\dot{\tau}_M$, G) is a strictly weaker topology than that induced on X_M by val($\dot{\tau}$, G). However we shall make frequent appeals to ω-absoluteness to overcome this. For example, if $\dot{A}, \dot{B}$ are both countable names in M and $1 \Vdash_{P_M} \mathrm{cl}_{\dot{\tau}_M} \dot{A} \cap \mathrm{cl}_{\dot{\tau}_M} \dot{B} \neq (=) \emptyset$, then $1 \Vdash_P \mathrm{cl}_{\dot{\tau}} \dot{A} \cap \mathrm{cl}_{\dot{\tau}} \dot{B} \neq (=) \emptyset$ where the second pair of closures are with respect to the larger topology.

5. Main Results

Let us begin by considering initially ω_1-compact spaces of countable tightness.

Theorem 5.1. *If P is a ccc poset, $1 \Vdash_P < X, \dot{\tau} >$ is initially ω_1-compact and $t = \omega$, and if M is an elementary submodel of $H(\theta)$ closed under ω-sequences, then $1 \Vdash_{P_M} < X_M, \dot{\tau}_M >$ has countable tightness.*

Proof: Suppose that $p \Vdash_{P_M} < X_M, \dot{\tau}_M >$ does not have countable tightness. By ω-absolutenss, we know that $p \Vdash < X_M, \dot{\tau}_M >$ is countably compact. Therefore, by Arhangel'skii's theorem we may choose P_M-names $\{ \dot{x}_\alpha : \alpha < \omega_1 \}$ so that each $x_\alpha \in M$ and $p \Vdash_{P_M} \{\dot{x}_\alpha : \alpha < \omega_1\}$ form a free sequence. But now, since $1 \Vdash_P < X, \dot{\tau} >$ is initially ω_1-compact, we may choose a condition $q \in P$ with $q < p$ and some $x \in X$ so that $q \Vdash x$ is a complete accumulation point of $\{ \dot{x}_\alpha : \alpha < \omega_1 \}$. But now, by countable tightness and the fact that P is *ccc* we may choose $\alpha_0 < \alpha_1 < \omega_1$ so that $q \Vdash x \in \mathrm{cl}_{\dot{\tau}}\{ \dot{x}_\beta : \beta < \alpha_0 \} \cap \mathrm{cl}_{\dot{\tau}}\{ \dot{x}_\beta : \alpha_0 < \beta < \alpha_1 \}$. This contradicts, by ω-absoluteness, that $p \Vdash \mathrm{cl}_{\dot{\tau}_M}\{ \dot{x}_\beta : \beta < \alpha_0 \} \cap \mathrm{cl}_{\dot{\tau}_M}\{ \dot{x}_\beta : \alpha_0 < \beta < \alpha_1 \} = \emptyset$. ∎

Proposition 5.2. *[Fr2] Suppose $< X, \dot{\tau} >$ is an initially ω_1-compact space with countable tightness and $\mathcal{F}$ is a countably complete filter on the closed subsets of X. If $H \in \mathcal{F}^+$ (i.e. $H \cap F \neq \emptyset$ for each $F \in \mathcal{F}$) then $clH' \in \mathcal{F}^+$ for some countable $H' \subset H$.*

Proof: Suppose that H is a counterexample and choose by recursion on $\alpha < \omega_1$ $h_\alpha \in H \cap \bigcap\{ F_\beta : \beta < \alpha\}$ and $F_\alpha \in \mathcal{F}$ so that $F_\alpha \cap \mathrm{cl}\{ h_\beta : \beta < \alpha\} = \emptyset$. But now by countable tightness $F = \bigcup_{\beta<\omega_1} \mathrm{cl}\{ h_\alpha : \alpha < \beta\}$ is closed and $\emptyset = \bigcap_{\beta<\omega_1} F \cap F_\beta$ which contradicts that X is initially ω_1-compact. ∎

The next result is our key tool for most of our preservation results.

MAIN LEMMA 5.3. *Let $P = Fn(I,2)$ and suppose that*

$$1 \Vdash_P < X, \dot{\tau} > \text{ is a completely regular initially } \omega_1\text{-compact space of countable tightness .}$$

Let M be an elementary submodel closed under ω-sequences , and suppose that $\dot{A}$, is a P_M-name of a subset of X_M such that

$$1 \Vdash_{P_M} \text{ if } B\,,\, C \text{ are uncountable subsets of } \dot{A} \text{ then } cl_{\tau_M} B \cap cl_{\tau_M} C \neq \emptyset\,.$$

Then

$$1 \Vdash_P clB \cap clC \neq \emptyset \text{ for each uncountable } B\,,\, C \subset \dot{A}\,.$$

PROOF: Let us first note that , by countable tightness and ω-absoluteness, if $p \in P_M$ and $\dot{B}$, $\dot{C}$ are P_M-names such that $p \Vdash_P \mathrm{cl}_{\tau_M} \dot{B} \cap \mathrm{cl}_{\tau_M} \dot{C} \neq \emptyset$, then $p \Vdash_P \mathrm{cl}_\tau \dot{B} \cap \mathrm{cl}_\tau \dot{C} \neq \emptyset$. Suppose indirectly that the conclusion of the lemma does not hold. Since $1 \Vdash_P < X, \dot{\tau} >$ is initially ω_1 -compact , we may choose complete accumulation points of some subset of size ω_1 of each of the two uncountable sets which are to have disjoint closure. Then since $1 \Vdash_P (< X, \dot{\tau} >$ is completely regular) , we may choose a $p \in P$ and a P-name $\dot{f}$ such that $p \Vdash_P \dot{f}$ is a continuous function from $< X, \dot{\tau} >$ into the unit interval such that $\dot{f}^{\leftarrow}(0) \cap \dot{A}$ and $\dot{f}^{\leftarrow}(1) \cap \dot{A}$ are uncountable . For each $a \in X$, choose (if possible) $p_a \in P$ such that $p_a < p$ and $p_a \Vdash (a \in \dot{A}$ and $\dot{f}(a) = 1)$. Let $B = \{a \in X_M : p_a$ was chosen$\}$; B is uncountable since $p \Vdash_P \dot{f}^{\leftarrow}(1) \cap \dot{A}$ is uncountable . Let us say that a family $S \subset P$ forms a Δ-system , if $S' = \{dom(s) \ : \ s \subset S\}$ forms a Δ-system of sets and all functions $s \in S$ agree on the root of S' . The common restriction to the root of S' will be called the root of S . By passing to an uncountable subset of B , if necessary, we may suppose that $\{p_a : a \in B\}$ form a Δ-system with root $p' \leq p$. Next, for each $a \in X_M$, choose (if possible) $r_a \in P$ so that $r_a \leq p'$ and $r_a \Vdash_P a \in \dot{A}$ and $\dot{f}(a) = 0$. Similarly assume that B' is an uncountable set so that $\{r_a : a \in B'\}$ forms a Δ-system with root r . By removing at most finitely many members of B we may also assume that $dom(p_a - p') \cap dom(r) = \emptyset$ for each $a \in B$. Let $B_0 = \{a \in B : p_a - p' \subset M\}$ and $B'_0 = \{a \in B' : r_a - r \subset M\}$.

Claim: B_0 and B'_0 cannot both be uncountable.

To see this note first that

$$\dot{C} = \{(a, p_a - p') : a \in B_0\} \quad \text{and} \quad \dot{C}' = \{(a, r_a - r) : a \in B'_0\}$$

are P_M-names such that

$$r \Vdash_P \dot{C} \subset \dot{f}^{\leftarrow}(1) \cap \dot{A} \quad \text{and} \quad r \Vdash_P \dot{C}' \subset \dot{f}^{\leftarrow}(0) \cap \dot{A}\,.$$

Therefore $r \Vdash_P \mathrm{cl}_\tau \dot{C} \cap \mathrm{cl}_\tau \dot{C}' = \emptyset$. But then, as noted at the beginning of the proof, it must be the case that $r \cap M \Vdash_{P_M} \mathrm{cl}_{\tau_M} \dot{C} \cap \mathrm{cl}_{\tau_M} \dot{C}' = \emptyset$. By our assumption on $\dot{A}$, $r \cap M \Vdash$ (either $\dot{C}$ or $\dot{C}'$ is countable) . But since $\{p_a - p' : a \in B_0\}$ and $\{r_a - r : a \in B'_0\}$ are Δ-systems (with (empty) root compatible with r) , it follows that either B_0 or B'_0 is countable .

Since the proofs are similar, we shall assume that B_0 is countable. By considering $B - B_0$, we may assume that for each $a \in B$, it is not the case that $(p \cap M) \cup p' \Vdash_P \dot{f}(a) = 1$ (note that it is the case that $(p \cap M) \Vdash_{P_M} a \in \dot{A}$). Now, for each $a \in B$, choose some $q_a \in P$ and $n_a \in \omega$ so that $q_a < (p \cap M) \cup p'$ and $dom(q_a) \supset dom(p_a)$ and $q_a \Vdash \dot{f}(a) < 1 - 1/n_a$. Choose an uncountable $C \subset B$, $n \in \omega$ and $q \in P$ so that $\{q_a : a \in C\}$ forms a Δ-system with root q and $n_a = n$ for all $a \in C$.

Now let $q' = q \cap M$ and choose $\{a_n : n < \omega\}$ any infinite subset of C. For each $n \in \omega$, let $p_n = p_{a_n} \cap M$ and $q_n = q_{a_n} \cap M$ (observe that $p_n \subset q_n$). Since M is closed under ω-sequences $\{a_n : n \in \omega\}$, $\{p_n : n < \omega\}$ and $\{q_n : n \in \omega\}$ are all in M. Let us call a sequence $\{< p'_n, q'_n >: n < \omega\}$ a Δ-system pair (for $\{a_n : n < \omega\}$, $\{p_n : n \in \omega\}$ and $\{q_n : n \in \omega\}$) if there is a P-name $\dot{f}$ such that :

(i) $1 \Vdash_P \dot{f}$ is a continuous real-valued function on $< X, \dot{\tau} >$;

(ii) for each $n \in \omega$ $p'_n \Vdash_p \dot{f}(a_n) = 1$, $q'_n \Vdash_P \dot{f}(a_n) = 0$;

(iii) for each $n \in \omega$, $p_n \subset p'_n \cap q'_n$, $q_n \subset q'$;

(iv) $\{p'_n : n \in \omega\}$ forms a Δ-system with root p';

(v) $\{q'_n : n \in \omega\}$ forms a Δ-system with root q'.

The sequence $\{< p_{a_n}, q_{a_n} >: n \in \omega\}$ constructed above witnesses the fact that there are Δ-system pairs for $\{a_n : n \in \omega\}$ which are not in M, hence there must be uncountably many in M. We shall inductively choose an uncountable sequence $\mathbf{p}_\alpha = \{< p^\alpha_n, q^\alpha_n >: n \in \omega\}$ for $\alpha < \omega_1$ of Δ-system pairs for $\{a_n : n \in \omega\}$ so that if

$$D = \bigcup \{dom(p_n) \cup dom(q_n) : n \in \omega\}$$

and for each $\alpha < \omega_1$

$$D_\alpha = \bigcup \{dom(p^\alpha_n - p_n) \cup dom(q^\alpha_n - q_n) : n \in \omega\}$$

then for each $\alpha < \beta < \omega_1$ $D_\alpha \cap D = \emptyset$ and $D_\alpha \cap D_\beta = \emptyset$. Indeed, suppose we have chosen $\mathbf{p}_\alpha$ for $\alpha < \beta$ so that each $\mathbf{p}_\alpha$ is a Δ-system pair which is a subset of M. Since M is closed under ω-sequences the entire sequence $< \mathbf{p}_\alpha : \alpha < \beta >$ is a member of M. The Δ-system pair constructed in the above paragraph would serve as the next member except that it is not a subset of M. However by absoluteness there must be such a Δ-system pair in M, so we may choose $\mathbf{p}_\beta$. For each $\alpha < \omega_1$ let q_α be the root of the Δ-system $\{q^\alpha_n : n \in \omega\}$ in the Δ-system pair $\mathbf{p}_\alpha$. By construction $\{q_\alpha : \alpha < \omega_1\}$ forms a Δ-system with root q'. For each $\alpha < \omega_1$, fix a P-name, $\dot{f}_\alpha$, of a continuous real-valued function which witnesses that $\mathbf{p}_\alpha$ is a Δ-system pair.

We are going to prove the following

Claim $q' \Vdash_P$ the closure of $\Pi_{\beta < \omega_1} \dot{f}_\alpha[\{a_n : n \in \omega\}]$ in $[0,1]^{\omega_1}$ contains a homeomorphic copy of 2^{ω_1}.

Note that since $\Vdash_P < X, \tau >$ is initially ω_1-compact and $[0,1]^{\omega_1}$ has weight ω_1, $q' \Vdash_P$ the image of $< X, \tau >$ under $\Pi_{\alpha < \omega_1} \dot{f}_\alpha$ is compact. However by **[2.3]** this is a contradiction since it means that $q' \Vdash < X, \tau >$ does not have countable tightness.

Now to establish the claim. If G is P-generic and $q' \in$ G then $L = \{\alpha < \omega_1 : q_\alpha \in \text{G}\}$ is uncountable . Furthermore, let $t \in Fn(L, 2)$ and $r \in$G be arbitrary For each $\alpha \in dom(t)$, there is an $n_\alpha \in \omega$ such that $[dom(p^\alpha_n - p') \cup dom(q^\alpha_n - q')] \cap dom(r) = \emptyset$.

Therefore there is an $n \in \omega$ such that p_n^α and q_n^α are both compatible with r for each $\alpha \in dom(t)$. Furthermore, since $D_\alpha \cap D_\beta = \emptyset$ for $\alpha < \beta < \omega_1$ we have that $[dom(p_n^\alpha - p_n) \cup dom(q_n^\alpha - q_n)] \cap [dom(p_n^\beta - p_n) \cup dom(q_n^\beta - q_n)] = \emptyset$. Therefore there is an extension r' of r such that for each $\alpha \in dom(t)$ $\quad r' \Vdash_P \dot{f}_\alpha(a_n) = t(\alpha)$. From this it follows that V[G]$\models$ the map $\Pi \dot{f}_\alpha$ $(\alpha \in L)$ takes $\{a_n : n \in \omega\}$ onto a dense subset of 2^L . ∎

By a very similar proof one can show the following.

PROPOSITION 5.4. *Let $P = Fn(I,2)$ and suppose that*

$$1 \Vdash_P < X, \dot{\tau} > \text{ is a completely regular initially } \omega_1\text{-compact space of countable tightness .}$$

Let M be an elementary submodel closed under ω-sequences , and suppose that $\dot{A}$, $\dot{C}$ are P_M-names of subsets of X_M such that

$$1 \Vdash_{P_M} \text{ if } B \text{ is an uncountable subset of } \dot{A} \text{ then } cl_{\tau_M} B \cap cl_{\tau_M} \dot{C} \neq \emptyset .$$

Then

$$1 \Vdash_P cl_\tau B \cap cl_\tau \dot{C} \neq \emptyset \text{ for each uncountable } B \subset \dot{A} .$$

THEOREM 5.5. *[CH] If $P = Fn(I,2)$ and $1 \Vdash_P < X, \dot{\tau} >$ is initially ω_1-compact and has countable tightness , then $1 \Vdash_P < X, \dot{\tau} >$ is compact .*

PROOF: Suppose indirectly that there is some $p \in P$ and some P-name $\dot{\mathcal{F}}$ such that $p \Vdash_P$ ($\dot{\mathcal{F}}$ is a maximal free filter on the closed subsets of the initially ω_1-compact space of countable tightness , $< X, \dot{\tau} >$) . By [5.2] , $p \Vdash_P \dot{\mathcal{F}}$ has a base of separable sets. Let M be an elementary submodel containing $< p, X, \dot{\tau}, \dot{\mathcal{F}} >$ which is closed under ω-sequences and has cardinality ω_1 . Let $\{\dot{F}_\alpha : \alpha < \omega_1\} \subset M$ be a listing of all P_M-names of countable subsets of X_M such that $p \Vdash_P \text{cl}\dot{F}_\alpha \in \dot{\mathcal{F}}$. By ω-absoluteness $p \Vdash_{P_M} \{ \text{cl}_{\tau_M} \dot{F}_\alpha : \alpha < \omega_1\}$ is a filter base for a countably complete filter on X_M . For each $\alpha \in \omega_1$, let $\dot{a}_\alpha \in M$ be a P_M-name such that $p \Vdash_{P_M} \dot{a}_\alpha \in \bigcap_{\beta < \alpha} \text{cl}_{\tau_M} \dot{F}_\beta$. Let $\dot{A}$ be the name $\{(\dot{a}_\alpha , p) : \alpha \in \omega_1\}$. Note that $p \Vdash_{P_M} \dot{A} - \text{cl}_{\tau_M} \dot{F}_\alpha$ is countable for all $\alpha \in \omega_1$. By elementarity, we have that for each $\beta < \alpha < \omega_1$, $p \Vdash_P \dot{a}_\alpha \in \text{cl}_\tau \dot{F}_\beta$.

Now let $\{\dot{b}_\alpha : \alpha \in \omega_1\}$ be any sequence of P_M-names and $q < p$ with $q \in P_M$ so that $q \Vdash \{\dot{b}_\alpha : \alpha \in \omega_1\}$ is an uncountable subset of $\dot{A}$. For convenience, let $\dot{B} = \{\dot{b}_\alpha : \alpha \in \omega_1\}$. Therefore $q \Vdash_P \dot{B} \in \{\text{cl}_\tau \dot{F}_\alpha : \alpha \in \omega_1\}^+$. By [5.2], $q \Vdash_P$ (there is some countable B' such that $B' \subset \dot{B}$ and $\text{cl}_\tau B' \in \{\text{cl}_\tau \dot{F}_\alpha : \alpha < \omega_1\}^+$) . Clearly then, (since P is *ccc*) there is some $\alpha \in \omega_1$ such that $q \Vdash_P \text{cl}_\tau \{b_\beta : \beta < \alpha\} \in \{\text{cl}_\tau \dot{F}_\beta : \beta < \omega_1\}^+$. Let $\dot{B}'$ be the name $\{(b_\beta , q) : \beta < \alpha\}$. But now $\dot{B}' \subset M$ hence, $\dot{B}' \in M$. Furthermore $M \models q \Vdash_P \text{cl}_\tau \dot{B}' \in \dot{\mathcal{F}}^+$, hence $q \Vdash_P \text{cl}_\tau \dot{B}' \in \dot{\mathcal{F}}^+$. Therefore $q \Vdash_P \text{cl}_\tau \dot{B}' \in \dot{\mathcal{F}}$ since $\dot{\mathcal{F}}$ is a maximal filter. It then follows that $q \Vdash_P \dot{A} - \text{cl}_\tau \dot{B}' \subset \dot{A} - \text{cl}_\tau \dot{B}$ is countable. By the Main Lemma [5.3] (with p replacing 1) , $p \Vdash_P$ (Uncountable subsets of $\dot{A}$ do not have disjoint closures) . But now let G be P-generic over V with $p \in$ G , and let $F_\alpha = \text{val}(\dot{F}_\alpha,\text{G})$, $\mathcal{F} = \text{val}(\dot{\mathcal{F}} ,\text{G})$ etc. . Since $< X, \tau >$ is initially ω_1-compact, $\mathcal{F}$ must be ω_2-complete. Hence $F = \bigcap_{\alpha<\omega_1} \mathrm{cl}_\tau\{a_\beta : \beta > \alpha\} \in \mathcal{F}$. But since A has at most one complete accumulation point, $|F| = 1$. Clearly this is a contradiction. ∎

A routine closure argument (which we omit) proves the following fact.

THEOREM 5.6. *If initially ω_1-compact spaces of countable tightness are compact then compact separable spaces of countable tightness have cardinality at most 2^{ω_1} .*

COROLLARY 5.7. *If G is $Fn(I,2)$-generic over a model of CH , V , then $V[G] \models$ separable compact spaces of countable tightness have cardinality at most 2^{ω_1} .*

Other interesting consequences of the Main Lemma are the following results. The first shows that compact spaces of countable tightness in the Cohen model have a property which might be called ω_1-sequential.

THEOREM 5.8. *[CH] If G is $Fn(I,2)$-generic over V then , in V[G] , if X is a compact space of countable tightness then :*

(a) *A set $F \subset X$ is closed iff F is countably compact and contains the complete accumulation point of any of its subsets of cardinality ω_1 which have unique complete accumulation points in X ;*

(b) *Each point of uncountable character is the unique complete accumulation point of some set of cardinality ω_1 ; and*

(c) *X is metrizable if it has a small diagonal*
([H] the diagonal $\Delta \subset X^2$ is <u>small</u> if for each uncountable $Y \subset X^2 - \Delta$, there is an uncountable $Y' \subset Y$ such that $\mathrm{cl}Y' \cap \Delta = \emptyset$).

PROOF: To prove a), suppose $x \in \overline{F}$ and that F is countably compact. Fix names $\dot{F}$ for F and $\dot{\tau}$ for the topology on X . Choose an elementary submodel M of cardinality ω_1 which is closed under ω-sequences and contains $\{x, \dot{F}, \dot{\tau}, X, I\}$. In V[G$\cap M$] , let F_M = val($\dot{F}_M$,G$\cap M$) . Note that with respect to the toplogy val($\dot{\tau}_M$,G$\cap M$) x will have character at most ω_1 in $\overline{F}_M$ (since $|\dot{\tau}_M| = \omega_1$) . Clearly we may as well assume that the character of x in $\overline{F}_M$ is uncountable. Furthermore F_M is countably compact so we may choose $\{x_\alpha : \alpha < \omega_1\} \subset F_M$ so that, in V[G$\cap M$] , x is the unique complete accumulation point of $\{x_\alpha : \alpha < \omega_1\}$. By the Main Lemma , x remains the unique complete accumulation point of $\{x_\alpha : \alpha < \omega_1\}$. Therefore if F satisfies the property in (a) , $x \in F$.

While (b) does not follow directly from (a) it does follow from the proof of (a) .

For (c) , X^2 has countable tightness **[M]** and the quotient space X^2/Δ has countable tightness by **[2.3]** . Now if X is not metrizable then the character of Δ in the quotient space is uncountable. Therefore by (b) X^2 does not have a small diagonal. ∎

We now turn our attention to finding points of small character. We begin by recalling the proof of the classical Čech-Pospíšil theorem.

PROPOSITION 5.9. *Let κ be a cardinal and X a compact space such that $\chi(x, X) \geq \kappa$ for all $x \in X$. Then there is a family $\{U_s : s \in {}^{<\kappa}2\}$ of open subsets of X so that for each $\alpha < \kappa$ and $s \in {}^{\alpha}2$;*

(i) *the family $\{U_{s|\beta} : \beta < \alpha\}$ has the finite intersection property; and* *(ii)* $U_{s\frown 0}$ *and* $U_{s\frown 1}$ *are chosen so that* $\emptyset = \overline{U_{s\frown 0}} \cap \overline{U_{s\frown 1}}$ *and* $\overline{U_{s\frown 0}} \cup \overline{U_{s\frown 1}} \subset U_s$.

PROOF: Character equals pseudocharacter in compact spaces hence no point is equal to the intersection of fewer than κ open sets. Choose the sets U_s by induction on the domain of s . Conditions i) and ii) guarantee that $K_s = \bigcap\{U_{s|\beta} : \beta \in dom(s)\}$ is not empty for each $s \in 2$. ∎

PROPOSITION 5.10. *If X is a compact space in which each point has character at least* κ *, then there is a subset* $Y \subset X$ *with* $|Y| \leq 2^{<\kappa}$ *such that* $|\overline{Y}| \geq 2^\kappa$.

PROOF: Fix the family $\{U_s : s \in {}^{<\kappa}2\}$ as in [5.9] . Let K_s be as defined in the proof of [5.9] . Choose Y so that $Y \cap K_s \neq \emptyset$ for each $s \in {}^{<\kappa}2$. For each $s \in {}^{\kappa}2$, $K_{s|\alpha}\alpha < \kappa$ is a descending sequence of closed sets each of which meets Y . Therefore $\overline{Y} \cap K_s \neq \emptyset$ for each $s \in {}^{\kappa}2$. Now observe that condition ii) in [5.9] guarantees that $K_s \cap K_t = \emptyset$ for $s,t \in {}^{\kappa}2$ with $s \neq t$. ∎

COROLLARY 5.11. *If* $2^{\omega_1} < 2^{\omega_2}$ *and X is a compact space in which every point has character at least* ω_2 *, then X contains an initially* ω_1 *-compact non-compact subspace.*

PROOF: Apply [5.10] with $\kappa = \omega_2$ to obtain a set $Y \subset X$ of cardinality 2^{ω_1} whose closure has cardinality 2^{ω_2} . Clearly there is an initially ω_1 -compact Y' such that $Y \subset Y' \subset \overline{Y}$ and $|Y'| = 2^{\omega_1}$. ∎

COROLLARY 5.12. *If* $V \models GCH$ *and G is* $Fn(\omega_2, 2)$*-generic over V then* $V[G] \models$ *Each compact space of countable tightness contains a point with character at most* ω_1 .

Let us try to improve on [5.11] by restricting to spaces with countable tightness. The next result has likely not appeared before but it is an easy consequence of Arhangel'skii's proof that there are free sequences in compact spaces of uncountable tightness.

THEOREM 5.13. *CH implies that there must be points of character at most* ω_1 *in each compact space with countable tightness.*

PROOF: Let $\{U_s : s \in {}^{<\omega_2}2\}$ be chosen as in [5.9] . Begin choosing an increasing sequence $\{s_\gamma : \gamma < \omega_1\} \subset {}^{<\omega_2}2$ and points $x_\gamma \in K_{s_\gamma}$ as follows. At stage $\alpha < \omega_1$ we let $s = \cup\{s_\gamma : \gamma < \alpha\}$ and ask if the closure of $\{x_\gamma : \gamma < \alpha\}$ meets each set K_t such that $s \subset t$. If it does we stop , if it does not choose $s_\alpha \supset s$ such that $K_{s_\alpha} \cap cl\{x_\gamma : \gamma < \alpha\} = \emptyset$ and choose x_α any point in K_{s_α} . Countable tightness guarantees this process will stop before we reach ω_1 since we are building a free sequence. On the other hand the process does not stop since if it did we would have a separable space with 2^{ω_1} distinct regular open sets. ∎

THEOREM 5.14. *If* $V \models GCH$ *and G is* $Fn(I, 2)$*-generic over V then* $V[G] \models$ *Each compact space of countable tightness contains a point with character at most* ω_1 .

PROOF: Let X be a set and let $\dot{\tau}$ be a $Fn(I, 2)$-name of a countably tight compact topology on X . Let M be an elementary submodel of our "sufficiently large" $H(\theta)$ so that M contains $\{X, \dot{\tau}\}$ and so that $|M| = \omega_2$ and M is closed under ω_1-sequences. We can now use "ω_1-absoluteness" to deduce that

$$V[G \cap M] \models < X_M, \text{val}(\dot{\tau}_M, G \cap M) > \text{ is an initially } \omega_1\text{-compact space of countable tightness.}$$

Therefore, by **[5.5]**, we have that $< X_M, \text{val}(\dot{\tau}_M, G \cap M) >$ is compact in $V[G \cap M]$. Next we use **[5.12]** to deduce that $V[G \cap M] \models < X_M, \text{val}(\dot{\tau}_M, G \cap M) >$ contains a point, say x, of character ω_1. But since M is closed under ω_1-sequences the base for x will actually be a member of $M[G \cap M]$, i.e. it will have a name in M. We can therefore deduce, by elementarity, that this base will be a base for x even in the full $V[G]$. ∎

We finish with a somewhat surprising reflection lemma, however I do not have any applications for it.

LEMMA 5.15. *Suppose V is a model of CH, $P = Fn(I, 2)$ and $1 \Vdash < X, \dot{\tau} >$ is a compact space of countable tightness. If M is an elementary submodel closed under ω-sequences and G is P-generic then*

$V[G \cap M] \models$ The Stone-Čech compactification of

$< X_M, val(\dot{\tau}_M, G \cap M) >$ has countable tightness.

Furthermore if we let $< K_M, \sigma >$ be this compactification from $V[G \cap M]$ then

$V[G] \models$ (there is a set K' such that $X_M \subset K' \subset X$, K' with the topology induced by val($\dot{\tau}$, G) is Lindelöf and there is a unique continuous one-to-one map from K' to K_M which is the identity on X_M).

PROOF: Let us begin by working in V[G $\cap M$]. Define K_M to be the Stone-Čech compactification of $< X_M, \text{val}(\dot{\tau}, G) >$. By countable tightness and ω-absoluteness each continuous real-valued function on X_M extends, in V[G], to one on clX_M; indeed if $f : X_M \to [0,1]$ is such that, in V[G], $\text{cl} f^{\leftarrow}(0) \cap \text{cl} f^{\leftarrow}(1) \neq \emptyset$, then this would be exhibited by a countable set. Now that we know the continuous real-valued functions are absolute upward, we can use them to lift any free sequence of K_M to one in X. Therefore K_M has countable tightness. An immediate consequence of this fact is that K_M will have character $\aleph_1$ since the set of complements of the closures of countable subsets of X_M which do not have a given point in their closure will form a base for that point.

Next, we know that X_M is countably compact, hence each zero-set ultrafilter on X_M is countably complete. Furthermore we claim that each ultrafilter of zero-sets will have a unique accumulation point in X. Indeed, if not, we can choose, by countable tightness, a countable subset of X_M whose closure will contain both the purported limit points. But then we can assume we are working with a separable space in which case there will only be $\aleph_1$ continuous real-valued functions. A routine diagonalization process allows us to pick a set A which will meet each member of the ultrafilter in a co-countable set and which will have both of the above accumulation points as complete accumulation points. Of course this contradicts the conclusion of **[5.3]**. To complete the proof of the claim we note that the hypotheses on A in **[5.3]** can be weakened to just the assumption that no two uncountable subsets can be completely separated (as is the case here) since this is what was proven. We then define the space K' and the map $g : K_M \to K'$ in the obvious way.

It remains to show that K' is Lindelöf. We first observe yet another consequence of the Main Lemma which is interesting in its own right.

Fact The G_δ-topologies on K' generated by

$$\text{val}(\dot{\tau}_M, G \cap M) \qquad \text{and} \qquad \text{val}(\dot{\tau}, G) \qquad \text{are identical.}$$

Indeed, suppose that we have a point $x \in K'$ which is in the G_δ-closure of a set F with respect to the smaller topology. Let $\{W_\alpha : \alpha \in \omega_1\}$ be a neighbourhood base for x with respect to val$(\dot{\tau}_M, G \cap M)$. Working in V[G∩M] fix a name $\dot{F}$ for F . Let $p \in G$ be arbitrary and for each $\alpha \in \omega_1$, choose a condition $p_\alpha \in Fn(I - M, 2)$ below p and a point $x_\alpha \in K_M$ so that

$$p_\alpha \Vdash g(x_\alpha) \in \dot{F} \cap \bigcap_{\beta<\alpha} W_\beta \ .$$

We may then find an uncountable set $J \subset \omega_1$ so that $\{p_\alpha : \alpha \in J\}$ form a Δ-system with root q . Since p was arbitrary we may assume that $q \in G$. But now it again follows by the Main Lemma that x will be in the G_δ-closure of each uncountable subset of $\{x_\alpha : \alpha \in J\}$ with respect to val$(\dot{\tau}, G)$. Furthermore G will meet $\{p_\alpha : \alpha \in J\}$ in an uncountable set since it forms a Δ-system whose root is in G . Therefore x is in the G_δ-closure of $F \cap \{x_\alpha : \alpha \in J\}$ with respect to either topology.

Finally let us suppose that $\mathcal{F}$ is a maximal countably complete filter of closed subsets of K' . Clearly we can view $\mathcal{F}$ as a filter on K_M under the identification given by the above mapping. We claim it suffices to find a point $x \in K_M$ which is the closure (with respect to val$(\dot{\tau}_M, G \cap M)$) of each member of $\mathcal{F}$. Indeed if x is such a point then the closure of each val$(\dot{\tau}_M, G \cap M)$ neighbourhood of x will be a member of $\mathcal{F}$. Since $\mathcal{F}$ is countably complete, it follows that x will be in the G_δ-closure of each member of $\mathcal{F}$ (in both topologies) . Therefore $\mathcal{F}$ will not be a free filter thus showing that K' is Lindelöf.

By **[5.2]** , $\mathcal{F}$ has a base of separable sets hence we may assume we are working in such a separable subspace. In V$[G \cap M]$, there are only $\aleph_1$ zero-sets $\{Z\alpha : \alpha \in \omega_1\}$ of this subspace. In V[G] we may inductively choose a set $J \subset \omega_1$ as follows. At stage α choose β_α to be the smallest member of ω_1 so that $Z_{\beta_\alpha} \subset \bigcap_{\gamma<\alpha} Z_{\beta_\gamma a}$ and $Z_{\beta_\alpha} \in \mathcal{F}$. Next we work in V$[G \cap M]$. Fix a name $\dot{J}$ for J and let $p \in G$ be arbitrary. For each $\alpha \in \omega_1$, choose if possible $p_\alpha \in Fn(I - M, 2)$ so that $p_\alpha < p$ and $p_\alpha \Vdash \alpha \in \dot{J}$. We now choose an uncountable subset D of ω_1 so that $\{p_\alpha : \alpha \in D\}$ forms a Δ-system with some root q . Since p was arbitrary, we may assume that $q \in G$. But now $D \in V[G \cap M]$ and it is easily checked that $\{Z_\alpha : \alpha \in D\}$ generates a maximal filter of zero-sets. Let x be the unique accumulation point of this filter. Clearly the closure of each neighbourhood of x is in $\mathcal{F}$ hence x is our desired point.

We finish with a question.

16: If in **[5.15]** , one assumes , in addition, that
$1 \Vdash$ (X is sequential) , does it follow that $X_M = K_M$?

REFERENCES

[Ar71] A. V. Arhangel'skii, *On bicompacta hereditarily satisfying Suslin's condition. Tightness and free sequences*, Soviet Math. Dokl. **12** (1971), 1253-1257.

[Ar78] ____________, *Structure and Classification of Topological spaces and cardinal invariants*, Uspehi Mat. Nauk. **33** (1978), 23-84.

[Ba86a] Z. Balogh, *On compact Hausdorff spaces of countable tightness*, preprint (1986).

[Ba86b] ________, *A note on closed pre-images of ω and the PFA*, preprint (1986).

[D80] A. Dow, *Absolute C-embedding of spaces with countable character and pseudocharacter conditions*, Can. J. Math. **34 4** (1980), 945-956.

[D88] ———, *An Introduction to Applications of Elementary Submodels in Topology*, Topology Proceedings (1988). to appear

[Fe76] V. V. Fedorčuk, *Fully closed mappings and the consistency of some theorems of general topology with the axioms of set theory*, Math. USSR **28** (1976), 1-26; Mat. Sbornik **99**, 3-33.

[Fr86a] D. Fremlin, *Consequences of Martin's Maximum*, preprint (1986).

[Fr86b] ———, *Perfect pre-images of ω and the PFA*, preprint (1986).

[Ju88] I. Juhász, *A Weakening of club , with Applications to Topology*, Math. Inst. Hung. Acad. Sci. preprint No. 51 (1988).

[Hu77] M. Hušek, *Topological spaces without κ-inaccessible diagonal*, Comm. Math. Univ. Carolinae **18** (1977), 777-788.

[Ma87] V. Malykin, *Bicompact Fréchet-Urysohn bez toček*, Mat. Zametki 41 (1987), 365-376.

[Ma72] ———, *On tightness and Suslin number in expX and in a product of spaces*, Sov. Math. Dokl. **13** (1972), 496-499.

[Ny86a] P. Nyikos, *Forcing compact non-sequential spaces of countable tightness*, preprint (1986).

[Ny86b] ———, *Progress on countably compact spaces*, to appear in the proceedings of the 1986 Prague Symposium (1986).

[Os76] A. Ostaszewski, *On countably compact perfectly normal spaces*, J. London Math. Soc. **(2) 14** (1976), 505-516.

Two remarks about analytic sets

Fons van Engelen[1]
Vrÿe Universiteit
Subfaculteit Wiskunde
De Boeleaan 1081
1081 HV Amsterdam
The Netherlands

Kenneth Kunen
Department of Mathematics
University of Wisconsin
Madison, Wisconsin 53706

Arnold W. Miller[2]
Department of Mathematics
University of Wisconsin
Madison, Wisconsin 53706

Abstract

In this paper we give two results about analytic sets. The first is a counterexample to a problem of Fremlin. We show that there exists ω_1 compact subsets of a Borel set with the property that no σ-compact subset of the Borel set covers them. In the second section we prove that for any analytic subset A of the plane either A can be covered by countably many lines or A contains a perfect subset P which does not have three collinear points.

In his book about Martin's Axiom [2] p. 61 D. H. Fremlin shows that, assuming MA, for any analytic set X and set $A \subset X$ of cardinality less than the continuum there exists a σ-compact set L such that $A \subset L \subset X$. (σ-compact means countable union of compact sets.) Here we show that the set A cannot be replaced by a family of compact sets of cardinality ω_1. This answers a question of Fremlin [2] p.67.

Let $\mathbf{Q}$ be the rationals and let $E = \mathbf{Q}^\omega$. E is a Π^0_3 set, or equivalently an $F_{\sigma\delta}$ set. For each $\alpha < \omega_1$ let C_α be a compact subset of $\mathbf{Q}$ which is homeomorphic to $\alpha + 1$ and let $H_\alpha = C_\alpha^\omega$.

Theorem 1 *There does not exists a σ-compact set L such that for every $\alpha < \omega_1$, $H_\alpha \subset L \subset E$.*

Lemma 2 *For every compact set $K \subset \mathbf{Q}$ there exists $\alpha < \omega_1$ such that for all $\beta > \alpha$, $C_\beta \setminus K$ is nonempty.*

proof

This follows easily from a well-known theorem of Sierpiński that every countable scattered space is isomorphic to an ordinal. For simplicity we sketch a proof here.

---

[1]Research partially supported by the Netherlands organization for the advancement of pure research
[2]Research partially supported by NSF grant MCS-8401711

Let $D(X)$ be the derivative operator, i.e. $D(X)$ is the set of nonisolated points of X. Then let $D^\alpha(X)$ be the usual α^{th} iterate of D, defined by induction as follows.

$$D^{\alpha+1}(X) = D(D^\alpha(X))$$

$$D^\lambda(X) = \cap_{\alpha<\lambda} D^\alpha(X) \text{ if } \lambda \text{ a limit ordinal}$$

Define the rank of any X (rank(X)) as the least α such that $D^\alpha(X)$ is empty.

Then the lemma follows easily from the following facts:

1. Every compact subset of **Q** has a countable rank.
2. If $X \subset Y$ then $D(X) \subset D(Y)$.
3. If $X \subset Y$ then rank(X)$\leq$rank(Y).
4. rank$(C_{\omega^\alpha+1}) = \alpha + 1$.

□

To prove the Theorem let $L = \bigcup_{n\in\omega} L_n$ where each L_n is compact. Let $K_n \subset \mathbf{Q}$ be the projection of L_n onto the nth coordinate. By the lemma there exists C_β which is not covered by any K_n. It follows that H_β is not covered by L.

□

We don't know whether the theorem is true for Σ_3^0 sets ($G_{\delta\sigma}$) or even for a set which is the union of a countable set and a Π_2^0 (G_δ).

Next we prove the following theorem:

Theorem 3 *Suppose that A is an analytic subset of the plane,* $\mathbf{R}^2$*, which cannot be covered by countably many lines. Then there exists a perfect subset P of A such that no three points of P are collinear.*

proof

A set is perfect iff it is homeomorphic to the Cantor space 2^ω. The proof we give is similar to the classical proof that uncountable analytic sets must contain a perfect subset. A subset A of a complete separable space X is analytic iff there exists a closed set $C \subset \omega^\omega \times X$ such that A is the projection of C, i.e.

$$A = p(C) = \{y \in X \mid \exists x \in \omega^\omega \ (x, y) \in C\}$$

Every Borel subset of X is analytic.

Let A be analytic subset of the plane $\mathbf{R}^2$ which cannot be covered by countably many lines. Let S be the unit square ($[0,1] \times [0,1]$) minus all lines of the form $x = r$ or $y = r$ for r a rational number. Without loss of generality we may assume that A is a subset of S. Since S is a complete separable space there exists $C \subset \omega^\omega \times S$ a closed set such that $A = p(C)$.

Give S the basis B_s for $s \in 4^{<\omega}$ described in figure 1. ($4^{<\omega}$ is the set of finite sequences of elements from $4 = \{0,1,2,3\}$)

For $y \in S$ define $y \restriction n = s$ iff $y \in B_s$ for s of length n and for $x \in \omega^\omega$ let $x \restriction n$ be the restriction of x to the set $n = \{0,1,2,\ldots,n-1\}$. Let

$$T = \{(x \restriction n, y \restriction n) \mid (x, y) \in C\}$$

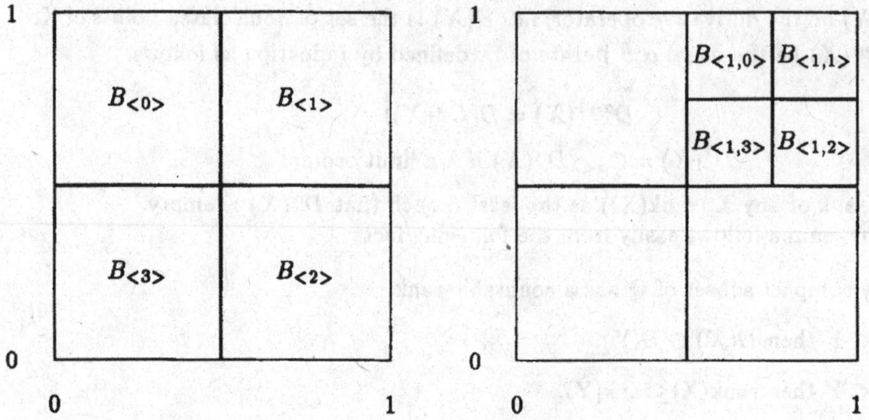

Figure 1: B_s for $s \in 4^{<\omega}$

Then

$$C = [T] = \{(x, y) \mid \forall n\, (x \restriction n, y \restriction n) \in T\}$$

and

$$A = \mathrm{p}[T]$$

For any $(s,t) \in T$ define

$$T_{(s,t)} = \{(\hat{s}, \hat{t}) \in T \mid (\hat{s} \subset s \wedge \hat{t} \subset t) \text{ or } (s \subset \hat{s} \wedge t \subset \hat{t})\}$$

Let

$$T' = \{(s,t) \in T \mid \mathrm{p}[T_{(s,t)}] \text{ cannot be covered by countably many lines } \}$$

Lemma 4 *For any $(s,t) \in T'$ $\mathrm{p}[T'_{(s,t)}]$ cannot be covered by countably many lines, where $T'_{(s,t)} = T_{(s,t)} \cap T'$.*

proof

By definition $\mathrm{p}[T_{(s,t)}]$ cannot be covered by countably many lines. For any x in $\mathrm{p}[T_{(s,t)}]$ but not in $\mathrm{p}[T'_{(s,t)}]$ there must be some $(\hat{s}, \hat{t})$ in $T_{(s,t)}$ but not in $T'_{(s,t)}$ such that x is in $\mathrm{p}[T_{(\hat{s},\hat{t})}]$. Since each such $\mathrm{p}[T_{(\hat{s},\hat{t})}]$ can be covered by countably many lines, $\mathrm{p}[T'_{(s,t)}]$ cannot be covered by countably many lines.
□

Lemma 5 *(Split and shrink) Suppose $(s_i, t_i) \in T'$ for $i = 0, 1, 2, \ldots, n$ are given with the properties that for $i \neq j$ B_{t_i} is disjoint from B_{t_j} and no line meets three or more of the B_{t_j}'s. Then there exists $(\hat{s}_i, \hat{t}_i) \in T'$ for $i = 1, 2, \ldots, n$ with $(\hat{s}_i, \hat{t}_i) \supset (s_i, t_i)$ and*

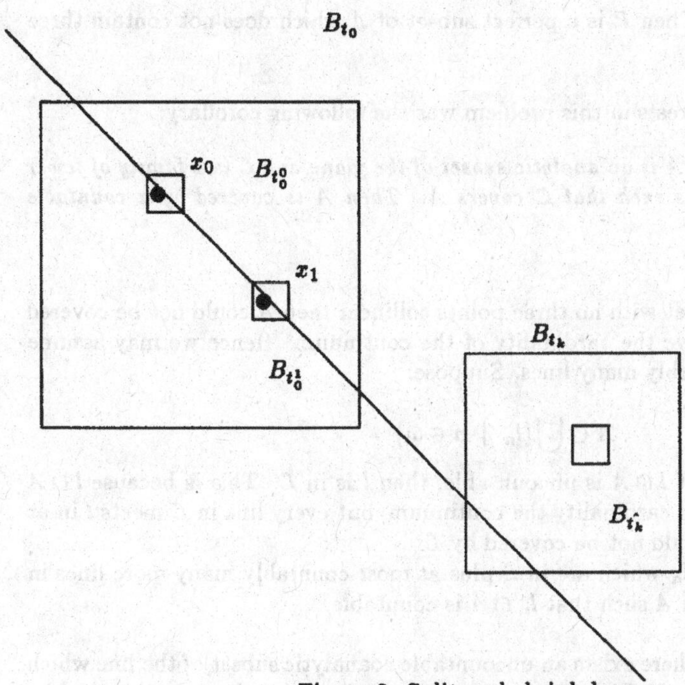

Figure 2: Split and shrink lemma

$(s_0^j, t_0^j) \in T'$ *for* $j = 0, 1$ *with* $(s_0^j, t_0^j) \supset (s_0, t_0)$ *and* $B_{t_0^0}$ *disjoint from* $B_{t_0^1}$ *and no line meets three or more of the*

$$B_{t_0^0}, B_{t_0^1}, B_{\hat{t}_1}, B_{\hat{t}_2}, B_{\hat{t}_3}, \ldots, B_{\hat{t}_n},$$

proof

Since $p[T'_{(s_0,t_0)}]$ cannot be covered by countably many lines we can find distinct elements of it x_0, x_1. Let l be the line containing x_0 and x_1. Since this line cannot cover $p[T'_{(s_i,t_i)}]$ for $i = 1, 2, \ldots, n$ we can find $(\hat{s}_i, \hat{t}_i) \supset (s_i, t_i)$ with each $B_{\hat{t}_i}$ a positive distance from the line l. Now choose $(s_0^j, t_0^j) \supset (s_0, t_0)$ in T' with $B_{t_0^j}$ a small enough neighborhood of x_j so as ensure that no line meets three or more of these squares. See figure 2.

□

Lemma 6 *Suppose* $(s_i, t_i) \in T'$ *for* $i = 0, 1, 2, \ldots, n$ *are given with the properties that for* $i \neq j$ B_{t_i} *is disjoint from* B_{t_j} *and no line meets three or more of the* B_{t_j}*'s. Then there exists* $(s_i^j, t_i^j) \in T'$ *for* $i = 0, 1, 2, \ldots, n$ *and* $j = 0, 1$ *with* $(s_i^j, t_i^j) \supset (s_i, t_i)$ *and* $B_{t_i^0}$ *disjoint from* $B_{t_i^1}$ *and no line meets three or more of the* $B_{t_i^j}$ *for* $i = 0, 1, 2, \ldots, n$ *and* $j = 0, 1$.

proof

Apply the split and shrink lemma iteratively $n + 1$ times.

□

To prove the theorem construct a subtree $T^* \subset T'$ with the property that $p[T^*] = P$ is perfect and for every $n \in \omega$ no line meets three or more of the B_t with $(s, t) \in T^*$ for

some s and t of length n. Then P is a perfect subset of A which does not contain three collinear points.
□

One of our original interests in this problem was the following corollary:

Corollary 7 *Suppose that A is an analytic subset of the plane and $\mathcal{L}$ is a family of fewer than continuum many lines such that $\mathcal{L}$ covers A. Then A is covered by a countable subfamily of lines from $\mathcal{L}$.*

proof

If A contains a perfect set with no three points collinear then A could not be covered by $\mathcal{L}$, since perfect sets have the cardinality of the continuum. Hence we may assume that A is covered by countably many lines. Suppose:

$$A \subset \bigcup \{l_n \mid n \in \omega\}$$

If l is any line such that $l \cap A$ is uncountable, then l is in $\mathcal{L}$. This is because $l \cap A$ is an analytic set, hence has cardinality the continuum, but every line in $\mathcal{L}$ meets l in at most one point, so $l \cap A$ could not be covered by $\mathcal{L}$.

So A is covered by the l_n which are in $\mathcal{L}$ plus at most countably many more lines in $\mathcal{L}$ which cover the points in A such that $l_n \cap A$ is countable.
□

Note that if V=L then there exists an uncountable coanalytic subset of the line which contains no perfect subsets. If this set is arranged around a circle then we see that the theorem cannot be generalized to include coanalytic sets.

However Dougherty, Jackson, and Kechris have proved the following result:

Theorem 8 *Suppose the axiom of determinacy and V=L[R] is true. Then every subset of the plane either can be covered by countably many lines or contains a perfect subset P with no three points collinear.*

Their proof uses a technique of Harrington (see Kechris and Martin [1]) to prove Silver's theorem that every coanalytic equivalence relation with uncountably many equivalence classes contains a perfect set of inequivalent points. They generalize this result and our result.

Is it true in Solovay's model [3] that every subset of the plane either can be covered by countably many lines or contains a perfect subset P with no three points collinear?

References

[1] A. S. Kechris and D. A. Martin, Infinite games and effective descriptive set theory, in **Analytic Sets**, ed. by C. A. Rogers et al, Academic Press, (1980), 404-470.

[2] D. H. Fremlin, **Consequences of Martin's Axiom**, Cambridge University Press, (1984).

[3] R.Solovay, A model of set-theory in which every set of reals is Lebesgue measurable, Annals of Mathematics, 92(1970), 1-56.

Saturated ideals obtained via restricted iterated collapse of huge cardinals

Frantisek Franek

Abstract.

A uniform method to define a (restricted iterated) forcing notion to collapse a huge cardinal to a small one to obtain models with various types of highly saturated ideals over small cardinals is presented. The method is discussed in great technical details in the first chapter, while in the second chapter the application of the method is shown on three different models: Model I with an $\aleph_1$-complete $\aleph_2$-saturated ideal over $\aleph_1$ that satisfies Chang's conjecture, Model II with an $\aleph_1$-complete $\aleph_3$-saturated ideal over $\aleph_3$, and Model III with an $\aleph_1$-complete $(\aleph_2, \aleph_2, \aleph_0)$-saturated ideal over $\aleph_1$.

Introduction

"There is no κ^+-complete κ^+-saturated ideal over κ (κ an uncountable cardinal)" is a straightforward generalization of the classical result of Ulam (see [U] or [J]) "there is no non-trivial σ-additive measure on $\aleph_1$". Solovay (see [S]) proved that if "there exists a κ-complete κ-saturated ideal over κ", then κ is a large cardinal (Mahlo). So if one wants to generalize the notion of saturated ideals to an ideal over a smaller cardinal κ, either completeness of such ideal must be less than κ, or saturatedness must be at least κ^+. From Solovay's work (see [S]) follows that the consistency strength of the existence of an $\aleph_1$-complete $\aleph_2$-saturated ideal over $\aleph_1$ is at least the existence of a measurable cardinal. Later improved by Mitchell (see [Mi]), the consistency strength is in fact at least the existence of a certain sequence of measurable cardinals. Until Kunen's paper [K_2], there was no model known with an $\aleph_1$-complete $\aleph_2$-saturated ideal over $\aleph_1$. Kunen used a collapse of huge cardinal to obtain such a model. Variations of his method were used by Magidor (see [M]), Laver (see [L]), and Forman-Laver (see [FL]) to obtain various saturated ideals over $\aleph_1$ or $\aleph_3$.

In this paper the author tried to unify all these variations. In Chapter 1 an exposition of the method (a restricted iterated forcing) with most of details worked out is presented. Only the general knowledge of forcing and iterated forcing is assumed. The aspects of restricted forcing (keeping forcing terms "small" so the resulting posets are not getting too "big"), extension and covering properties of elementary embeddings (i.e. when an elementary embedding $j{:}V \to M$ can be extended to one from a generic extension of V to a generic extension of M, and when a subset of a generic extension of M is a set from the generic extension of M) are the main thrust of the first chapter. Also the circumstances which give rise to a particular ideal (which can be made saturated depending on the forcing used) are discussed as well.

Starting with an elementary embedding $j{:}V \to M$ with the critical point κ, restricted iterated forcing is used to obtain a poset $B = P * Q$ so that B is a regular suborder of $j(P)$, and in any generic extension of V via B there is an ideal $\mathcal{I}$ over κ so that $\wp(\kappa)/\mathcal{I}$ can be embedded into Boolean completion of $j(P)/B$, and hence it inherits the saturatedness of $j(P)/B$. The extension of the elementary embedding j to one from $V[G]$ to $M[H]$ can satisfy (if the circumstances are right) the "transfer" property, i.e. for every object X from $V[G]$ of certain size $j''X \in M[H]$, and of course $j(X) \in M[H]$. In many situations $j''X$ becomes a "subobject" of $j(X)$, lending itself to prove properties like Chang's conjecture, or the graph one proven by Forman-Laver.

In Chapter 2 three different models for various saturated ideals are produced via restricted iterated collapse of a huge cardinal. In the first chapter the author tried to set up the machinery of restricted iterated forcing so that only certain properties of the forcing to be iterated must be checked for all pieces to fit together to get the posets P, Q, and B so that B can be regularly embedded into $j(P)$. Some additional properties of the model V^B then follow from properties of P and/or $j(P)/B$.

Lately the field has been quite active by efforts of Forman, Magidor, and Shelah (see [FMS]) who obtained a model where MM (Martin's Maximum Axiom) holds by collapsing "just" a supercompact cardinal to $\aleph_1$. Some of the consequences of MM are that $2^{\aleph}_0 = \aleph_2$ and the non-stationary ideal over $\aleph_1$ is $\aleph_2$-saturated. Forman, Magidor, and Shelah (private communication) using a forcing similar to the one used to produce a model where MM holds, obtained a model where GCH holds and the non-stationary ideal over $\aleph_1$ is "somewhere" $\aleph_2$-saturated (i.e. the restriction of the ideal to a stationary subset of $\aleph_1$ is $\aleph_2$-saturated). It seems at the moment that huge cardinals can give rise to some "fancy" saturated ideals, while supercompact cardinals can give similar results as far as $\aleph_2$-saturatedness is concerned.

Let us mention an interesting open problem. Although a supercompact cardinal is enough to get an $\aleph_1$-complete $\aleph_2$-saturated ideal over $\aleph_1$, and a huge cardinal cardinal is enough to get an $\aleph_1$-complete $\aleph_3$-saturated ideal over $\aleph_3$, a model with an $\aleph_1$-complete $\aleph_2$-saturated ideal over $\aleph_2$ is not known. Note that if $\aleph_\omega < \kappa < \aleph_{\omega_1}$, then there is no an $\aleph_1$-complete $\aleph_2$-saturated ideal over κ (see [F_2]). For completeness, Woodin (see [W]) obtained a model of ZFC with an $\aleph_1$-complete ideal over $\aleph_1$ which has a dense set of size $\aleph_1$ via the axiom of determinacy. He can now obtain (private communication) enough of determinacy by collapsing a huge cardinal to $\aleph_1$ to use similar construction to get a model of ZFC with an $\aleph_1$-complete $\aleph_1$-dense ideal over $\aleph_1$, but it is a completely different approach from the one presented in this paper.

The motivation of the author to undertake writing of this paper was twofold: first is the scarcity of published literature in this area (no wonder due to the enormous technicality of the subject), and second the non-existence of expository literature in the topic allowing non-experts to understand and use the methods. The author sincerely hopes that this paper will succeed in at least partially filling both gaps.

Notation and basic definitions.

For all basic notations about sets see [J] and [K_1], about forcing and iterated forcing see [J],[K_1], and [B]. For definitions and properties of large cardinals see [MK] and [SRK].

We are using as standard set-theoretical notation as possible. To distinguish formulas from text, they are enclosed between " ", e.g. "$x \in X$". WLOG abbreviates "without loss of generality", $\emptyset$ denotes the empty set. Lower case Greek letters are reserved for ordinal numbers. *Ord* denotes the class of all ordinal numbers. If X is a set, then $|X|$ denotes its size (cardinality). If f is a function, $dom(f)$ denotes its domain, while $rng(f)$ denotes its range. If X is a set, then $f''X$ denotes the range of the function $f|X$ (f restricted to X). For a set X, $\wp(X)$ denotes the power set of X, while $[X]^{<\gamma}$ denotes the set of all subsets of X of size $< \gamma$, and $[X]^{\leq\gamma}$ denotes the set of all subsets of X of size $\leq \gamma$, and $[X]^{\gamma}$ denotes the set of all subsets of X of size γ. If X and Y are sets, then ${}^XY$ denotes the set of all functions from X into Y, and for an ordinal γ, ${}^{<\gamma}X = \bigcup\{{}^{\alpha}X : \alpha < \gamma\}$, while ${}^{\leq\gamma}X = \bigcup\{{}^{\alpha}X : \alpha \leq \gamma\}$. If V is the set universe, the cumulative hierarchy of sets $\langle V_\alpha : \alpha \in Ord\rangle$ is defined by $V_0 = \emptyset$, $V_{\alpha+1} = V_\alpha \cup \wp(V_\alpha)$, and $V_\alpha = \bigcup\{V_\beta : \beta < \alpha\}$ for α limit. Then $V = \bigcup\{V_\alpha : \alpha \in Ord\}$. A set X is said to have rank α, if $X \in V_{\alpha+1} - V_\alpha$.

Let P be a poset (i.e. a set partially ordered by $\leq$). Let $p \in P$, and let $D \subset P$. Then $p << D$ iff $p \leq d$ for every $d \in D$. D is dense in P iff for any $p \in P$ there is $d \in D$ so that $d \leq p$, while D is said to be dense below p iff for any $p' \leq p$ there is $d \in D$ so that $d \leq p'$. $p, p' \in P$ are compatible, we shall denote it by $p \mathbin{\text{⊐⊏}} q$, if there is $q \in P$ so that $q \leq p, p'$. $p \supset\subset q$ will denote that p and q are incompatible. D is an antichain, iff D consists of pairwise incompatible elements. P satisfies the κ-c.c. iff for every $X \in [P]^\kappa$ there are $p, q \in X$ so that they are compatible in P. P satisfies the (κ, κ, μ)-c.c. iff for any $X \in [P]^\kappa$ there is $Y \in [X]^\kappa$ so that for any $Z \in [Y]^\mu$ there is $z \in P$ so that $z << Z$. P satisfies the $(\kappa, \kappa, <\mu)$-c.c. iff P satisfies the (κ, κ, γ)-c.c. for every $\gamma < \mu$. D is directed iff every two elements of D are compatible. D is centered iff for any $D_0 \in [D]^{<\omega}$ there is $p \in P$ so that $p << D_0$. P is κ-centered iff P is a union of κ centered posets. P is said to be λ-closed ($<\lambda$-closed) if for every $\xi \leq \lambda$ ($\xi < \lambda$) and every descending sequence $\langle p_\alpha : \alpha \leq \xi \rangle$ of elements of P there is $p \in P$ so that $p \leq p_\alpha$ for every $\alpha \leq \xi$.

$\mathcal{I}$ is a κ-complete ideal over λ iff (i) $\mathcal{I} \subset \wp(\lambda)$; and (ii) if $X \subset Y \subset \lambda$, and $Y \in \mathcal{I}$, then $X \in \mathcal{I}$; and (iii) if $\{X_\alpha : \alpha < \xi\}$, and $\xi < \kappa$, then $\bigcup\{X_\alpha : \alpha < \xi\} \in \mathcal{I}$; (iv) $\emptyset \in \mathcal{I}$; and (v) $\lambda \notin \mathcal{I}$; and (vi) for every $\alpha \in \lambda$, $\{\alpha\} \in \mathcal{I}$. (Note: usually a κ-complete ideal is defined as one satisfying (i)-(iv), a proper ideal is one satisfying (v), non-principal ideal as one satisfying (vi). Since we shall deal only with proper, non-principal ideals, we included these properties right in the definition.) $\mathcal{I}^+ = \{X \subset \kappa : X \notin \mathcal{I}\}$. $\mathcal{I}$ is κ-saturated iff $\{X_\alpha : \alpha < \kappa\} \subset \mathcal{I}^+$, then $X_\alpha \cap X_\beta \notin \mathcal{I}$ for some $\alpha \neq \beta < \kappa$. $\mathcal{I}$ is (κ, κ, μ)-saturated iff for every $X \in [\mathcal{I}^+]^\kappa$ there is $Y \in [X]^\kappa$ so that for every $Z \in [Y]^\mu$, $\bigcap Z \notin \mathcal{I}$. $\mathcal{I}$ is κ-centered iff $\mathcal{I} = \bigcup\{\mathcal{I}^+_\alpha : \alpha < \kappa\}$, and each $\mathcal{I}^+_\alpha$ is centered, i.e. whenever $X_0, \ldots, X_n \in \mathcal{I}^+_\alpha$, then $\bigcap\{X_i : i \leq n\} \notin \mathcal{I}$.

If $M \subset V$ is a transitive model of ZFC, then $\mathcal{U}$ is a non-principal M-κ-complete M-ultrafilter over λ iff (i) $\mathcal{U} \subset \wp(\lambda) \cap \mathcal{M}$; and (ii) if $X \subset Y \subset \lambda$, and $X \in \mathcal{Y}$, and $Y \in M$, then $Y \in \mathcal{U}$; and (iii) if $\{X_\alpha : \alpha < \xi\} \in M$, and $\xi < \kappa$, then $\bigcap\{X_\alpha : \alpha < \xi\} \in \mathcal{U}$; and (iv) if $X \subset \lambda$, $X \in M$ and $X \notin \mathcal{U}$, then $\lambda - X \in \mathcal{U}$; and (v) for every $\alpha \in \lambda$, $\{\alpha\} \notin \mathcal{U}$.

If P is a forcing notion (i.e. a poset), $p \in P$, then $p \Vdash^V_P$ "ϕ" (read p forces over V that ϕ) means that for any G P-generic over V, so that $p \in G$, $V[G] \models$ "ϕ". Symbol 1_P denotes the greatest element of P (if it exists). $\Vdash^V_P$ "ϕ" means that $p \Vdash^V_P$ "ϕ" for any $p \in P$, which is equivalent to $1_P \Vdash^V_P$ "ϕ" if P has the greatest element, and also it is equivalent to $V[G] \models$ "ϕ" for any G P-generic over V. If P and Q are posets, $P \simeq Q$ denotes that they are isomorphic. $P \subset\subset Q$ denotes that P is a complete suborder (see Def. 13). $P \times Q$ is a poset of ordered pairs $\langle p, q \rangle$, $p \in P$, $q \in Q$ ordered by $\langle p, q \rangle \leq \langle p', q' \rangle$ iff $p \leq p'$ and $q \leq q'$. Then $P \subset\subset P \times Q$, as well as $Q \subset\subset P \times Q$.

As much as possible we shall adhere to denoting forcing terms (i.e. names, see Def. 5) with $\circ$ accent. E.g. $\mathring{X} \in V^P$ denotes a V^P-term. When forcing with P, every object X from the ground model V has a canonical name by which V is embedded into $V[G]$. For simplicity we shall use the same notation for the canonical name for X as for X itself. The notion of iterated forcing $P * \mathring{Q}$ represents a poset of pairs $\langle p, \mathring{q} \rangle$ so that $p \in P$, and $\Vdash^V_P$ "$\mathring{q} \in \mathring{Q}$ & $\mathring{Q}$ is a poset", with the order defined by $\langle p, \mathring{q} \rangle \leq \langle p', \mathring{q}' \rangle$ iff $p \leq p'$ and $p' \Vdash^V_P$ "$\mathring{q} \leq \mathring{q}'$". It is standard (see e.g. [J], [K₁]) that if $P \subset\subset Q$, then there is $(Q/P) \in V^P$ so that $P * (Q/P) \simeq Q$. A sequence $\langle P_\alpha : \alpha \leq \kappa \rangle$ is a forcing iteration iff for every $\alpha < \kappa$, there is $\mathring{R}_\alpha \in V^{P_\alpha}$ so that $P_{\alpha+1} = P_\alpha * \mathring{R}_\alpha$. For α limit $p \in P_\alpha$ iff p is an α-sequence so that $p|\beta \in P_\beta$ and $p|\beta \Vdash^V_{P_\beta}$ "$p(\beta) \in \mathring{R}_\beta$" and $supp(p) = \{\beta \in \alpha : p|\beta \Vdash^V_{P_\beta}$ "$p(\beta) \neq 1_{\mathring{R}_\beta}$"$\}$ satisfies some specified properties. For example, if at every limit stage we require that only the sequences with finite support are taken, it is called finite support iteration. Different ideals for support may be used (see [K_1]).

Let B be a Boolean algebra, and let ϕ be a formula (using V^B-terms). The symbol $\|\phi\|_B$ denotes the Boolean value of ϕ (see [J]). Symbols O_B, 1_B denote the least and greatest elements of B. $Comp(P)$ for a poset P denotes its Boolean completion. Let $p \in P$. Then $p \Vdash^V_P$ "ϕ" iff $p \leq \|\phi\|_{Comp(P)}$.

Let formula ϕ define a set. Let P be a poset. Let M be a model of ZFC. Then ϕ^P denotes the V^P-term for the set ϕ defines in a generic extension of V via P, and ϕ^M denotes the set ϕ defines in M.

An uncountable regular cardinal α is inaccessible (we shall abbreviate it by inacc.) iff $2^\lambda < \alpha$ for every $\lambda < \alpha$.

If M is a model of ZFC, $j: V \to M$ is an elementary embedding iff for any formula $\phi(X_0, \ldots, X_n)$ with $n+1$ free variables and no constants, and any $A_0, \ldots, A_n \in V$, $V \models$ "$\phi(A_0, \ldots, A_n)$" iff

$M \models$ "$\phi(j(A_0), \ldots, j(A_n))$". An ordinal κ is the critical point of j iff $j(\alpha) = \alpha$ for all $\alpha < \kappa$, and $j(\kappa) > \kappa$ (such κ must be at least a measurable cardinal - see [J]). If $V_\alpha = M_\alpha$ for all $\alpha \leq \kappa$, then $j(x) = x$ for every $x \in V_\kappa$.

Chapter 1.

Def. 1: Let $j{:}V \to M$ be an elementary embedding with critical point κ. Let $M \subset V$ and let j be definable in V. j is **huge** if ${}^{j(\kappa)}M \subset M$ (where ${}^{j(\kappa)}M$ is defined in V).

Def. 2: Let ρ be an ordinal. $X \subset Ord$ is **ρ-Easton** if $|X \cap \gamma| < \gamma$ for all regular $\gamma > \rho$.

Lemma 3: Let λ be Mahlo, $\rho \geq \omega$, and let $X \subset \lambda$ be ρ-Easton. Then X is bounded below λ, i.e. $|X| < \lambda$.

Proof: $A = \{\gamma : \rho \leq \gamma < \lambda \ \& \ \lambda$ regular$\}$ is stationary in λ since λ is Mahlo. For every $\gamma \in A$ define $f(\gamma)$ as the least ν so that $X \cap \gamma \subset \nu$. Since X is ρ-Easton, f is regressive and so by Fodor's theorem there are a stationary $B \subset A$ and $\sigma < \lambda$ so that $X \cap \gamma \subset f(\gamma) = \sigma$ for all $\gamma \in B$. Hence $X \subset \sigma$. □

Lemma 4: Let $j{:}V \to M$ be huge with critical point κ. For all $\alpha \leq j(\kappa)$, all $\rho \geq \omega$, and all $X \subset j(\kappa)$

(4.1) "α is a cardinal" iff $M \models$ "α is a cardinal";
(4.2) "α is regular" iff $M \models$ "α is regular";
(4.3) "α is weakly inaccessible" iff $M \models$ "α is weakly inaccessible";
(4.4) "α is inaccessible" iff $M \models$ "α is inaccessible";
(4.5) "α is Mahlo" iff $M \models$ "α is Mahlo";
(4.6) "X is ρ-Easton and $X \in M$" iff $M \models$ "X is ρ-Easton".
(4.6) "X is a ρ-Easton subset of $j(\kappa)$" iff $M \models$ "X is a ρ-Easton subset of $j(\kappa)$".

Proof: Left to the reader. □

Def. 5: Let P be a poset. Then $V_0^P = \emptyset$, $V_{\alpha+1}^P = V_\alpha^P \cup \wp(V_\alpha^P \times P)$, $V_\alpha^P = \bigcup\{V_\beta^P : \beta < \alpha\}$ for α limit, and $V^P = \bigcup\{V_\alpha^P : \alpha \in Ord\}$.

Lemma 6: Let P be a poset so that $P \in V_\lambda$, λ a regular cardinal. Then

(6.1) $V_\alpha^P \in V_\lambda$ for every $\alpha < \lambda$;
(6.2) $V_{\alpha+n}^P \subset V_{\alpha+3n}$ for every $\alpha \geq \lambda$, α limit, and every $n \in \omega$.

Proof: Left to the reader. □

Lemma 7: Let P be a poset. If $X \in V^P$ and $X \in V_\alpha$, then $X \in V_\alpha^P$.

Proof: Left to the reader. □

Lemma 8: Let P be a poset. If $X \in V^P$, then $X \cap V_\alpha \in V_{\alpha+1}^P$.

Proof: Left to the reader. □

Lemma 9: The maximum principle.
Let P be a poset, A an antichain in P (i.e. a set of mutually incompatible elements of P). For each $a \in A$, let $\mathring{X}_a \in V_{\alpha_a}^P$, α_a ordinals. Then there is a $\mathring{X} \in V_{\alpha+1}^P$, $\alpha = \bigcup\{\alpha_a : a \in A\}$, so that $a \Vdash_P$ "$\mathring{X} = \mathring{X}_a$" for every $a \in A$.

Proof: Define $\mathring{X}$ by $\langle \mathring{Y}, p \rangle \in \mathring{X}$ iff for some $a \in A$, $\mathring{Y} \in dom(\mathring{X}_a)$, $p \leq a$ and $p \Vdash_P$ "$\mathring{Y} \in \mathring{X}_a$". Then $dom(\mathring{X}) = \bigcup\{dom(\mathring{X}_a) : a \in A\} \subset \bigcup\{V_{\alpha_a}^P : a \in A\} = V_\alpha^P$. Now, to prove that $a \Vdash_P$ "$\mathring{X} = \mathring{X}_a$" for each $a \in A$ is fairly standard (see e.g. [K1]) and hence left to the reader. □

Lemma 10: Let P be a poset, $p\in P$, $\mathring{X}\in V^P$ and $\alpha+n\geq 1$, where $\alpha=0$, or α is a limit ordinal, and $n\in\omega$. Let $p \Vdash_P$ "$\mathring{X}$ has rank at most $\alpha+n$". Then there are $q\leq p$ and $\mathring{Y}\in V^P_{\alpha+2n}$ so that $q \Vdash_P$ "$\mathring{X}=\mathring{Y}$".

Proof: By contradiction. Let α, n be the least such that the negation holds.

(1) If $n=0$ (so α is limit), then $p \Vdash_P$ "$(\exists\beta<\alpha)(\mathring{X}$ has rank at most $\beta)$". There are $q\leq p$ and $\beta<\alpha$ so that $q\Vdash_P$ "$\mathring{X}$ has rank at most β". By the minimality of α, n there are $\bar{q}\leq q$ and $\mathring{Y}\in V^P_\beta$ such that $\bar{q}\Vdash_P$ "$\mathring{X}=\mathring{Y}$", a contradiction.

(2) So $n\geq 1$.

Let $t=\langle\mathring{x},q\rangle\in\mathring{X}$. Define $D_t=\{r\leq q:$ (r incompatible with p) or ($r\leq p$ and for some $\mathring{z}\in V^P_{\alpha+2n-2}$, $r\Vdash_P$ "$\mathring{z}=\mathring{x}$")$\}$. We claim that D_t is dense below q.

Let $q'\leq q$. We are to show that there is $q''\leq q'$ so that $q''\in D_t$. If q' is incompatible with p, then q' is in D_t and we are done. If on the other hand q' is compatible with p, then there is $\bar{q}\leq p,q'$. Then $\bar{q}\Vdash_P$ "$\mathring{x}\in\mathring{X}$ and has rank at most $\alpha+n-1$". By the minimality of α, n there are $q''\leq\bar{q}$ and $\mathring{z}\in V^P_{\alpha+2n-2}$ so that $q''\Vdash_P$ "$\mathring{x}=\mathring{z}$". Therefore $q''\in D_t$ and we are done. The claim is proven.

Let A_t be a maximal antichain in D_t. For every $a\in A_t$ there is $\mathring{z}_a\in V^P_{\alpha+2n-2}$ so that $a\Vdash_P$ "$\mathring{x}=\mathring{z}_a$" whenever a is compatible with p. By Lemma 9 there is $\mathring{R}_t\in V^P_{\alpha+2n-1}$ so that $a\Vdash_P$ "$\mathring{R}_t=\mathring{z}_a$" for every $a\in A_t$, hence $a\Vdash_P$ "$\mathring{R}_t=\mathring{x}$" whenever $a\in A_t$ si compatible with p. Define $\mathring{Y}=\{\langle\mathring{R}_t,q\rangle:\langle\mathring{x},q\rangle\in\mathring{X}\}$. Then $\mathring{Y}\in V^P_{\alpha+2n}$.

Let G be P-generic over V so that $p\in G$. If $t=\langle\mathring{x},q\rangle\in\mathring{X}$ and $q\in G$, then some $a\in A_t$ is in G (and hence compatible with p) and since $a\Vdash_P$ "$\mathring{R}_t=\mathring{x}$", $(\mathring{x})^G=(\mathring{R}_t)^G$. Thus $(\mathring{X})^G=\{(\mathring{x})^G: t=\langle\mathring{x},q\rangle\in\mathring{X}\ \&\ q\in G\}=\{(\mathring{x})^G:\langle\mathring{R}_t,q\rangle\in\mathring{Y}\ \&\ q\in G\}=\{(\mathring{R}_t)^G:\langle\mathring{R}_t,q\rangle\in\mathring{Y}\ \&\ q\in G\}=(\mathring{Y})^G$. Therefore there is some $q\leq p$ so that $q\Vdash_P$ "$\mathring{X}=\mathring{Y}$" and $\mathring{Y}\in V^P_{\alpha+2}$, a contradiction. □

Lemma 11: Let P be a λ-c.c. poset, λ a regular cardinal, $p\in P$ and $\mathring{X}\in V^P$. Let $p\Vdash_P$ "$\mathring{X}$ has rank at most λ". Then there is $\mathring{Y}\in V^P_\lambda$ so that $p\Vdash_P$ "$\mathring{X}=\mathring{Y}$".

Proof: By Lemma 10, $D=\{q\leq p:(\exists\mathring{Y}\in V^P_\lambda)(q\Vdash_P$ "$\mathring{X}=\mathring{Y}$"$)\}$ is dense below p. Let A be a maximal antichain in D. For every $a\in A$ there is $\mathring{X}_a\in V^P_\lambda$ be so that $a\Vdash_P$ "$\mathring{X}=\mathring{X}_a$". Since λ is regular and $|A|<\lambda$ (as P satisfies the λ-c.c.), there is $\beta<\lambda$ so that $\mathring{X}_a\in V^P_\beta$ for each $a\in A$. By Lemma 9, there is $\mathring{Y}\in V^P_{\beta+1}\subset V^P_\lambda$ so that $a\Vdash_P$ "$\mathring{Y}=\mathring{X}_a=\mathring{X}$" for every $a\in A$. Hence $p\Vdash_P$ "$\mathring{Y}=\mathring{X}$". □

Lemma 12: Let P be a λ-c.c. poset, λ a regular cardinal. Let $p\in P$, $X\in V$, $\mathring{Y}\in V^P$, and $p\Vdash_P$ "$\mathring{Y}\subset X\ \&\ |\mathring{Y}|\leq\lambda$". Then there are $q\leq p$ and $\hat{Y}\in V^P$ so that $|\hat{Y}|\leq\lambda$ and $q\Vdash_P$ "$\mathring{Y}=\hat{Y}$".

Proof: Let $f:\rho\to X$ be a bijection. There are $\mathring{g}\in V^P$, $\xi\leq\lambda$ and $q\leq p$ so that $q\Vdash_P$ "$\mathring{g}:\xi\to\mathring{Y}$". Hence $q\Vdash_P$ "$(\exists\beta\in\rho)(\mathring{g}(\alpha)=f(\beta))$", for any $\alpha<\xi$. Let $D_\alpha=\{r\leq q:(\exists\beta\in\rho)(r\Vdash_P$ "$\mathring{g}(\alpha)=f(\beta)$"$)\}$. Then D_α is dense below q. Let A_α be a maximal antichain in D_α. Define $\mathring{h}\in V^P$ by $\langle\langle\alpha,f(\beta)\rangle,r\rangle\in\mathring{h}$ iff $r\in A_\alpha$ and $r\Vdash_P$ "$\mathring{g}(\alpha)=f(\beta)$". Since $|A_\alpha|\leq\lambda$ for every $\alpha<\xi$, and λ is a regular cardinal in V, $|\mathring{h}|\leq\lambda$. It is left to the reader to check that (1) $\Vdash_P$ "$\mathring{h}\subset\xi\times X$", (2) $q\Vdash_P$ "$\mathring{h}$ is a function", (3) $q\Vdash_P$ "$dom(\mathring{h})=\xi$", (4) $q\Vdash_P$ "$\mathring{h}\subset\mathring{g}$", and so $q\Vdash_P$ "$\mathring{h}=\mathring{g}$".

Now define $\hat{Y}\in V^P$ by $\hat{Y}=\{\langle f(\beta),r\rangle:(\exists\alpha<\xi)(\langle\langle\alpha,f(\beta)\rangle,r\rangle\in\mathring{h})\}$. Thus $|\hat{Y}|\leq\lambda$ and it is left to the reader to check that (5) $q\Vdash_P$ "$\mathring{h}''\xi\subset\hat{Y}$", (9) $q\Vdash_P$ "$\hat{Y}\subset\mathring{h}''\xi$". Thus $q\Vdash_P$ "$\hat{Y}=\mathring{h}''\xi=\mathring{g}''\xi=\mathring{Y}$". □

Def. 13: P,Q be posets. A mapping $i:P\to Q$ is a complete (regular) embedding of P into Q, iff

(13.1) for every $p,q \in P$, if $p \leq q$ in P, then $i(p) \leq i(q)$ in Q;

(13.2) for every $p,q \in P$, if $p \| q$ in P, then $i(p) \| i(q)$ in Q;

(13.3) for every $q \in Q$ there is $p \in P$ so that whenever $p' \in P$ and $p' \| p$ in P, then $i(p') \| q$ in Q (we shall denote this relationship between p and q by $p \prec_P q$ in Q).

$P \subset\subset Q$ denotes that P is a complete suborder of Q, i.e. the identity is a complete embedding of P into Q.

Note: (13.3) can be replaced by: for every A, a maximal antichain (a set of mutually incompatible elements) in P, $i''A$ is a maximal antichain in Q.

Def. 14: Let P be a poset. We shall say that P is separative if for every $p,q \in P$, whenever $p \not\leq q$, then there is $p' \in P$ so that $p' \leq p$ and $p' \perp q$.

Lemma 15: Let P,Q be posets such that $P \subset\subset Q$ and P is separative. Let $p, p_1, p_2 \in P$, $q \in Q$. Let $p \prec_P q$ and $q \leq p_1, p_2$ in Q. Then $p \leq p_1, p_2$ in P.

Proof: Assume that $p \not\leq p_1$ in P. Then by separativeness of P, there is $P_3 \in P$ such that $p_3 \leq p$ and $p_3 \perp p_1$ in P. Thus $p_3 \| p$ in P. Since $p \prec_P q$, $p_3 \| q$ in Q. Therefore $p_3 \| p_1$ in Q, and since $P \subset\subset Q$, $p_3 \| p_1$ in P, a contradiction. Thus $p \leq p_1$, and by the same argument $p \leq p_2$. □

Lemma 16: Let $j: V \to M$ be an elementary embedding. Let P be a poset. Then $j|P: P \to j(P)$ satisfies (13.1) and (13.2).

Proof: Left to the reader. □

Lemma 17: Let P,Q be posets such that $P \subset\subset Q$. Let $\phi(x_1, ..., x_n)$ be an upward absolute formula. Let $\mathring{X}_1, ..., \mathring{X}_n \in V^P$. Let $p \in P$, $q \in Q$ and $q \leq p$ in Q. Let $p \Vdash_P$ "$\phi(\mathring{X}_1, ..., \mathring{X}_n)$". Then $q \Vdash_Q$ "$\phi(\mathring{X}_1, ..., \mathring{X}_n)$".

Proof: Let $q_1 \leq q$ in Q. Let G be Q-generic over V so that $q_1 \in G$. Since $p \Vdash_P$ "$\phi(\mathring{X}_1, ..., \mathring{X}_n)$", $V[G \cap P] \models$ "$\phi(\mathring{X}_1^{G \cap P}, ..., \mathring{X}_n^{G \cap P})$". Since each $\mathring{X}_i \in V^P$, $\mathring{X}_i^{G \cap P} = \mathring{X}_i^G$, and by upward absolutness of ϕ, $V[G] \models$ "$\phi(\mathring{X}_1^G, ..., \mathring{X}_n^G)$". Thus for some $q_2 \in G$ $q_2 \Vdash_Q$ "$\phi(\mathring{X}_1, ..., \mathring{X}_n)$". Then $q_2 \| q_1$ in Q, and so there is $q_3 \in Q$, $q_3 \leq q_2, q_1$ and $q_3 \Vdash_Q$ "$\phi(\mathring{X}_1, ..., \mathring{X}_n)$". Hence $q \Vdash_Q$ "$\phi(\mathring{X}_1, ..., \mathring{X}_n)$". □

Lemma 18: Let P,Q be posets such that $P \subset\subset Q$. Let $\phi(x_1, ..., x_n)$ be a downward absolute formula. Let $\mathring{X}_1, ..., \mathring{X}_n \in V^P$. Let $p \in P$ and let $p \Vdash_Q$ "$\phi(\mathring{X}_1, ..., \mathring{X}_n)$". Then $p \Vdash_P$ "$\phi(\mathring{X}_1, ..., \mathring{X}_n)$".

Proof: Let $p_1 \leq p$ in P. Then $p_1 \leq p$ in Q. Let G be Q-generic over V so that $p_1 \in G$. Since $p_1 \Vdash_Q$ "$\phi(\mathring{X}_1, ..., \mathring{X}_n)$", $V[G] \models$ "$\phi(\mathring{X}_1^G, ..., \mathring{X}_n^G)$". Since each $\mathring{X}_i \in V^P$, $\mathring{X}_i^{G \cap P} = \mathring{X}_i^G$, and by downward absolutness of ϕ, $V[G \cap P] \models$ "$\phi(\mathring{X}_1^{G \cap P}, ..., \mathring{X}_n^{G \cap P})$". Thus for some $p_2 \in G \cap P$ $p_2 \Vdash_P$ "$\phi(\mathring{X}_1, ..., \mathring{X}_n)$". Then $p_2 \| p_1$ in P, and so there is $p_3 \in P$, $p_3 \leq p_2, p_1$ and $p_3 \Vdash_P$ "$\phi(\mathring{X}_1, ..., \mathring{X}_n)$". Hence $p \Vdash_P$ "$\phi(\mathring{X}_1, ..., \mathring{X}_n)$". □

Lemma 19: Let P,Q be posets such that $P \subset\subset Q$. Let $\phi(x_1, ..., x_n)$ be a downward absolute formula. Let $\mathring{X}_1, ..., \mathring{X}_n \in V^P$. Let $p \in P$, $q \in Q$, $p \prec_P q$, and let $q \Vdash_Q$ "$\phi(\mathring{X}_1, ..., \mathring{X}_n)$". Then $p \Vdash_P$ "$\phi(\mathring{X}_1, ..., \mathring{X}_n)$".

Proof: Let $p_1 \leq p$ in P. Then $p_1 \| p$ in P, and so $p_1 \| q$ in Q. Let $q_1 \in Q$ be such that $q_1 \leq p_1, q$ in Q. Let G be Q-generic over V so that $q_1 \in G$. Since $q_1 \Vdash_Q$ "$\phi(\mathring{X}_1, ..., \mathring{X}_n)$", $V[G] \models$ "$\phi(\mathring{X}_1^G, ..., \mathring{X}_n^G)$". Since each $\mathring{X}_i \in V^P$, $\mathring{X}_i^{G \cap P} = \mathring{X}_i^G$, and by downward absolutness of ϕ, $V[G \cap P] \models$ "$\phi(\mathring{X}_1^{G \cap P}, ..., \mathring{X}_n^{G \cap P})$". Thus for some $p_2 \in G \cap P$ $p_2 \Vdash_P$ "$\phi(\mathring{X}_1, ..., \mathring{X}_n)$". Then $p_2 \| p_1$ in P, and so there is $p_3 \in P$, $p_3 \leq p_2, p_1$ and $p_3 \Vdash_P$ "$\phi(\mathring{X}_1, ..., \mathring{X}_n)$". Hence $p \Vdash_P$ "$\phi(\mathring{X}_1, ..., \mathring{X}_n)$". □

Def. 20: Let P, Q be posets so that $P \subset\subset Q$. Let $\mathring{R} \in V^P$ so that $\Vdash_P$ "$\mathring{R}$ *is a poset*". Define $Q \otimes_P \mathring{R} = \{\langle p, \mathring{q}\rangle : q \in Q, \mathring{r} \in V^P, q \Vdash_Q \text{"}\mathring{r} \in \mathring{R}\text{"}\}$, $\langle q_1, \mathring{r}_1\rangle \leq \langle q_2, \mathring{r}_2\rangle$ iff $q_1 \leq q_2$ in Q, and $q_1 \Vdash_Q$ "$\mathring{r}_1 \leq \mathring{r}_2$ *in* $\mathring{R}$".

Lemma 21: Let P, Q be posets such that $P \subset\subset Q$. Let $\mathring{R} \in V^P$ so that $\Vdash_P$ "$\mathring{R}$ *is a poset*". Then $Q \otimes_P \mathring{R}$ is dense in $Q * \mathring{R}$.

Proof: Obviously $Q \otimes_P \mathring{R}$ is a suborder of $Q * \mathring{R}$. Let $\langle q, \mathring{r}\rangle \in Q * \mathring{R}$. Then $q \in Q$, $\mathring{r} \in V^Q$ and $q \Vdash_Q$ "$\mathring{r} \in \mathring{R}$". Let G be Q-generic over V so that $q \in G$. Since $\mathring{R} \in V^P$ and $G \cap P$ is P-generic over V, $\mathring{R}^G = \mathring{R}^{G \cap P} \in V[G \cap P]$. Thus $\mathring{r}^G \in V[G \cap P]$, and so there is $\mathring{r}_1 \in V^P$ such that $\mathring{r}_1^{G \cap P} = \mathring{r}^G$. Thus for some $q_1 \in G$, $q_1 \Vdash_Q$ "$\mathring{r}_1 = \mathring{r}$". Then $q_1 \supset\!\subset q$ in Q, and so there is $q_3 \in Q$ such that $q_2 \leq q_1, q$ in Q and $q_2 \Vdash_Q$ "$\mathring{r}_1 = \mathring{r}$". Then $\langle q_2, \mathring{r}_1\rangle \in Q \otimes_P \mathring{R}$ and $\langle q_2, \mathring{r}_1\rangle \leq \langle q, \mathring{r}\rangle$ in $Q * \mathring{R}$. □

Lemma 22: Let P, Q, R be posets so that $P \subset\subset Q \subset\subset R$. Let $\mathring{S} \in V^P$ and let $\Vdash_P$ "$\mathring{S}$ *is s poset*". Let Q be separative. Then $Q \otimes_P \mathring{S} \subset\subset R \otimes_P \mathring{S}$.

Proof: Obviously $Q \otimes_P \mathring{S} \subset R \otimes_P \mathring{S}$.
First we verify that (13.1) holds. Let $\langle q_1, \mathring{s}_1\rangle \leq \langle q_2, \mathring{s}_2\rangle$ in $Q \otimes_P \mathring{S}$. Then $q_1 \leq q_2$ in Q and $q_1 \Vdash_Q$ "$\mathring{s}_1 \leq \mathring{s}_2$ *in* $\mathring{S}$". By Lemma 17 $q_1 \Vdash_R$ "$\mathring{s}_1 \leq \mathring{s}_2$ *in* $\mathring{S}$". As $q_1 \leq q_2$ in R, $\langle q_1, \mathring{s}_1\rangle \leq \langle q_1, \mathring{s}_1\rangle$ in $R \otimes_P \mathring{S}$.
Next we verify that (13.2) holds. Let $\langle q_1, \mathring{s}_1\rangle \supset\!\subset \langle q_2, \mathring{s}_2\rangle$ in $R \otimes_P \mathring{S}$. There is $\langle r, \mathring{s}\rangle \in R \otimes_P \mathring{S}$ so that $\langle r, \mathring{s}\rangle \leq \langle q_1, \mathring{s}_1\rangle, \langle q_2, \mathring{s}_2\rangle$ in $R \otimes_P \mathring{S}$. Thus $r \leq q_1, q_2$ in R and $r \Vdash_R$ "$\mathring{s} \leq \mathring{s}_1, \mathring{s}_2$ *in* $\mathring{S}$". Since $Q \subset\subset R$, there is $q \in Q$ so that $q \prec_Q r$. By Lemma 5 $q \leq q_1, q_2$ in Q. Since "$s \leq s_1$ *in* S" is an absolute formula, by Lemma 19 $q \Vdash_Q$ "$\mathring{s} \leq \mathring{s}_1, \mathring{s}_2$ *in* $\mathring{S}$". By Lemma 17 $q \Vdash_R$ "$\mathring{s} \leq \mathring{s}_1, \mathring{s}_2$ *in* $\mathring{S}$". So $\langle q, \mathring{s}\rangle \in Q \otimes_P \mathring{S}$, and thus $\langle q, \mathring{s}\rangle \leq \langle q_1, \mathring{s}_1\rangle, \langle q_2, \mathring{s}_2\rangle$ in $Q \otimes_P \mathring{S}$.
Verify that (13.3) holds. Let $\langle r, \mathring{s}\rangle \in R \otimes_P \mathring{S}$. Then $\mathring{s} \in V^P$, $r \in R$ and $r \Vdash_R$ "$\mathring{s} \in \mathring{S}$". Since $Q \subset\subset R$, there is $q \in Q$ such that $q \prec_Q r$. Since "$x \in X$" is absolute, by Lemma 19 $q \Vdash_Q$ "$\mathring{s} \in \mathring{S}$", and by Lemma 17 $q \Vdash_R$ "$\mathring{s} \in \mathring{S}$". So $\langle q, \mathring{s}\rangle \in Q \otimes_P \mathring{S}$. We shall show that $\langle q, \mathring{s}\rangle \prec \langle r, \mathring{s}\rangle$: let $\langle q_1, \mathring{s}_1\rangle \in Q \otimes_P \mathring{S}$ so that $\langle q_1, \mathring{s}_1\rangle \supset\!\subset \langle q, \mathring{s}\rangle$ in $Q \otimes_P \mathring{S}$. Thus there is $\langle q_2, \mathring{s}_2\rangle \in Q \otimes_P \mathring{S}$ such that $\langle q_2, \mathring{s}_2\rangle \leq \langle q_1, \mathring{s}_1\rangle, \langle q, \mathring{s}\rangle$ in $Q \otimes_P \mathring{S}$. Hence $q_2 \leq q_1, q$ in Q, and $q \Vdash_Q$ "$\mathring{s}_2 \leq \mathring{s}_1, \mathring{s}$ *in* $\mathring{S}$". Since $q_2 \supset\!\subset q$ in Q, $q_2 \supset\!\subset r$ in R. Let $r_1 \in R$ so that $r_1 \leq r, q_2$ in R. Since "$s_2 \leq s_1, s$ *in* S" is absolute, by Lemma 17 $r_1 \Vdash_R$ "$\mathring{s}_2 \leq \mathring{s}_1, \mathring{s}$ *in* $\mathring{S}$". Then $\langle r_1, \mathring{s}_2\rangle \in R \otimes_P \mathring{S}$ and $\langle r_1, \mathring{s}_2\rangle \leq \langle r, \mathring{s}\rangle, \langle q_1, \mathring{s}_1\rangle$ in $R \otimes_P \mathring{S}$. Hence $\langle q_1, \mathring{s}_1\rangle \supset\!\subset \langle r, \mathring{s}\rangle$ in $R \otimes_P \mathring{S}$. □

Lemma 23: Let P, Q be posets so that $P \subset\subset Q$. Let $\mathring{S}, \mathring{R} \in V^P$ so that $\Vdash_P$ "$\mathring{R} \subset\subset \mathring{S}$". Then $Q \otimes_P \mathring{R} \subset\subset Q \otimes_P \mathring{S}$.

Proof: Obviously $Q \otimes_P \mathring{R}$ is a suborder of $Q \otimes_P \mathring{S}$.
Let's verify (13.2). Let $\langle q_1, \mathring{r}_1\rangle, \langle q_2, \mathring{r}_2\rangle \in Q \otimes_P \mathring{S}$ be so that $\langle q_1, \mathring{r}_1\rangle \supset\!\subset \langle q_2, \mathring{r}_2\rangle$ in $Q \otimes_P \mathring{S}$. There is $\langle q_3, \mathring{r}_3\rangle \in Q \otimes_P \mathring{S}$ so that $\langle q_3, \mathring{r}_3\rangle \leq \langle q_1, \mathring{r}_1\rangle, \langle q_2, \mathring{r}_2\rangle$ in $Q \otimes_P \mathring{S}$. Hence $q_3 \leq q_1, q_2$ in Q and $q_3 \Vdash_Q$ "$\mathring{s}_3 \leq \mathring{r}_1, \mathring{r}_2$ *in* $\mathring{S}$". So $q_3 \Vdash_Q$ "$\mathring{r}_1 \supset\!\subset \mathring{r}_2$ *in* $\mathring{S}$". And thus $q_3 \Vdash_Q$ "$\mathring{r}_1 \supset\!\subset \mathring{r}_2$ *in* $\mathring{R}$". It follows that $q_3 \Vdash_Q$ "$(\exists\, \mathring{r}_3)(\mathring{r}_3 \leq \mathring{r}_1, \mathring{r}_2$ *in* $\mathring{R})$". There is $\mathring{r}_3 \in V^Q$ so that $q_3 \Vdash_Q$ "$\mathring{r}_3 \leq \mathring{r}_1, \mathring{r}_2$ *in* $\mathring{R}$". Then $\langle q_3, \mathring{r}_3\rangle \in Q * \mathring{R}$. By Lemma 21 there is $\langle q_4, \mathring{r}_4\rangle \in Q \otimes_P \mathring{R}$ so that $\langle q_4, \mathring{r}_4\rangle \leq \langle q_3, \mathring{r}_3\rangle$ in $Q * \mathring{R}$. $q_4 \Vdash_Q$ "$\mathring{r}_4 \leq \mathring{r}_1, \mathring{r}_2$ *in* $\mathring{R}$". Therefore $\langle q_4, \mathring{r}_4\rangle \leq \langle q_1, \mathring{r}_1\rangle, \langle q_2, \mathring{r}_2\rangle$ in $Q \otimes_P \mathring{R}$, and so $\langle q_1, \mathring{r}_1\rangle \supset\!\subset \langle q_2, \mathring{r}_2\rangle$ in $Q \otimes_P \mathring{R}$.
Let's verify (13.3). Let $\langle q, \mathring{s}\rangle \in Q \otimes_P \mathring{S}$. Then $q \Vdash_Q$ "$\mathring{s} \in \mathring{S}$". So $q \Vdash_Q$ "$(\exists \mathring{r} \in \mathring{R})(\mathring{r} \prec_{\mathring{R}} \mathring{s})$", since

$q \Vdash_Q$ "$\mathring{R} \subset\subset \mathring{S}$". Thus there is $\mathring{r} \in V^Q$ such that $q \Vdash_Q$ "$\mathring{r} \prec_{\mathring{R}} \mathring{s}$", and so $\langle q,\mathring{r}\rangle \in Q*\mathring{R}$. By Lemma 21 there is $\langle q_1,\mathring{r}_1\rangle \in Q\otimes_P\mathring{R}$ so that $\langle q_1,\mathring{r}_1\rangle \le \langle q,\mathring{r}\rangle$ in $Q*\mathring{R}$. We shall show that $\langle q_1,\mathring{r}_1\rangle$ is the element in $Q\otimes_P\mathring{R}$ we are looking for: let $\langle q_2,\mathring{r}_2\rangle \in Q\otimes_P\mathring{R}$ so that $\langle q_2,\mathring{r}_2\rangle \sqsupset\!\sqsubset \langle q_1,\mathring{r}_1\rangle$ in $Q\otimes_P\mathring{R}$. Then there is $\langle q_3,\mathring{r}_3\rangle \in Q\otimes_P\mathring{R}$ so that $\langle q_3,\mathring{r}_3\rangle \le \langle q_2,\mathring{r}_2\rangle, \langle q_1,\mathring{r}_1\rangle$ in $Q\otimes_P\mathring{R}$. Then $q_3 \le q_1, q_2$ in Q and $q_3 \Vdash_Q$ "$\mathring{r}_3 \le \mathring{r}_1,\mathring{r}_2$ *in* $\mathring{R}$". Also $q_3 \Vdash_Q$ "$\mathring{r}_1 \le \mathring{r} \prec_{\mathring{R}} \mathring{s}$". Thus $q_3 \Vdash_Q$ "$\mathring{r}_3 \sqsupset\!\sqsubset \mathring{s}$ *in* $\mathring{S}$", and so $q_3 \Vdash_Q$ "$(\exists\, \mathring{s}_1)(\mathring{s}_1 \le \mathring{r}_3,\mathring{s}$ *in* $\mathring{S})$". Thus there is $\mathring{s}_1 \in V^Q$ so that $q_3 \Vdash_Q$ "$\mathring{s}_1 \le \mathring{r}_3,\mathring{s}$ *in* $\mathring{S}$". It follows that $\langle q_3,\mathring{s}_1\rangle \in Q*\mathring{S}$. By Lemma 21 there is $\langle q_4,\mathring{s}_2\rangle \in Q\otimes_P\mathring{S}$ so that $\langle q_4,\mathring{s}_2\rangle \le \langle q_3,\mathring{s}_1\rangle$. Then $q_4 \Vdash_Q$ "$\mathring{s}_2 \le \mathring{r}_3,\mathring{s}$ *in* $\mathring{S}$". Thus $\langle q_4,\mathring{s}_2\rangle \le \langle q_2,\mathring{r}_2\rangle, \langle q,\mathring{s}\rangle$ in $Q\otimes_P\mathring{S}$. Hence It follows that $\langle q_2,\mathring{r}_2\rangle \sqsupset\!\sqsubset \langle q,\mathring{s}\rangle$ in $Q\otimes_P\mathring{S}$. □

Lemma 24: Let P, Q be posets. Let $\mathring{R}\in V^P$ and let $\Vdash_P$ "$\mathring{R}$ *is a separative poset*". Let Q be separative. Then $Q\otimes_P\mathring{R}$ is separative.

Proof: We shall prove that $Q*\mathring{R}$ is separative; since $Q\otimes_P\mathring{R}$ is dense in $Q*\mathring{R}$ it follows that $Q\otimes_P\mathring{R}$ must also be separative, too.
Let $\langle q_1,\mathring{r}_1\rangle \not\le \langle q_2,\mathring{r}_2\rangle$ in $Q*\mathring{R}$. There are two possible cases:
(i) $q_1 \not\le q_2$ in Q. Then by separativness of Q, there is $q_3\in Q$ such that $q_3 \le q_1$ and $q_3 \supset\subset q_2$ in Q. Then $\langle q_3,\mathring{r}_1\rangle \in Q*\mathring{R}$, $\langle q_3,\mathring{r}_1\rangle \le \langle q_1,\mathring{r}_1\rangle$, and $\langle q_3,\mathring{r}_1\rangle \supset\subset \langle q_2,\mathring{r}_2\rangle$ in $Q*\mathring{R}$.
(ii) $q_1 \le q_2$ in Q. Then $q_1 \not\Vdash_Q$ "$\mathring{r}_1 \le \mathring{r}_2$ *in* $\mathring{R}$". So there is $q_3\in Q$, $q_3 \le q_1$ in Q, so that $q_3 \Vdash_Q$ "$\mathring{r}_1 \not\le \mathring{r}_2$ *in* $\mathring{R}$", and so $q_3 \Vdash_Q$ "$(\exists\, \mathring{r}_3)(\mathring{r}_3 \le \mathring{r}_1$ *and* $\mathring{r}_3 \supset\subset \mathring{r}_2$ *in* $\mathring{R})$". So for some $\mathring{r}_3 \in V^Q$, $q_3 \Vdash_Q$ "$\mathring{r}_3 \le \mathring{r}_1$ *and* $\mathring{r}_3 \supset\subset \mathring{r}_2$ *in* $\mathring{R}$". Thus, $\langle q_3,\mathring{r}_3\rangle \in Q*\mathring{R}$. Hence $\langle q_3,\mathring{r}_3\rangle \le \langle q_1,\mathring{r}_1\rangle$, and $\langle q_3,\mathring{r}_3\rangle \supset\subset \langle q_2,\mathring{r}_2\rangle$ in $Q*\mathring{R}$. Hence $Q*\mathring{R}$ is separative. □

Lemma 25: Let $M\subset V$ be a transitive model of ZFC.

(25.1) Let $P\in M$ be a poset and ϕ a restricted formula with n free variable and no constants. Let $p\in P$ and let $\mathring{X}_1, \dots ,\mathring{X}_n\in M^P$. Then $p \Vdash_P$ "$\phi(\mathring{X}_1, \dots ,\mathring{X}_n)$" iff $p \Vdash^M_P$ "$\phi(\mathring{X}_1, \dots ,\mathring{X}_n)$".

(25.2) If $P\in M_\lambda$ and ${}^{<\lambda}M_\lambda\subset M$ and P satisfies the λ-c.c. in V, then "G is P-generic over V" iff "G is P-generic over M".

Proof:

(25.1) Let $p \Vdash^M_P$ "$\phi(\mathring{X}_1, \dots ,\mathring{X}_n)$" and assume that $p \not\Vdash_P$ "$\phi(\mathring{X}_1, \dots ,\mathring{X}_n)$". Then for some $q \le p$, $q \Vdash_P$ "$\neg\phi(\mathring{X}_1, \dots ,\mathring{X}_n)$". Choose G, a P-generic filter over V (and so over M as well) so that $q\in G$. Then $V[G]\models$ "$\neg\phi(X_1, \dots, X_n)$" and so (as ϕ is restricted and $X_1, \dots, X_n\in M[G]$) $M[G]\models$ "$\neg\phi(X_1, \dots, X_n)$", where $X_i = (\mathring{X}_i)^G$ for $i = 1, \dots, n$. Since $p\in G$, $M[G]\models$ "$\phi(X_1, \dots, X_n)$", a contradiction. Hence $p \Vdash_P$ "$\phi(\mathring{X}_1, \dots ,\mathring{X}_n)$".
The opposite direction: let $p \Vdash_P$ "$\phi(\mathring{X}_1, \dots ,\mathring{X}_n)$". Assume that $p \not\Vdash^M_P$ "$\phi(\mathring{X}_1, \dots ,\mathring{X}_n)$". Then for some $q \le p$, $q \Vdash^M_P$ "$\neg\phi(\mathring{X}_1, \dots ,\mathring{X}_n)$". By the same argument as above, $q \Vdash_P$ "$\neg\phi(\mathring{X}_1, \dots ,\mathring{X}_n)$", a contradiction as $q \le p$.

(25.2) One direction is easy and so is left to the reader. For the other direction assume that G is P-generic over M. Let D be dense in P. Let A be a maximal antichain in D. Since $|A| < \lambda$, $A\in M$ and so $G\cap A \neq \emptyset$. Hence $G\cap D \neq \emptyset$. □

Lemma 26: Let $j{:}V\to M$ be an elementary embedding, $M\subset V$, and let j be definable in V. Let $P\in V$ be a poset. Let G be P-generic over V and let H be $j(P)$-generic over M so that if $p\in G$, then $j(p)\in H$. Then

(26.1) there is an elementary embedding $\bar{j}{:}V[G]\to M[H]$ definable in $V[H]$ and extending j;

(26.2) if ${}^\lambda M\subset M$ and $j(P)$ satisfies the λ-c.c. in V, λ regular, then if $X\in V$, $Y\in V[H]$, $Y\subset X$, $|Y| < \lambda$, and $Y\subset M[H]$, then $Y\in M[H]$;

(26.3) if $M_\alpha = V_\alpha$ for all $\alpha \leq \lambda$, λ a limit ordinal, and if $P \in V_\lambda$, then $V[H]_\alpha = M[H]_\alpha$ for all $\alpha \leq \lambda$.

Proof:

(26.1) Define $\hat{\jmath}((\mathring{X})^G) = (j(\mathring{X}))^H$.
$\hat{\jmath}$ is well-defined, because if $(\mathring{X})^G = (\mathring{Y})^G$, then for some $p \in G$, $p \Vdash_P$ "$\mathring{X} = \mathring{Y}$", and so $j(p) \Vdash^M_{j(P)}$ "$j(\mathring{X}) = j(\mathring{Y})$" by elementarity of j. Since $j(p) \in H$, $\hat{\jmath}((\mathring{X})^G) = (j(\mathring{X}))^H = (j(\mathring{Y}))^H = \hat{\jmath}((\mathring{Y})^G)$. $\hat{\jmath}$ is elementary, for if $V[G] \models$ "$\phi((\mathring{X}_1)^G, \ldots, (\mathring{X}_n)^G)$", then $p \Vdash_P$ "$\phi(\mathring{X}_1, \ldots, \mathring{X}_n)$" for some $p \in G$, and so $j(p) \Vdash^M_{j(P)}$ "$\phi(j(\mathring{X}_1), \ldots, j(\mathring{X}_n))$". Since $j(p) \in H$, $M[H] \models$ "$\phi((j(\mathring{X}_1))^H, \ldots, (j(\mathring{X}_n))^H)$", so $M[H] \models$ "$\phi(\hat{\jmath}((\mathring{X}_1)^G), \ldots, \hat{\jmath}((\mathring{X}_n)^G))$".
$\hat{\jmath}$ extends j, for if $X \in V$, then $(X)^G = X$ and so $\hat{\jmath}(X) = j(X)$.

(26.2) Let $\mathring{Y} \in V^{j(P)}$ so that $Y = (\mathring{Y})^H$. By Lemma 12 we can assume WLOG that $|\mathring{Y}| \leq \lambda$. Let $\langle \mathring{y}, q \rangle \in \mathring{Y}$. Then $(\mathring{y})^H \in (\mathring{Y})^H = Y \subset M[H]$, so there is $\hat{y} \in M^{j(P)}$ such that $(\hat{y})^H = (\mathring{y})^H$. Define $\hat{Y} = \{\langle \hat{y}, q \rangle : \langle \mathring{y}, q \rangle \in \mathring{Y}\}$. Then $\hat{Y} \subset M^{j(P)} \subset M$ and $|\hat{Y}| = |\mathring{Y}| \leq \lambda$, so $\hat{Y} \in {}^\lambda M \subset M$, hence $\hat{Y} \in M$. Since $\hat{Y} \in V^{j(P)}$, $\hat{Y} \in M^{j(P)}$ and thus $(\hat{Y})^H \in M[H]$. $(\hat{Y})^H = \{(\hat{y})^H : \langle \hat{y}, q \rangle \in \hat{Y}\} = \{(\mathring{y})^H : \langle \mathring{y}, q \rangle \in \mathring{Y}\} = (\mathring{Y})^H = Y$. Hence $Y \in M[H]$.

(26.3) will be proven by induction:

(i) $(V[H])_0 = (M[H])_0 = \emptyset$.

(ii) Assume that $(V[H])_\alpha = (M[H])_\alpha$ and $\alpha < \lambda$.
Let $(\mathring{X})^H \in (V[H])_{\alpha+1}$. Then $(\mathring{X})^H \subset (V[H])_\alpha = (M[H])_\alpha$. By Lemma 10, we can assume WLOG that $\mathring{X} \in V^P_{\alpha+2}$, and by Lemma 6, $V^P_{\alpha+2} \subset V_\lambda = M_\lambda$. Hence $(\mathring{X})^H \in M[H]$ and so $(\mathring{X})^H \in (M[H])_{\alpha+1}$. Thus $(V[H])_{\alpha+1} \subset (M[H])_{\alpha+1}$, and so $(V[H])_{\alpha+1} = (M[H])_{\alpha+1}$.

(iii) if $(V[H])_\beta = (M[H])_\beta$ for all $\beta < \alpha \leq \lambda$, α limit, then $(V[H])_\alpha = (M[H])_\alpha$. □

Lemma 27: Let $j{:}V \to M$ be huge with critical point κ. Let $P \in V_\kappa$ be a poset. Then

(27.1) $j(p) = p$ for all $p \in P$;

(27.2) $j(P) = P$;

(27.3) G is P-generic over V iff G is P-generic over M;

(27.4) for any G P-generic over V, there is a nearly huge $\hat{\jmath}{:}V[G] \to M[G]$ extending j.

Proof: (27.1) and (27.2) are easy and so they are left to the reader to prove.

(27.3) follows directly from Lemma 25.

(27.4) follows from Lemma 26. □

Properties 28: Let $C(\gamma, \delta)$, $\gamma < \delta$ cardinals, define in V a poset.

(28.1) $C(\gamma, \delta) \subset V_\delta$ for all cardinals $\gamma < \delta$, δ inacc.;

(28.2) $C(\gamma, \tau) = \{s \cap V_\tau : s \in C(\gamma, \delta)\}$, for all cardinals $\gamma < \tau \leq \delta$, τ, δ inacc.;

(28.3) for every $s \in C(\gamma, \delta)$, $s \leq s \cap V_\tau$ in $C(\gamma, \delta)$, for all cardinals $\gamma < \tau \leq \delta$, τ, δ inacc.;

(28.4) for every $s \in C(\gamma, \delta)$, $s \cap V_\tau \prec_{C(\gamma,\tau)} s$ in $C(\gamma, \delta)$, for all cardinals $\gamma < \tau \leq \delta$, τ, δ inacc.;

(28.5) $C(\gamma, \tau) \subset\subset C(\gamma, \delta)$, for all cardinals $\gamma < \tau \leq \delta$, τ, δ inacc.;

(28.6) $C(\gamma, \delta)$ is separative for all cardinals $\gamma \leq \delta$, δ inacc.

Properties 29: Let $I = \langle I_\alpha : \alpha \leq \kappa, \alpha \text{ limit} \rangle$ be a sequence such that:

(29.1) I_α is an ideal on α containing all finite subsets of α, for all limit $\alpha \leq \kappa$;

(29.2) $I_\alpha \subset I_\beta$ for all limit $\alpha \leq \beta \leq \kappa$;

(29.3) if α is inaccessible, than $x \in I_\alpha$ implies that $|x| < \alpha$.

Lemma 30: Let $|\{\alpha \in \kappa : \alpha \text{ inaccessible }\}| = \kappa$. Let $I = \langle I_\alpha : \alpha \text{ limit}, \alpha \leq \kappa \rangle$ be a sequence of ideals satisfying (29.1) - (29.3). Let C define a poset and satisfy (28.1) - (28.6). Then there exists an (iterated forcing) sequence $\langle P_\alpha : \alpha \leq \kappa \rangle$ such that

(30.1) $P_0 = C(\omega_0, \kappa)$;

(30.2) $P_{\alpha+1} = P_\alpha * \{\emptyset\}$ whenever $\alpha < \kappa$ is not inaccessible;

(30.3) $P_{\alpha+1} = P_\alpha \otimes_{P_\alpha \upharpoonright V_\alpha} C(\alpha,\kappa)^{P_\alpha \upharpoonright V_\alpha}$ whenever $\alpha < \kappa$ is inaccessible, where $C(\alpha,\kappa)^{P_\alpha \upharpoonright V_\alpha}$ denotes $C(\alpha,\kappa)$ as defined in the extension by $P_\alpha \upharpoonright V_\alpha$;
(this is a sound definition since it follows from (30.7) - (30.10) that $P_\alpha \upharpoonright V_\alpha \subset\subset P_\alpha$, see (3*) below) where for any β inacc. such that $\alpha \leq \beta \leq \kappa$ we define $P_\alpha \upharpoonright V_\beta = \{p \upharpoonright V_\beta : p \in P_\alpha\}$, and $p \upharpoonright V_\beta$ is defined as follows: $(p \upharpoonright V_\beta)(0) = p(0) \cap V_\beta$, if $p(\xi) = \emptyset$, then $(p \upharpoonright V_\beta)(\xi) = \emptyset$, and when $p(\xi) \neq \emptyset$ (hence ξ is inacc.), then $(p \upharpoonright V_\beta)(\xi) \in V_\beta^{P_\xi \upharpoonright V_\xi}$ so that $\Vdash_{P_\xi \upharpoonright V_\xi}$ "$(p \upharpoonright V_\beta)(\xi) = p(\xi) \cap V_\beta^{P_\xi \upharpoonright V_\xi}$" (such $P_\xi \upharpoonright V_\xi$-name in V_β exists by Lemma 11 as ξ is inacc.). The ordering is defined by $p \upharpoonright V_\beta \leq q \upharpoonright V_\beta$ in $P_\alpha \upharpoonright V_\beta$ if $p \upharpoonright V_\beta \leq q \upharpoonright V_\beta$ in P_α (this is a sound definition since by (30.7) $P_\alpha \upharpoonright V_\beta \subset P_\alpha$)

(30.4) For $\alpha \leq \kappa$ limit, P_α consists of all limits of conditions of $\langle P_\beta : \beta < \alpha \rangle$ with support from I_α;

And furthermore the sequence $\langle P_\alpha : \alpha \leq \kappa \rangle$ satisfies:

(30.5) P_α is separative for every $\alpha \leq \kappa$;

(30.6) $P_\alpha \subset V_\kappa$ for every $\alpha < \kappa$;

for every inacc. $\alpha \leq \kappa$, any inacc. β such that $\alpha \leq \beta \leq \kappa$:

(30.7) $P_\alpha \upharpoonright V_\beta \subset P_\alpha$;

(30.8) $P_\alpha \upharpoonright V_\beta$ is separative;

(30.9) $p \not\perp q$ in $P_\alpha \upharpoonright V_\beta$ iff $p \not\perp q$ in P_α for any $p, q \in P_\alpha \upharpoonright V_\beta$;

(30.10) $p \upharpoonright V_\beta \prec_{P_\alpha \upharpoonright V_\beta} p$ in P_α, for any $p \in P_\alpha$;

(30.11) $p \upharpoonright V_\beta \geq p$ in P_α, for any $p \in P_\alpha$;

(30.12) $\{p | supp(p) : p \in P_\alpha \upharpoonright V_\alpha\} \subset V_\alpha$.

Proof:

(1*) For any $\alpha \leq \kappa$, any $\gamma < \alpha$, and any β such that $\gamma < \beta \leq \kappa$, $(p|\gamma) \upharpoonright V_\beta = (p \upharpoonright V_\beta)|\gamma$, for any $p \in P_\alpha$. $((p|\gamma)\upharpoonright V_\beta)(0) = (p|\gamma)(0) \cap V_\beta = p(0) \cap V_\beta = (p \upharpoonright V_\beta)(0) = ((p \upharpoonright V_\beta)|\gamma)(0)$. If $p(\xi) = \emptyset$, then $(p|\gamma)(\xi) = \emptyset$ and so $((p|\gamma)\upharpoonright V_\beta)(\xi) = \emptyset = (p \upharpoonright V_\beta)(\xi) = ((p \upharpoonright V_\beta)|\gamma)(\xi)$. If $p(\xi) \neq \emptyset$, then ξ is inacc. and $\xi < \gamma$. Then $((p|\gamma)\upharpoonright V_\beta)(\xi) \in V_\beta^{P_\xi \upharpoonright V_\xi}$ so that $\Vdash_{P_\xi \upharpoonright V_\xi}$ "$((p|\gamma)\upharpoonright V_\beta)(\xi) = (p|\gamma)(\xi) \cap V_\beta^{P_\xi \upharpoonright V_\xi}$", hence $\Vdash_{P_\xi \upharpoonright V_\xi}$ "$((p|\gamma)\upharpoonright V_\beta)(\xi) = p(\xi) \cap V_\beta^{P_\xi \upharpoonright V_\xi}$", thus $((p|\gamma)\upharpoonright V_\beta)(\xi)$ and $(p \upharpoonright V_\beta)(\xi)$ are names for the same object.

(2*) For any $\alpha \leq \kappa$, any $\alpha < \gamma \leq \beta \leq \kappa$, γ, β inacc., $p \upharpoonright V_\gamma = (p \upharpoonright V_\beta)\upharpoonright V_\gamma = (p \upharpoonright V_\gamma)\upharpoonright V_\beta$, for any $p \in P_\alpha$. $(p \upharpoonright V_\gamma)(0) = p(0) \cap V_\gamma$. $((p \upharpoonright V_\gamma)\upharpoonright V_\beta)(0) = (p \upharpoonright V_\gamma)(0) \cap V_\beta = (p(0) \cap V_\gamma) \cap V_\beta = p(0) \cap V_\gamma$. $((p \upharpoonright V_\beta)\upharpoonright V_\gamma)(0) = (p \upharpoonright V_\beta)(0) \cap V_\gamma = (p(0) \cap V_\beta) \cap V_\gamma = p(0) \cap V_\gamma$. Thus $(p \upharpoonright V_\gamma)(0) = ((p \upharpoonright V_\gamma)\upharpoonright V_\beta)(0) = ((p \upharpoonright V_\beta)\upharpoonright V_\gamma)(0)$. If $p(\xi) = \emptyset$, then $(p \upharpoonright V_\gamma)(\xi) = \emptyset$, $((p \upharpoonright V_\gamma)\upharpoonright V_\beta)(\xi) = \emptyset$, and $((p \upharpoonright V_\beta)\upharpoonright V_\gamma)(\xi) = \emptyset$. Thus $(p \upharpoonright V_\gamma)(\xi) = ((p \upharpoonright V_\gamma)\upharpoonright V_\beta)(\xi) = ((p \upharpoonright V_\beta)\upharpoonright V_\gamma)(\xi)$. If $p(\xi) \neq \emptyset$, then ξ is inacc. and $\xi \leq \gamma$, and so $((p \upharpoonright V_\gamma)(\xi) \in V_\gamma^{P_\xi \upharpoonright V_\xi}$ so that $\Vdash_{P_\xi \upharpoonright V_\xi}$ "$(p \upharpoonright V_\gamma)(\xi) = p(\xi) \cap V_\gamma^{P_\xi \upharpoonright V_\xi}$", and $((p \upharpoonright V_\beta)(\xi) \in V_\beta^{P_\xi \upharpoonright V_\xi}$ so that $\Vdash_{P_\xi \upharpoonright V_\xi}$ "$(p \upharpoonright V_\beta)(\xi) = p(\xi) \cap V_\beta^{P_\xi \upharpoonright V_\xi}$", therefore $((p \upharpoonright V_\beta)\upharpoonright V_\gamma)(\xi) \in V_\gamma^{P_\xi \upharpoonright V_\xi}$ so that $\Vdash_{P_\xi \upharpoonright V_\xi}$ "$((p \upharpoonright V_\beta)\upharpoonright V_\gamma)(\xi) = (p \upharpoonright V_\beta)(\xi) \cap V_\gamma^{P_\xi \upharpoonright V_\xi}$", and hence $\Vdash_{P_\xi \upharpoonright V_\xi}$ "$((p \upharpoonright V_\beta)\upharpoonright V_\gamma)(\xi) = (p(\xi) \cap V_\beta^{P_\xi \upharpoonright V_\xi}) \cap V_\gamma^{P_\xi \upharpoonright V_\xi} = p(\xi) \cap V_\gamma^{P_\xi \upharpoonright V_\xi}$". So $(p \upharpoonright V_\gamma)(\xi)$ and $((p \upharpoonright V_\beta)\upharpoonright V_\gamma)(\xi)$ are names for the same object. Similarly for $((p \upharpoonright V_\gamma)\upharpoonright V_\beta)$.

(3*) For any $\alpha \leq \kappa$, any inacc. β such that $\alpha < \beta \leq \kappa$, $P_\alpha \upharpoonright V_\beta \subset\subset P_\alpha$.
To verify (13.1) notice that by (30.7) and by the definiton of the order on $P_\alpha \upharpoonright V_\beta$, $P_\alpha \upharpoonright V_\beta$ is a suborder of P_α.
(13.2) is in fact (30.9).
(13.3) follows from (30.10).

(4*) For any $\alpha \leq \kappa$, any inacc. β and inacc. γ such that $\alpha < \gamma \leq \beta \leq \kappa$, $P_\alpha{\restriction}V_\gamma \subset\subset P_\alpha{\restriction}V_\beta$.
Let $p \in P_\alpha{\restriction}V_\gamma$. Then $p = q{\restriction}V_\gamma$ for some $q \in P_\alpha$. By (30.7) $p \in P_\alpha$, so $p{\restriction}V_\beta \in P_\alpha{\restriction}V_\beta$. By (2*) $p{\restriction}V_\beta = p$, so $p \in P_\alpha{\restriction}V_\beta$. Hence $P_\alpha{\restriction}V_\gamma \subset P_\alpha{\restriction}V_\beta$, and by the definition (see 30.3) it is a suborder.
Let $p, q \in P_\alpha{\restriction}V_\beta$ so that $p \supset\!\subset q$ in $P_\alpha{\restriction}V_\beta$, then $p \supset\!\subset q$ in P_α by (3*) as $P_\alpha{\restriction}V_\beta \subset\subset P_\alpha$, and so $p \supset\!\subset q$ in $P_\alpha{\restriction}V_\gamma$ by (3*) as $P_\alpha{\restriction}V_\gamma \subset\subset P_\alpha$.
Let $p \in P_\alpha{\restriction}V_\beta$. Consider $p{\restriction}V_\gamma$. Let $p' \in P_\alpha{\restriction}V_\gamma$. Let $p' \supset\!\subset p{\restriction}V_\gamma$ in $P_\alpha{\restriction}V_\gamma$. Then $p' \supset\!\subset p$ in P_α by (30.10), hence $p' \supset\!\subset p$ in $P_\alpha{\restriction}V_\beta$ as by (3*) $P_\alpha{\restriction}V_\beta \subset\subset P_\alpha$.

(5*) For any $\alpha \leq \kappa$, any inacc. β and inacc. γ such that $\alpha < \gamma \leq \beta \leq \kappa$, if $p, q \in P_\alpha$ so that $p \leq q$, then $p{\restriction}V_\beta \leq q{\restriction}V_\beta$.
If not, then $p{\restriction}V_\beta \not\leq q{\restriction}V_\beta$. By (30.8) there is $p_1 \in P_\alpha{\restriction}V_\beta$ so that $p_1 \leq p{\restriction}V_\beta$ and $p_1 \supset\!\subset q{\restriction}V_\beta$. Since $p_1 \supset\!\subset p{\restriction}V_\beta$, $p_1 \supset\!\subset p$ in P_α, by (30.10). Thus $p_1 \supset\!\subset q$ in P_α. By (30.11) $q \leq q{\restriction}V_\beta$ and so $p_1 \supset\!\subset q{\restriction}V_\beta$, a contradiction.

The proper proof will be conducted by induction over the level of iteration.

Define $P_0 = C(\omega_0, \kappa)$. Then $P_0 \subset V_\kappa$ and is separative by (28.1), (28.6). Hence (30.1),(30.5), and (30.6) are satisfied.

For $\alpha < \kappa$ not inacc. define $P_{\alpha+1} = P_\alpha * \{\emptyset\}$. Since by the induction hypothesis $P_\alpha \subset V_\kappa$ and is separative, so is $P_{\alpha+1}$. Thus (30.2),(30.5), and (30.6) are satisfied.

For $\alpha < \kappa$, α inacc. define $P_{\alpha+1} = P_\alpha \otimes_{P_\alpha{\restriction}V_\alpha} C(\alpha,\kappa)^{P_\alpha{\restriction}V_\alpha}$.

Let's explain this part a bit more. Let $\mathring{C} \in V^{P_\alpha{\restriction}V_\alpha}$ be a name for $C(\alpha,\kappa)$ as defined in $V^{P_\alpha{\restriction}V_\alpha}$. The forcing conditions of $P_{\alpha+1}$ are $\langle p, \mathring{s}\rangle$ such that $p \in P_\alpha$, $\mathring{s} \in V^{P_\alpha{\restriction}V_\alpha}$, and $p \Vdash_{P_\alpha}$ "$\mathring{s} \in \mathring{C}$". By the induction hypothesis (30.10), $p{\restriction}V_\alpha \prec_{P_\alpha{\restriction}V_\alpha} p$, and so by Lemma 19 (as "$x \in X$" is absolute), $p{\restriction}V_\alpha \Vdash_{P_\alpha{\restriction}V_\alpha}$ "$\mathring{s} \in \mathring{C}$". Thus, $p{\restriction}V_\alpha \Vdash_{P_\alpha{\restriction}V_\alpha}$ "$\mathring{s}$ has rank at most κ", and because $|P_\alpha{\restriction}V_\alpha| \leq \alpha$ (follows from (30.12) as α is inacc.), and $\alpha < \kappa$, by Lemma 11 there is $\mathring{s}_1 \in V_\kappa^{P_\alpha{\restriction}V_\alpha}$ so that $p{\restriction}V_\alpha \Vdash_{P_\alpha{\restriction}V_\alpha}$ "$\mathring{s}_1 = \mathring{s}$". By Lemma 17 (as "$x = y$" is absolute), $p{\restriction}V_\alpha \Vdash_{P_\alpha}$ "$\mathring{s}_1 = \mathring{s}$". By (30.10) $p \Vdash_{P_\alpha}$ "$\mathring{s}_1 = \mathring{s}$". Hence $\langle p, \mathring{s}_1\rangle \in P_\alpha \otimes_{P_\alpha{\restriction}V_\alpha} C(\alpha,\kappa)^{P_\alpha{\restriction}V_\alpha}$ and $\langle p, \mathring{s}_1\rangle = \langle p, \mathring{s}\rangle$. Thus we can restrict our conditions to those with the second coordinate from $V_\kappa^{P_\alpha{\restriction}V_\alpha}$, and so $P_{\alpha+1} \subset V_\kappa$. By Lemma 24 it also is separative as P_α is by the induction hypothesis, and $C(\alpha,\kappa)^{P_\alpha{\restriction}V_\alpha}$ is separative in the generic extension via $P_\alpha{\restriction}V_\alpha$ by (28.6). Hence (30.3), (30.5), and (30.6) are satisfied.

Let's look at the limit case.
Let $\alpha < \kappa$, α limit. Consider the set A of all limits of $\langle P_\xi : \xi < \alpha\rangle$ (full limits in Kunen's terminology [K1], or inverse limits in Baumgartner's terminology [B]). If $a \in A$, then $a|\xi \in P_\xi$ for every $\xi < \alpha$ and so $a(\xi) \in V_\kappa$. Hence $a \in V_\kappa$, and so $A \subset V_\kappa$. Since P_α contains all limits with support from I_α, $P_\alpha \subset A$, thus $P_\alpha \subset V_\kappa$ and so (30.6) holds.
To show that P_α for $\alpha \leq \kappa$, α limit, is separative, consider $p, q \in P_\alpha$ so that $p \not\leq q$. Since $(\forall \xi < \alpha)(p|\xi \leq q|\xi)$ implies that $p \leq q$, there is $\xi < \alpha$ so that $p|\xi \not\leq q|\xi$. Take such ξ. Since P_ξ is separative by the induction hypothesis (30.5), there is $t \in P_\xi$ so that $t \leq p|\xi$ and $t \supset\!\subset q|\xi$ in P_ξ. Define r by $r(\eta) = t(\eta)$ for all $\eta < \xi$, and $r(\eta) = p(\eta)$ for all $\xi \leq \eta < \alpha$. Then $supp(r) \subset supp(t) \cup supp(p)$, hence $supp(r) \in I_\alpha$, and so $r \in P_\alpha$. Clearly $r \leq p$, and since $r|\xi = t$, $r \supset\!\subset q$ in P_α. Thus (30.5) holds.
To prove (30.12) : let $p \in P_\alpha$. Then $(p{\restriction}V_\alpha)(0) = p(0) \cap V_\alpha \in V_\alpha$ by (28.1). If $p(\xi) = \emptyset$, then $(p{\restriction}V_\alpha)(\xi) = \emptyset \in V_\alpha$. On the other hand if $p(\xi) \neq \emptyset$, then ξ is inacc. and $\xi < \alpha$. Then $(p{\restriction}V_\alpha)(\xi) \in V_\alpha^{P_\xi{\restriction}V_\xi}$. Thus $(p{\restriction}V_\alpha) \subset V_\alpha$. Since α is inacc., $|supp(p{\restriction}V_\alpha)| < \alpha$ and so $(p{\restriction}V_\alpha)|supp(p{\restriction}V_\alpha) \in V_\alpha$.
We have proven everything but (30.7) - (30.11). So let's assume that $\alpha \leq \kappa$ is inacc. We shall discuss it in three steps, (A), (B), and (C). Let β be inacc. so that $\alpha \leq \beta \leq \kappa$.

(A) Case that α is the least inacc. (and so $\alpha < \kappa$).
$p \in P_\alpha$ iff $p(0) \in C(\omega_0, \kappa)$ and $p(\xi) = \emptyset$ for all $0 < \xi < \alpha$.
$p \in P_\alpha \upharpoonright V_\beta$ iff $p(0) \in C(\omega_0, \beta)$ and $p(\xi) = \emptyset$ for all $0 < \xi < \alpha$ (it follows from (28.2)).
Verify (30.7): follows from $C(\omega_0,\beta) \subset C(\omega_0,\kappa)$, which follows from (28.3).
Verify (30.8): follows from (28.6).
Verify (30.9): follows from (28.5).
Verify (30.10): follows from (28.4).
Verify (30.11): follows from (28.3).

(B) Case that α is a successor inacc., i.e. α has an immediate inacc. predecessor γ (and so $\alpha < \kappa$).
Let $\mathring{C} \in V^{P_\gamma \upharpoonright V_\gamma}$ be a name for $C(\gamma,\kappa)$ as defined in $V^{P_\gamma \upharpoonright V_\gamma}$. Let $\mathring{C}_\beta \in V^{P_\gamma \upharpoonright V_\gamma}$ be a name for $C(\gamma,\beta)$ as defined in $V^{P_\gamma \upharpoonright V_\gamma}$.
$p \in P_\alpha$ iff $supp(p) \subset \gamma$ & $p|\gamma \in P_\gamma$, or $\gamma \in supp(p) \subset \gamma{+}1$ & $p|\gamma \in P_\gamma$ & $p(\gamma) \in V_\kappa^{P_\gamma \upharpoonright V_\gamma}$ & $p|\gamma \Vdash_{P_\gamma}$ "$p(\gamma) \in \mathring{C}$".

(*) $p \in P_\alpha \upharpoonright V_\beta$ iff $supp(p) \subset \gamma$ & $p|\gamma \in P_\gamma \upharpoonright V_\beta$, or $\gamma \in supp(p) \subset \gamma{+}1$ & $p|\gamma \in P_\gamma \upharpoonright V_\beta$ & $p(\gamma) \in V_{\beta+1}^{P_\gamma \upharpoonright V_\gamma}$ & $p|\gamma \Vdash_{P_\gamma}$ "$p(\gamma) \in \mathring{C}_\beta$".

The direction from right to left is easy, as any p satisfying the right hand side must be in P_α, and since $p \upharpoonright V_\beta = p$, p must be in $P_\alpha \upharpoonright V_\beta$. Now, the opposite direction. Let $p = q \upharpoonright V_\beta$ for some $q \in P_\alpha$. There are two possibilities:
(i) $supp(q) \subset \gamma$. Then $supp(p) \subset \gamma$ as well. $p|\gamma = (q \upharpoonright V_\beta)|\gamma = (q|\gamma) \upharpoonright V_\beta$. Thus the first part of the right hand side condition is satisfied.
(ii) $\gamma \in supp(q) \subset \gamma{+}1$. Then $q|\gamma \in P_\gamma$, $q(\gamma) \in V_\kappa^{P_\gamma \upharpoonright V_\gamma}$ and $q|\gamma \Vdash_{P_\gamma}$ "$q(\gamma) \in \mathring{C}$". $p(\gamma) = q(\gamma) \cap V_\beta \in V_{\beta+1}^{P_\gamma \upharpoonright V_\gamma}$. By Lemma 19, using (30.10), $(q|\gamma) \upharpoonright V_\gamma \Vdash_{P_\gamma \upharpoonright V_\gamma}$ "$q(\gamma) \in \mathring{C}$". $(q|\gamma) \upharpoonright V_\gamma \Vdash_{P_\gamma \upharpoonright V_\gamma}$ "$p(\gamma) = q(\gamma) \cap V_\beta^{P_\gamma \upharpoonright V_\gamma}$ & $q(\gamma) \in \mathring{C}$". By (28.2), $(q|\gamma) \upharpoonright V_\gamma \Vdash_{P_\gamma \upharpoonright V_\gamma}$ "$p(\gamma) \in \mathring{C}_\beta$", as β is inacc. in $V^{P_\gamma \upharpoonright V_\gamma}$ since $|P_\gamma \upharpoonright V_\gamma| \leq \gamma$, and $\gamma < \beta$, and β is inacc. in V. By Lemma 17, $(q|\gamma) \upharpoonright V_\gamma \Vdash_{P_\gamma}$ "$p(\gamma) \in \mathring{C}_\beta$". Using (30.11), $q|\gamma \Vdash_{P_\gamma}$ "$p(\gamma) \in \mathring{C}_\beta$". By Lemma 19, using (30.10), $(q|\gamma) \upharpoonright V_\beta \Vdash_{P_\gamma \upharpoonright V_\beta}$ "$p(\gamma) \in \mathring{C}_\beta$". By Lemma 17, $(q|\gamma) \upharpoonright V_\beta \Vdash_{P_\gamma}$ "$p(\gamma) \in \mathring{C}_\beta$". $p|\gamma = (q \upharpoonright V_\beta)|\gamma = (q|\gamma) \upharpoonright V_\beta$ by (1*), hence $p|\gamma \Vdash_{P_\gamma}$ "$p(\gamma) \in \mathring{C}_\beta$".
Verify (30.7): follows immediately from (*).
Verify (30.8): Let $p, q \in P_\alpha \upharpoonright V_\beta$ so that $p \not\leq q$. There are two possible cases:
(i) $p|\gamma \not\leq q|\gamma$. Then there is $t \in P_\gamma \upharpoonright V_\beta$ so that $t \leq p|\gamma$ and $t \supset\!\subset q|\gamma$, by (30.8). Define s so that $s|\gamma = t$, $s(\xi) = p(\xi)$ for all $\gamma \leq \xi < \alpha$. Then (by (*)) $s \in P_\alpha \upharpoonright V_\beta$, and $s \leq p$, and $s \supset\!\subset q$.
(ii) $p|\gamma \leq q|\gamma$. Then $\gamma \in supp(p) \subset \gamma{+}1$, and $p|\gamma, q|\gamma \in P_\gamma \upharpoonright V_\beta$, $p(\gamma), q(\gamma) \in V_{\beta+1}^{P_\gamma \upharpoonright V_\gamma}$. Clearly, $p|\gamma \not\Vdash_{P_\gamma}$ "$p(\gamma) \leq q(\gamma)$ *in* $\mathring{C}_\beta$". So there is $t \in P_\gamma$, $t \leq p|\gamma$ so that $t \Vdash_{P_\gamma}$ "$p(\gamma) \not\leq q(\gamma)$ *in* $\mathring{C}_\beta$". By Lemma 19, $t \upharpoonright V_\gamma \Vdash_{P_\gamma \upharpoonright V_\gamma}$ "$p(\gamma) \not\leq q(\gamma)$ *in* $\mathring{C}_\beta$". By (28.6) $t \upharpoonright V_\gamma \Vdash_{P_\gamma \upharpoonright V_\gamma}$ "$(\exists s \in \mathring{C}_\beta)(s \leq p(\gamma)$ & $s \supset\!\subset q(\gamma))$". Thus there is $\mathring{s} \in V_{\beta+1}^{P_\gamma \upharpoonright V_\gamma}$ so that $t \upharpoonright V_\gamma \Vdash_{P_\gamma \upharpoonright V_\gamma}$ "$\mathring{s} \in \mathring{C}_\beta$ & $\mathring{s} \leq p(\gamma)$ & $\mathring{s} \supset\!\subset q(\gamma)$". By Lemma 17, $t \upharpoonright V_\gamma \Vdash_{P_\gamma}$ "$\mathring{s} \in \mathring{C}_\beta$ & $\mathring{s} \leq p(\gamma)$ & $\mathring{s} \supset\!\subset q(\gamma)$". By (5*), Lemma 15, using (30.8), $t \upharpoonright V_\gamma \leq t \upharpoonright V_\beta \leq p|\gamma$. By (3*) $t \upharpoonright V_\gamma \in P_\gamma \upharpoonright V_\gamma \subset\subset P_\gamma \upharpoonright V_\beta$. Define r so that $r|\gamma = t \upharpoonright V_\gamma$, $r(\gamma) = \mathring{s}$, and $s(\xi) = \emptyset$ for all $\gamma < \xi < \alpha$. Then $r \in P_\alpha \upharpoonright V_\beta$, $r \leq p$ and $r \supset\!\subset q$.
Verify (30.9): it suffices to show it from right to left. Let $p, q \in P_\alpha \upharpoonright V_\beta$, so that $p \not\supset\!\subset q$ in P_α. Then for some $r \in P_\alpha$ $r \leq p, q$ in P_α. There are two possibilities:
(i) $supp(r) \subset \gamma$. Then $supp(p), supp(q) \subset \gamma$ as well. $r|\gamma \leq p|\gamma, q|\gamma$ in P_γ. By (5*) $(r|\gamma) \upharpoonright V_\beta \leq (p|\gamma) \upharpoonright V_\beta = p|\gamma$, $(q|\gamma) \upharpoonright V_\beta = q|\gamma$, hence $(r \upharpoonright V_\beta)|\gamma \leq p|\gamma, q|\gamma$ in $P_\gamma \upharpoonright V_\beta$. Since $supp(r) \subset \gamma$, $supp(r \upharpoonright V_\beta) \subset \gamma$, and so $r \upharpoonright V_\beta \leq p, q$ in $P_\alpha \upharpoonright V_\beta$.

(ii) $\gamma\in supp(r)\subset\gamma+1$. Then $r|\gamma\in P_\gamma$, $r(\gamma)\in V_\kappa^{P_\gamma\upharpoonright V_\gamma}$ so that $r|\gamma \Vdash_{\overline{P_\gamma}}$ "$r(\gamma)\in\mathring{C}$". Thus $r|\gamma \Vdash_{\overline{P_\gamma}}$ "$r(\gamma)\le p(\gamma), q(\gamma)$ *in* $\mathring{C}$". By (*), $r|\gamma \Vdash_{\overline{P_\gamma}}$ "$p(\gamma), q(\gamma)\in\mathring{C}_\beta$". By Lemma 19, using (30.10), $(r|\gamma)\upharpoonright V_\gamma \Vdash_{\overline{P_\gamma\upharpoonright V_\gamma}}$ "$r(\gamma)\le p(\gamma), q(\gamma)$ *in* $\mathring{C}$ & $p(\gamma), q(\gamma)\in\mathring{C}_\beta$". By (28.2) and (28.3), $(r|\gamma)\upharpoonright V_\gamma \Vdash_{\overline{P_\gamma\upharpoonright V_\gamma}}$ "$(\exists t\in\mathring{C}_\beta)(t\le p(\gamma), q(\gamma)$ *in* $\mathring{C})$". Thus there is $\mathring{t}\in V_{\beta+1}^{P_\gamma\upharpoonright V_\gamma}$ so that $(r|\gamma)\upharpoonright V_\gamma \Vdash_{\overline{P_\gamma\upharpoonright V_\gamma}}$ "$\mathring{t}\in\mathring{C}_\beta$ & $\mathring{t}\le p(\gamma), q(\gamma)$ *in* $\mathring{C}$". By Lemma 17, $(r|\gamma)\upharpoonright V_\gamma \Vdash_{\overline{P_\gamma}}$ "$\mathring{t}\in\mathring{C}_\beta$ & $\mathring{t}\le p(\gamma), q(\gamma)$ *in* $\mathring{C}$". By (30.11), $r|\gamma \Vdash_{\overline{P_\gamma}}$ "$\mathring{t}\in\mathring{C}_\beta$ & $\mathring{t}\le p(\gamma), q(\gamma)$ *in* $\mathring{C}$". Since $r|\gamma\le p|\gamma, q|\gamma$, by Lemma 15, using (30.8) $(r|\gamma)\upharpoonright V_\beta\le p|\gamma, q|\gamma$. By Lemma 19, $(r|\gamma)\upharpoonright V_\beta \Vdash_{\overline{P_\gamma\upharpoonright V_\beta}}$ "$\mathring{t}\in\mathring{C}_\beta$ & $\mathring{t}\le p(\gamma), q(\gamma)$ *in* $\mathring{C}$", and so by Lemma 17, using (30.11), $(r|\gamma)\upharpoonright V_\beta \Vdash_{\overline{P_\gamma}}$ "$\mathring{t}\in\mathring{C}_\beta$ & $\mathring{t}\le p(\gamma), q(\gamma)$ *in* $\mathring{C}$". Define s so that $s|\gamma=(r|\gamma)\upharpoonright V_\beta$, $s(\gamma)=\mathring{t}$, and $s(\xi)=\emptyset$ for all $\gamma<\xi<\alpha$. Then $s\in P_\alpha\upharpoonright V_\beta$ by (*), $s\le p, q$. Thus $p\sqsupset\!\subset q$ in $P_\alpha\upharpoonright V_\beta$.

Verify (30.10): Let $q=p\upharpoonright V_\beta$. Let $p'\in P_\alpha\upharpoonright V_\beta$ so that $p'\sqsupset\!\subset q$ in $P_\alpha\upharpoonright V_\beta$. There is $r\in P_\alpha\upharpoonright V_\beta$ such that $r\le p', q$. There are two possibilities:

(i) $supp(r)\subset\gamma$. Then $supp(p'), supp(q)\subset\gamma$. $r|\gamma\le p'|\gamma, q|\gamma$ in $P_\gamma\upharpoonright V_\beta$. $q|\gamma=(p\upharpoonright V_\beta)|\gamma=(p|\gamma)\upharpoonright V_\beta$, so $p'|\gamma\sqsupset\!\subset(p|\gamma)\upharpoonright V_\beta$ in $P_\gamma\upharpoonright V_\beta$. By (30.10) $p'|\gamma\sqsupset\!\subset p|\gamma$ in P_γ, and so $p'\sqsupset\!\subset p$ in P_α.

(ii) $\gamma\in supp(r)\subset\gamma+1$. Then $r|\gamma, p'|\gamma, q|\gamma\in P_\gamma\upharpoonright V_\beta$, $r(\gamma), p'(\gamma), q(\gamma)\in V_{\beta+1}^{P_\gamma\upharpoonright V_\gamma}$, and $r|\gamma\le p'|\gamma, q|\gamma$, and $r|\gamma \Vdash_{\overline{P_\gamma}}$ "$r(\gamma)\le p'(\gamma), q(\gamma)$ *in* $\mathring{C}$ & $r(\gamma), p'(\gamma), q(\gamma)\in\mathring{C}_\beta$". By Lemma 19, using (30.10), $(r|\gamma)\upharpoonright V_\gamma \Vdash_{\overline{P_\gamma\upharpoonright V_\gamma}}$ "$r(\gamma)\le p'(\gamma), q(\gamma)$ *in* $\mathring{C}$ & $r(\gamma), p'(\gamma), q(\gamma)\in\mathring{C}_\beta$". Since $q(\gamma)=p(\gamma)\cap V_\beta$, $(r|\gamma)\upharpoonright V_\gamma \Vdash_{\overline{P_\gamma\upharpoonright V_\gamma}}$ "$r(\gamma)\le p'(\gamma), p(\gamma)\cap V_\beta^{P_\gamma\upharpoonright V_\gamma}$ *in* $\mathring{C}$ & $r(\gamma), p'(\gamma)\in\mathring{C}_\beta$". By (28.5), $(r|\gamma)\upharpoonright V_\gamma \Vdash_{\overline{P_\gamma\upharpoonright V_\gamma}}$ "$p'(\gamma)\sqsupset\!\subset p(\gamma)$ *in* $\mathring{C}$". Hence $(r|\gamma)\upharpoonright V_\gamma \Vdash_{\overline{P_\gamma\upharpoonright V_\gamma}}$ "$(\exists t\in\mathring{C})(t\le p'(\gamma), p(\gamma))$". Thus there is $\mathring{t}\in V_\kappa^{P_\gamma\upharpoonright V_\gamma}$ so that $(r|\gamma)\upharpoonright V_\gamma \Vdash_{\overline{P_\gamma\upharpoonright V_\gamma}}$ "$\mathring{t}\in\mathring{C}$ & $\mathring{t}\le p'(\gamma), p(\gamma)$". By Lemma 17, $(r|\gamma)\upharpoonright V_\gamma \Vdash_{\overline{P_\gamma}}$ "$\mathring{t}\in\mathring{C}$ & $\mathring{t}\le p'(\gamma), p(\gamma)$". By (30.11), $r|\gamma \Vdash_{\overline{P_\gamma}}$ "$\mathring{t}\in\mathring{C}$ & $\mathring{t}\le p'(\gamma), p(\gamma)$". $r|\gamma\le p'|\gamma, q|\gamma$ in $P_\gamma\upharpoonright V_\beta$, $q|\gamma=(p\upharpoonright V_\beta)|\gamma=(p|\gamma)\upharpoonright V_\beta$. Thus $r|\gamma\sqsupset\!\subset(p|\gamma)\upharpoonright V_\beta$ in $P_\gamma\upharpoonright V_\beta$, and so by (30.10), $r|\gamma\sqsupset\!\subset p|\gamma$ in P_γ. Let $t\in P_\gamma$ so that $t\le r|\gamma, p|\gamma$. Define s so that $s|\gamma=t$, $s(\gamma)=\mathring{t}$, $s(\xi)=\emptyset$ for all $\gamma<\xi<\alpha$. Then $s\le p, p'$ and so $p'\sqsupset\!\subset p$ in P_α.

Verify (30.11): Let $p\in P_\alpha$. There are two possibilities:

(i) $supp(p)\subset\gamma$. $p|\gamma\in P_\gamma$, by (30.11) $p|\gamma\le(p|\gamma)\upharpoonright V_\beta=(p\upharpoonright V_\beta)|\gamma$, thus $p\le p\upharpoonright V_\beta$.

$\gamma\in supp(p)\subset\gamma+1$. Then $p|\gamma\in P_\gamma$, $p(\gamma)\in V_\kappa^{P_\gamma\upharpoonright V_\gamma}$ so that $p|\gamma \Vdash_{\overline{P_\gamma}}$ "$p(\gamma)\in\mathring{C}$". By Lemma 19, using (30.10), $(p|\gamma)\upharpoonright V_\gamma \Vdash_{\overline{P_\gamma\upharpoonright V_\gamma}}$ "$p(\gamma)\in\mathring{C}$". By (28.2), $(p|\gamma)\upharpoonright V_\gamma \Vdash_{\overline{P_\gamma\upharpoonright V_\gamma}}$ "$p(\gamma)\le p(\gamma)\cap V_\beta^{P_\gamma\upharpoonright V_\gamma}$ *in* $\mathring{C}$". So, $(p|\gamma)\upharpoonright V_\gamma \Vdash_{\overline{P_\gamma\upharpoonright V_\gamma}}$ "$p(\gamma)\le q(\gamma)$ *in* $\mathring{C}$", since $q(\gamma)=p(\gamma)\cap V_\beta$. By Lemma 17, $(p|\gamma)\upharpoonright V_\gamma \Vdash_{\overline{P_\gamma}}$ "$p(\gamma)\le q(\gamma)$ *in* $\mathring{C}$". Using (30.11), $p|\gamma \Vdash_{\overline{P_\gamma}}$ "$p(\gamma)\le q(\gamma)$ *in* $\mathring{C}$". Again by (30.11), $p|\gamma\le(p|\gamma)\upharpoonright V_\beta=(p\upharpoonright V_\beta)|\gamma=q|\gamma$. Thus $p\le q=p\upharpoonright V_\beta$.

(C) Case $\alpha\le\kappa$, is a limit inacc., i.e. there is a cofinal sequence of inacc. cardinals bellow α.
By (29.3), P_α contains only direct limits.

Verify (30.7): Let $p\in P_\alpha$. Then there is an inacc. $\gamma<\alpha$ so that $supp(p)\subset\gamma$. Then $p|\gamma\in P_\gamma$ and $(p\upharpoonright V_\beta)|\gamma=(p|\gamma)\upharpoonright V_\beta\in P_\gamma\upharpoonright V_\beta\subset\subset P_\gamma$ by (3*). Thus $(p\upharpoonright V_\beta)|\gamma\in P_\gamma$, and since $supp(p\upharpoonright V_\beta)\subset supp(p)\subset\gamma$, $p\upharpoonright V_\beta\in P_\alpha$. Thus $P_\alpha\upharpoonright V_\beta\subset P_\alpha$.

Verify (30.8): Let $p, q\in P_\alpha\upharpoonright V_\beta$ so that $p\not\le q$. There is an inacc. $\gamma<\alpha$ so that $supp(p), supp(q)\subset\gamma$. Then $p|\gamma\not\le q|\gamma$ in $P_\gamma\upharpoonright V_\beta$. By the induction hypothesis (30.8), there is $t\in P_\gamma\upharpoonright V_\beta$ so that $t\le p|\gamma$ and $t\supset\!\subset q|\gamma$. Define s so that $s|\gamma=t$, $s(\xi)=\emptyset$ for all $\gamma\le\alpha$. Then $s\in P_\alpha\upharpoonright V_\beta$, $s\le p$, and $s\supset\!\subset q$ in $P_\alpha\upharpoonright V_\beta$.

Verify (30.9): It suffices to prove right-to-left direction. Let $p,q \in P_\alpha \upharpoonright V_\beta$ so that $p \supset\!\subset q$ in P_α. There is $t \in P_\alpha$ so that $t \leq p,q$ in P_α. Then there is an inacc. $\gamma < \alpha$ so that $supp(t), supp(p), supp(q) \subset \gamma$. Hence $t|\gamma \leq p|\gamma, q|\gamma$ in P_γ. Since $p|\gamma, q|\gamma \in P_\gamma \upharpoonright V_\beta$, by the induction hypothesis (30.9)$p|\gamma \supset\!\subset q|\gamma$ in P_γ. So there is $r \in P_\gamma$ so that $r \leq p|\gamma, q|\gamma$. Define s so that $s|\gamma = r$, $s(\xi) = \emptyset$ for all $\gamma \leq \xi < \alpha$. Then $s \in P_\alpha$ and $s \leq p,q$.
Verify (30.10): Let $p \in P_\alpha$. There is an inacc. $\gamma < \alpha$ so that $supp(p) \subset \gamma$. Then $p|\gamma \in P_\gamma$. By the induction hypothesis (30.10) $(p|\gamma)\upharpoonright V_\beta \prec_{P_\alpha \upharpoonright V_\beta} p|\gamma$. Let $p' \in P_\alpha \upharpoonright V_\beta$ so that $p' \supset\!\subset p \upharpoonright V_\beta$ in $P_\alpha \upharpoonright V_\beta$. Since $(p\upharpoonright V_\beta)|\gamma = (p|\gamma)\upharpoonright V_\beta$, $p'|\gamma \supset\!\subset (p \upharpoonright V_\beta)|\gamma$ in $P_\gamma \upharpoonright V_\beta$, and so $p'|\gamma \supset\!\subset p|\gamma$ in P_γ. Since $supp(p) \subset \gamma$, $p' \supset\!\subset p$ in P_α.
Verify (30.11): Let $p \in P_\alpha$, there is an inacc. $\gamma < \alpha$ so that $supp(p) \subset \gamma$. Then $p|\gamma \in P_\gamma$ and by the induction hypothesis (30.11) $p|\gamma \leq (p|\gamma)\upharpoonright V_\beta$ in P_γ. Then $p|\gamma \leq (p \upharpoonright V_\beta)|\gamma$ in P_γ. Since $supp(p), supp(p \upharpoonright V_\beta) \subset \gamma$, $p \leq p \upharpoonright V_\beta$ in P_α. □

Def. 31: If C, I, κ, and $\langle P_\alpha : \alpha \leq \kappa \rangle$ are as in Lemma 30, we shall call $\langle P_\alpha : \alpha \leq \kappa \rangle$ the <u>(I,κ)-iteration of C in V.</u>

Properties 32: Let $C(\gamma,\delta)$ define a poset in V ($\gamma \leq \delta$ cardinals). Let λ be cardinal.
(32.1) For every transitive $M \subset V$ such that $M_\alpha = V_\alpha$ for all $\alpha \leq \lambda$, $C(\gamma,\delta)^V = C(\gamma,\delta)^M$ whenever $\delta \leq \lambda$, δ inacc. in V as well as in M.

Properties 33: Let $I = \langle I_\alpha : \alpha \leq \kappa,\ \alpha \text{ limit}\rangle$. Let $j{:}V \to M$ be an elementary embedding with critical point κ. Let $j(I) = \langle \hat{I}_\alpha : \alpha \leq j(\kappa),\ \alpha \text{ limit}\rangle$.
(33.1) $I_\alpha = \hat{I}_\alpha$ for all $\alpha \leq \kappa$, α limit.

Lemma 34: Let $\langle P_\alpha : \alpha \leq \kappa \rangle$ be the (I,κ)-iteration of C in V. Let $j{:}V \to M$ be a huge elementary embedding with critical point κ. Let $I = \langle I_\alpha : \alpha \leq \kappa,\ \alpha \text{ limit}\rangle$ satisfy (33.1) with respect to j. Let C satisfy (32.1) with respect to $j(\kappa)$. Let $\langle \hat{P}_\alpha : \alpha \leq j(\kappa)\rangle = j(\langle P_\alpha : \alpha \leq \kappa \rangle)$. Then $\langle \hat{P}_\alpha : \alpha \leq j(\kappa)\rangle$ is the $(j(I), j(\kappa))$-iteration of C in V, and $P_\kappa * C(\kappa, j(\kappa))^{P_\kappa} \subset\subset j(P_\kappa)$ in V.

Proof: Let $\langle R_\alpha : \alpha \leq j(\kappa)\rangle$ be the $(j(I), j(\kappa))$-iteration of C in V. By induction we shall prove that
(1) $R_\alpha = \hat{P}_\alpha$ for every $\alpha \leq j(\kappa)$;
(2) $P_\alpha \upharpoonright V_\beta = R_\alpha \upharpoonright V_\beta = \hat{P}_\alpha \upharpoonright M_\beta$, for every inacc. $\alpha \leq \kappa$, and every inacc. β so that $\alpha \leq \beta \leq \kappa$.

Let $\alpha = 0$. $R_0 = C(\omega_0, j(\kappa))^V$. By (32.1) $C(\omega_0, j(\kappa))^V = C(\omega_0, j(\kappa))^M = \hat{P}_0$. Thus $R_0 = \hat{P}_0$.

Assume that $\alpha < j(\kappa)$ is not inacc. in V (and hence in M, by Lemma 4). Then $R_{\alpha+1} = R_\alpha * \{\emptyset\}$, while $\hat{P}_{\alpha+1} = \hat{P}_\alpha * \{\emptyset\}$. By the induction hypothesis (1) $R_\alpha = \hat{P}_\alpha$, hence $R_{\alpha+1} = \hat{P}_{\alpha+1}$.

Let $\alpha < j(\kappa)$ be inacc. in V (and hence in M, by Lemma 4). Let $\mathring{C} \in V^{R_\alpha \upharpoonright V_\alpha}$ be a name for $C(\alpha, j(\kappa))$ as defined in $V^{R_\alpha \upharpoonright V_\alpha}$. Let Let $\mathring{D} \in M^{\hat{P}_\alpha \upharpoonright M_\alpha}$ be a name for $C(\alpha, j(\kappa))$ as defined in $M^{\hat{P}_\alpha \upharpoonright M_\alpha}$. Then $R_{\alpha+1} = R_\alpha \otimes_{R_\alpha \upharpoonright V_\alpha} \mathring{C}$ as defined in V, and $\hat{P}_{\alpha+1} = \hat{P}_\alpha \otimes_{\hat{P}_\alpha \upharpoonright M_\alpha} \mathring{D}$ as defined in M. Thus
$\langle p, \mathring{q}\rangle \in R_{\alpha+1}$ iff $p \in R_\alpha$ & $\mathring{q} \in V_{j(\kappa)}^{R_\alpha \upharpoonright V_\alpha}$ & $p \Vdash^{V}_{R_\alpha}$ "$\mathring{q} \in \mathring{C}$".
$\langle p, \mathring{q}\rangle \in \hat{P}_{\alpha+1}$ iff $p \in \hat{P}_\alpha$ & $\mathring{q} \in M_{j(\kappa)}^{\hat{P}_\alpha \upharpoonright M_\alpha}$ & $p \Vdash^{M}_{\hat{P}_\alpha}$ "$\mathring{q} \in \mathring{D}$".
Let $\langle p, \mathring{q}\rangle \in R_{\alpha+1}$.
Then $p \Vdash^{V}_{R_\alpha}$ "$\mathring{q} \in \mathring{C}$", so $p \upharpoonright V_\alpha \Vdash^{V}_{R_\alpha \upharpoonright V_\alpha}$ "$\mathring{q} \in \mathring{C}$", by Lemma 19, using (30.10). Let $p_1 \leq p \upharpoonright V_\alpha$ in $R_\alpha \upharpoonright V_\alpha = \hat{P}_\alpha \upharpoonright M_\alpha$. $p_1 \Vdash^{V}_{R_\alpha \upharpoonright V_\alpha}$ "$\mathring{q} \in \mathring{C}$". Let G be $R_\alpha \upharpoonright V_\alpha$-generic over V (and hence $\hat{P}_\alpha \upharpoonright M_\alpha$-generic over M by Lemma 25, as $|R_\alpha \upharpoonright V_\alpha| = |\hat{P}_\alpha \upharpoonright M_\alpha| \leq \alpha < \kappa$) so that $p_1 \in G$. Then $V[G] \models$ "$\mathring{q}^G \in C(\alpha, j(\kappa))$", and so $\mathring{q}^G \in C(\alpha, j(\kappa))^{V[G]}$. By (26.3) $V[G]_\xi = M[G]_\xi$ for all $\xi \leq j(\kappa)$, hence by (32.1), $\mathring{q}^G \in C(\alpha, j(\kappa))^{M[G]}$. So $M[G] \models$ "$\mathring{q}^G \in C(\alpha, j(\kappa))$", and so for some $p_2 \in G$,

$p_2 \Vdash^{M}_{\hat{P}_\alpha\upharpoonright M_\alpha}$ "$\mathring{q}\in\mathring{D}$". Since $p_2 \sqsupset p_1$ in $\hat{P}_\alpha\upharpoonright M_\alpha$, there is p_3 so that $p_3 \le p_1$ in $\hat{P}_\alpha\upharpoonright M_\alpha$ and $p_3 \Vdash^{M}_{\hat{P}_\alpha\upharpoonright M_\alpha}$ "$\mathring{q}\in\mathring{D}$". Thus, $p\upharpoonright V_\alpha \Vdash^{M}_{\hat{P}_\alpha\upharpoonright M_\alpha}$ "$\mathring{q}\in\mathring{D}$". $p\upharpoonright V_\alpha = p\upharpoonright M_\alpha$, hence by Lemma 17, using (30.11), $p \Vdash^{M}_{\hat{P}_\alpha}$ "$\mathring{q}\in\mathring{D}$". Now it follows that $\langle p,\mathring{q}\rangle\in\hat{P}_{\alpha+1}$.

Let $\langle p,\mathring{q}\rangle\in\hat{P}_{\alpha+1}$.

Then $p \Vdash^{M}_{\hat{P}_\alpha}$ "$\mathring{q}\in\mathring{D}$". By Lemma 19, using (30.10), $p\upharpoonright M_\alpha \Vdash^{M}_{\hat{P}_\alpha\upharpoonright M_\alpha}$ "$\mathring{q}\in\mathring{D}$". Let $p_1 \le p\upharpoonright M_\alpha$ in $\hat{P}_\alpha\upharpoonright M_\alpha = R_\alpha\upharpoonright V_\alpha$. Then $p_1 \Vdash^{M}_{\hat{P}_\alpha\upharpoonright M_\alpha}$ "$\mathring{q}\in\mathring{D}$". Let G be $\hat{P}_\alpha\upharpoonright M_\alpha$-generic over M (and hence $R_\alpha\upharpoonright V_\alpha$-generic over V by Lemma 25, as $|R_\alpha\upharpoonright V_\alpha| = |\hat{P}_\alpha\upharpoonright M_\alpha| \le \alpha < \kappa$) so that $p_1\in G$. Then $M[G]\models$ "$\mathring{q}^G\in C(\gamma,j(\kappa))$", and so $\mathring{q}^G\in C(\gamma,j(\kappa))^M[G]$. By Lemma 26 $V[G]_\xi = M[G]_\xi$ for all $\xi \le j(\kappa)$,and so by (32.1) $C(\gamma,j(\kappa))^M[G] = C(\gamma,j(\kappa))^V[G]$. Hence $\mathring{q}^G\in C(\gamma,j(\kappa))^V[G]$ and so $V[G]\models$ "$\mathring{q}^G\in C(\gamma,j(\kappa))$". Therefore there is $p_2\in G$ so that $p_2 \Vdash^{V}_{R_\alpha\upharpoonright V_\alpha}$ "$\mathring{q}\in\mathring{C}$". Since $p_2 \sqsupset p_1$ in $R_\alpha\upharpoonright V_\alpha$, there is $p_3 \le p_1, p_2$ in $R_\alpha\upharpoonright V_\alpha$ and so $p_3 \Vdash^{V}_{R_\alpha\upharpoonright V_\alpha}$ "$\mathring{q}\in\mathring{C}$". Thus $p\upharpoonright M_\alpha \Vdash^{V}_{R_\alpha\upharpoonright V_\alpha}$ "$\mathring{q}\in\mathring{C}$", and thus $p \Vdash^{V}_{R_\alpha}$ "$\mathring{q}\in\mathring{C}$", by Lemma 17, using (30.11) and the fact that $p\upharpoonright M_\alpha = p\upharpoonright V_\alpha$. It follows that $\langle p,\mathring{q}\rangle\in R_{\alpha+1}$.

Let $\alpha \le j(\kappa)$ be limit.

Let's prove (1) first.

$p\in R_\alpha$ iff $supp(p)\in\hat{I}_\alpha$ & $p(\xi) = \emptyset$ if $\xi < \alpha$ not inacc. in V, and $p(\xi)\in V^{R_\xi\upharpoonright V_\xi}_{j(\kappa)}$ if $\xi < \alpha$ is inacc. in V, and $p|\xi\in R_\xi$ for every $\xi < \alpha$.

$p\in\hat{P}_\alpha$ iff $supp(p)\in\hat{I}_\alpha$ & $p(\xi) = \emptyset$ if $\xi < \alpha$ not inacc. in M, and $p(\xi)\in M^{\hat{P}_\xi\upharpoonright M_\xi}_{j(\kappa)}$ if $\xi < \alpha$ is inacc. in M, and $p|\xi\in\hat{P}_\xi$ for every $\xi < \alpha$.

Since $\xi < \alpha$ is inacc. in V iff $\xi < \alpha$ is inacc. in M (by Lemma 4), and since $M^{\hat{P}_\xi\upharpoonright M_\xi}_{j(\kappa)} = V^{R_\xi\upharpoonright V_\xi}_{j(\kappa)}$ for every inacc. $\xi < \alpha$ (as $\hat{P}_\xi\upharpoonright M_\xi = R_\xi\upharpoonright V_\xi$ by the induction hypothesis (2)), and since $R_\xi = \hat{P}_\xi$ for every $\xi < \alpha$ by the induction hypothesis (1), then $R_\alpha = \hat{P}_\alpha$.

Let's prove (2) for inacc. $\alpha \le \kappa$, inacc. β so that $\alpha \le \beta \le \kappa$.

(A) α is the least inacc. (and so $\alpha < \kappa$). Then
$p\in P_\alpha\upharpoonright V_\beta$ iff $supp(p) = \{0\}$ & $p(0)\in C(\omega_0,\beta)^V$ (by (28.2)),
$p\in R_\alpha\upharpoonright V_\beta$ iff $supp(p) = \{0\}$ & $p(0)\in C(\omega_0,\beta)^V$ (by (28.2)),
hence $P_\alpha\upharpoonright V_\beta = R_\alpha\upharpoonright V_\beta$. By (1) $R_\alpha\upharpoonright V_\beta = \hat{P}\alpha\upharpoonright M_\beta$, since $V_\beta = M_\beta$.

(B) α has an immediate inacc. predecessor γ (and so $\alpha < \kappa$).
Let $\mathring{C}_\beta\in V^{P_\gamma\upharpoonright V_\gamma} = V^{R_\gamma\upharpoonright V_\gamma}$ be a name for $C(\gamma,\beta)$ as defined in $V^{R_\gamma\upharpoonright V_\gamma}$. Let $p\in P_\alpha\upharpoonright V_\beta$. Then there are two possibilities (see (*) in the proof of Lemma 30):
(i) $supp(p)\subset\gamma$ and $p|\gamma\in P_\gamma\upharpoonright V_\beta$. Since $P_\gamma\upharpoonright V_\beta = R_\gamma\upharpoonright V_\beta$ by the induction hypothesis (1), $p\in R_\alpha\upharpoonright V_\beta$.
(ii) $\gamma\in supp(p)\subset\gamma{+}1$, $p|\gamma\in P_\gamma\upharpoonright V_\beta$, $p(\gamma)\in V^{P_\gamma\upharpoonright V_\gamma}_{\beta+1}$ so that $p|\gamma \Vdash_{P_\gamma}$ "$p(\gamma)\in\mathring{C}_\beta$". By Lemma 19, using (30.10), $(p|\gamma)\upharpoonright V_\gamma \Vdash_{P_\gamma\upharpoonright V_\gamma}$ "$p(\gamma)\in\mathring{C}_\beta$". Thus $(p|\gamma)\upharpoonright V_\gamma \Vdash_{R_\gamma\upharpoonright V_\gamma}$ "$p(\gamma)\in\mathring{C}_\beta$", as $R_\gamma\upharpoonright V_\gamma = P_\gamma\upharpoonright V_\gamma$ by the induction hypothesis. By Lemma 17, using (30.11), $p|\gamma \Vdash_{R_\gamma\upharpoonright V_\gamma}$ "$p(\gamma)\in\mathring{C}_\beta$". Hence $p\in R_\alpha\upharpoonright V_\beta$ (by (*) in the proof of Lemma 30).
On the other hand, let $p\in R_\alpha\upharpoonright V_\beta$. Then there are two possibilities (see (*) in the proof of Lemma 30):
(i) $supp(p)\subset\gamma$ and $p|\gamma\in R_\gamma\upharpoonright V_\beta$. Since $R_\gamma\upharpoonright V_\beta = P_\gamma\upharpoonright V_\beta$ by the induction hypothesis (1), $p\in P_\alpha\upharpoonright V_\beta$.
(ii) $\gamma\in supp(p)\subset\gamma{+}1$, $p|\gamma\in R_\gamma\upharpoonright V_\beta$, $p(\gamma)\in V^{R_\gamma\upharpoonright V_\gamma}_{\beta+1}$ so that $p|\gamma \Vdash_{R_\gamma}$ "$p(\gamma)\in\mathring{C}_\beta$". By Lemma 19, using (30.10), $(p|\gamma)\upharpoonright V_\gamma \Vdash_{R_\gamma\upharpoonright V_\gamma}$, "$p(\gamma)\in\mathring{C}_\beta$". Thus $(p|\gamma)\upharpoonright V_\gamma \Vdash_{P_\gamma\upharpoonright V_\gamma}$, "$p(\gamma)\in\mathring{C}_\beta$", as $P_\gamma\upharpoonright V_\gamma = R_\gamma\upharpoonright V_\gamma$ by the induction hypothesis. By Lemma 17, using (30.11), $p|\gamma \Vdash_{P_\gamma\upharpoonright V_\gamma}$

"$p(\gamma)\in\mathring{C}_\beta$". Hence $p\in P_\alpha\upharpoonright V_\beta$ (by (*) in the proof of Lemma 30).

Thus $P_\alpha\upharpoonright V_\beta = R_\alpha\upharpoonright V_\beta$. Since $V_\beta = M_\beta$, by (1) $R_\alpha\upharpoonright V_\beta = \hat{P}_\alpha\upharpoonright M_\beta$.

(C) $\alpha\leq\kappa$ has a cofinal sequence of inacc. cardinals below.

Then $M\models$ "$(\forall X\in M_\kappa)(X\in\hat{I}_\alpha \Rightarrow |X|<\alpha)$". Since $M_\kappa = V_\kappa$, and using Lemma 4, $V\models$ "$(\forall X\in V_\kappa)(X\in\hat{I}_\alpha \Rightarrow |X|<\alpha)$". Hence both, P_α and R_α contain only direct limits with the same support.

Let $p\in P_\alpha\upharpoonright V_\beta$. Then for some inacc. $\gamma<\alpha$ $supp(p)\subset\gamma$ and $p|\gamma\in P_\gamma\upharpoonright V_\beta$. Hence $p|\gamma\in R_\gamma\upharpoonright V_\beta$ by the induction hypothesis, and so $p\in R_\alpha\upharpoonright V_\beta$.

The proof that $p\in R_\alpha\upharpoonright V_\beta \Rightarrow p\in P_\alpha\upharpoonright V_\beta$ is identical.

Thus (1) and (2) are proven.

By (30.6), if $p\in P_\kappa$, then $p(\xi)\in V_\kappa$ for every $\xi<\kappa$, and so $p\upharpoonright V_\kappa = p$. Hence $P_\kappa\upharpoonright V_\kappa = P_\kappa$, and by (1) and (2), $R_\kappa\upharpoonright V_\kappa = \hat{P}_\kappa\upharpoonright V_\kappa = P_\kappa$.

$(R_\kappa\upharpoonright V_\kappa)\otimes_{R_\kappa\upharpoonright V_\kappa} C(\kappa,j(\kappa))^{R_\kappa\upharpoonright V_\kappa} \subset\subset R_\kappa\otimes_{R_\kappa\upharpoonright V_\kappa} C(\kappa,j(\kappa))^{R_\kappa\upharpoonright V_\kappa} = R_{\kappa+1}$ by Lemma 22 as R_κ is separative (by (30.5)), and $R_\kappa\upharpoonright V_\kappa \subset\subset R_\kappa$ by (3*) in the proof of Lemma 30. As we have proven (see above) that $R_\kappa\upharpoonright V_\kappa = P_\kappa$, it follows that $(R_\kappa\upharpoonright V_\kappa)\otimes_{R_\kappa\upharpoonright V_\kappa} C(\kappa,j(\kappa))^{R_\kappa\upharpoonright V_\kappa} = P_\kappa\otimes_{P_\kappa} C(\kappa,j(\kappa))^{P_\kappa} = P_\kappa * C(\kappa,j(\kappa))^{P_\kappa}$.

Since $j(P_\kappa) = \hat{P}_{j(\kappa)} = R_{j(\kappa)}$ and contains only direct limits (by (29.3)) as $j(\kappa)$ is inacc. in both, M and V by Lemma 4). $R_{\kappa+1}\subset\subset R_{j(\kappa)}$. Hence $P_\kappa * C(\kappa,j(\kappa))^{P_\kappa}\subset\subset j(P_\kappa)$.□

To simplify the notation, we shall fix it for the rest of this chapter.

Let $j:V\to M$, κ, $I = \langle I_\alpha : \alpha \text{ limit} \leq\kappa\rangle$, $\hat{I} = j(I) = \langle \hat{I}_\alpha : \alpha \text{ limit} \leq j(\kappa)\rangle$, $\mathcal{P} = \langle P_\alpha : \omega\leq\alpha\leq\kappa\rangle$, $\hat{\mathcal{P}} = j(\mathcal{P}) = \langle \hat{P}_\alpha : \alpha\leq j(\kappa)\rangle$, and C be as in Lemma 34.

Let P denote P_κ. Then $|P|\leq\kappa$. Let G_1 be P-generic over V. Let Q denote $C(\alpha,j(\kappa))^{V[G_1]}$. Let $\mathring{Q}$ be a V^P-term for Q. Let B denote $P*\mathring{Q}$. Then by Lemma 34, B can be completely embedded in $j(P)$ and so $j(P) = B * j(P)/B$. Let G_2 be Q-generic over $V[G_1]$, and G_3 $j(P)/B$-generic over $V[G]$ $(G = G_1 * G_2)$, then $H_1 = G * G_3$ is $j(P)$-generic over V (and hence over M). If $p\in P$, then $supp(p)\in I_\kappa$ and so $supp(p) = supp(j(p))$. Thus $j(p)(\alpha) = p(\alpha)$ for $\omega\leq\alpha\leq\kappa$, and $j(p)(\alpha) = 1_{P_\alpha}$ for $\kappa<\alpha\leq j(\kappa)$. Hence $p\in G_1$ iff $j(p)\in H_1$. By Lemma 26 there is an elementary $\hat{\jmath}:V[G_1]\to M[H_1]$ definable in $V[H_1]$ and extending j, so that $(V[H_1])_\alpha = (M[H_1])_\alpha$ for every $\alpha\leq j(\kappa)$. Then $\hat{\jmath}(Q) = C(j(\kappa), j(j(\kappa)))^{M[H_1]}$.

Lemma 35: assume that

(35.1) j is huge;

(35.2) P satisfies the κ-c.c. in V;

(35.3) $V[G_1]\models$ "$|Q|\leq j(\kappa)$";

(35.4) for every directed $A\subset\hat{\jmath}''Q$ of size $\leq j(\kappa)$ and $A\in M[H_1]$, there is a $q\in\hat{\jmath}(Q)$ so that $q \ll A$.

Then there is a so-called <u>master condition</u> $q_m\in\hat{\jmath}(Q)$ so that if H_2 is $\hat{\jmath}(Q)$-generic over $V[H_1]$ and $q_m\in H_2$, then $j(p)\in H = H_1 * H_2$ whenever $p\in G$. Therefore, there is an elementary embedding $i:V[G]\to M[H]$ definable in $V[H]$ extending $\hat{\jmath}$ so that if $V[G_1]\models$ "Q satisfies the $j(\kappa)$-c.c.", then if $X\in V$, $Y\in V[H]$, $Y\subset X$, $|Y|\leq j(\kappa)$, $Y\subset M[H]$, then $Y\in M[H]$.

Proof: $G_2\in V[H_1]$ and $|G_2|\leq j(\kappa)$ by (35.3). Let $G_2 = \{e_\alpha : \alpha<j(\kappa)\}$. Since $\hat{\jmath}$ is definable in $V[H_1]$, $\hat{\jmath}''G_2 = \{\hat{\jmath}(e_\alpha) : \alpha<j(\kappa)\}\in V[H_1]$, and $\hat{\jmath}''G_2\subset\hat{\jmath}''Q$. Since ${}^{j(\kappa)}M\subset M$, and since P satisfies the κ-c.c. in V by (35.2), $j(P)$ satisfies the $j(\kappa)$-c.c. in M, and by hugeness of j, in V as well, by (26.2) $\hat{\jmath}''G_2\in M[H_1]$. Since $\hat{\jmath}''G_2$ is directed, there is a $q_m\in\hat{\jmath}(Q)$ so that $q_m \ll \hat{\jmath}''G_2$ by (35.4). Let H_2 be $\hat{\jmath}(Q)$-generic over $V[H_1]$ (and hence also over $M[H_1]$) so that $q_m\in H_2$. Then, if $\langle p,q\rangle\in G = G_1 * G_2$, $\hat{\jmath}(\langle p,q\rangle) = \langle\hat{\jmath}(p),\hat{\jmath}(q)\rangle$, and $\hat{\jmath}(p)\in H_1$ and $\hat{\jmath}(q)\geq q_m$, and so $\hat{\jmath}(q)\in H_2$. Thus $\hat{\jmath}(\langle p,q\rangle)\in H$. By Lemma 26 there is an elementary embedding $i:V[G]\to M[H]$ definable in $V[H]$ extending j (and also $\hat{\jmath}$). If $V[G_1]\models$ "Q satisfies the $j(\kappa)$-c.c.", then B satisfies the $j(\kappa)$-c.c., and so if $X\in V$, $Y\in V[H]$, $|Y|\leq j(\kappa)$, $Y\subset X$, and $Y\subset M[H]$, then $Y\in M[H]$ by (26.2). □

Lemma 36: Assume that

(36.1) j is huge;

(36.2) P satisfies the κ-c.c. in V;

(36.3) $V[G_1]\models$ "$|Q|\leq j(\kappa)$";

(36.4) $V[G_1]\models$ "Q is κ-closed";

(36.5) $V[G]\models$ "$|\wp(\kappa)|=\kappa^+$";

(36.6) for every directed $A\subset\hat{\jmath}''Q$ of size $\leq j(\kappa)$ and so that $A\in M[H_1]$, there is a $q\in\hat{\jmath}(Q)$ so that $q<<A$.

Then $V[H_1]\models$ "$(\exists\mathcal{U})(\mathcal{U}$ is a non-principal $V[G]-\kappa$-complete $V[G]$-ultrafilter over $j(\kappa))$".
(In fact $\Vdash^{V[G]}_{j(P)/B}$ "$(\exists\mathcal{U})(\mathcal{U}$ is a non-principal $V[G]-\kappa$-complete $V[G]$-ultrafilter over $j(\kappa))$", since G_3 was chosen arbitrarily.).

Proof: Apply Lemma 35 to obtain a master condition $q_m\in\hat{\jmath}(Q)$, and H_2 $\hat{\jmath}(Q)$-generic over $V[H_1]$ so that $q_m\in H_2$, and an elementary $i{:}V[G]\to M[H]$ definable in $V[H]$ and extending $\hat{\jmath}$ (where $H=H_1*H_2$). In $V[H]$ define for $X\in V[G]\cap\wp(j(\kappa))$: $X\in\mathcal{W}$ iff $\bigcup(i''j(\kappa))\in i(X)$.
It is easy to check that $\mathcal{W}$ is a non-principal $V[G]-\kappa$-complete $V[G]$-ultrafilter over $j(\kappa)$ in $V[H]$. Hence $q_m \Vdash^{V[H_1]}_{\hat{\jmath}(Q)}$ "$(\exists\mathcal{W})(\mathcal{W}$ is a non-principal $V[G]-\kappa$-complete $V[G]$-ultrafilter over $j(\kappa))$". Let $\mathring{\mathcal{W}}\in V[H_1]^{\hat{\jmath}(Q)}$ so that $q_m \Vdash^{V[H_1]}_{\hat{\jmath}(Q)}$ "$\mathring{\mathcal{W}}$ is a non-principal $V[G]-\kappa$-complete $V[G]$-ultrafilter over $j(\kappa)$". Now, $V[G]\models$ "$|\wp(\kappa)|=\kappa^+$", so $M[H]\models$ "$|\wp(j(\kappa))|=j(\kappa)^+$" by the elementarity of i. Hence $M[H]\models$ "$|\wp(j(\kappa))\cap V[G]|\leq j(\kappa)^+$", and so $V[H]\models$ "$|\wp(j(\kappa))\cap V[G]|\leq j(\kappa)^+$". Since Q is κ-closed, $\hat{\jmath}(Q)$ is $j(\kappa)$-closed, and so $(j(\kappa)^+)^{V[H]}=(j(\kappa)^+)^{V[H_1]}$. Thus $V[H_1]\models$ "$|\wp(j(\kappa))\cap V[G]|\leq j(\kappa)^+$". Let $\{K_\alpha:\alpha<j(\kappa)^+\}=\wp(j(\kappa))\cap V[G]$ in $V[H_1]$. In $V[H_1]$ let $\langle s_\alpha:\alpha<j(\kappa)^+\rangle$ be a descending sequence of elements of $\hat{\jmath}(Q)$ so that each s_α decides "$K_\alpha\in\mathring{\mathcal{W}}$". In $V[H_1]$ define $\mathcal{U}$ by:
if $X\in\wp(j(\kappa))\cap V[G]$, then $X\in\mathcal{U}$ iff $(\exists\alpha<j(\kappa)^+)(s_\alpha \Vdash^{V[H_1]}_{\hat{\jmath}(Q)}$ "$X\in\mathring{\mathcal{W}})$".
It is left to the reader to verify that $\mathcal{U}$ is a non-principal $V[G]-\kappa$-complete $V[G]$-ultrafilter over $j(\kappa)$ in $V[H_1]$. □

Lemma 37: Assume that

(37.1) j is huge;

(37.2) P satisfies the κ-c.c. in V;

(37.3) $V[G_1]\models$ "$|Q|\leq j(\kappa)$";

(37.4) $V[G_1]\models$ "Q is $<\kappa$-closed";

(37.5) $V[G]\models$ "$|\wp(\kappa)|=j(\kappa)$";

(37.6) for every directed $A\subset\hat{\jmath}''Q$ of size $\leq j(\kappa)$ and so that $A\in M[H_1]$, there is a $q\in\hat{\jmath}(Q)$ so that $q<<A$.

Then $V[H_1]\models$ "$(\exists\mathcal{U})(\mathcal{U}$ is a non-principal $V[G]-\kappa$-complete $V[G]$-ultrafilter over $\kappa)$".
(In fact $\Vdash^{V[G]}_{j(P)/B}$ "$(\exists\mathcal{U})(\mathcal{U}$ is a non-principal $V[G]-\kappa$-complete $V[G]$-ultrafilter over $\kappa)$", since G_3 was chosen arbitrarily.).

Proof: So similar to the proof of Lemma 36, that it is left to the reader. □

Lemma 38: If $\Vdash^{V[G]}_{j(P)/B}$ "$(\exists\mathcal{U})(\mathcal{U}$ is a non-principal $V[G]-\kappa$-complete $V[G]$-ultrafilter over $\lambda)$", then $V[G]\models$ "$(\exists\mathcal{I})(\mathcal{I}$ is a κ-complete ideal over λ so that $\wp(\lambda)/\mathcal{I}$ can be embedded into $Comp(j(P)/B))$".

Proof: Let $\mathring{\mathcal{U}}$ be a $V[G]^{j(P)/B}$-term so that $\Vdash^{V[G]}_{j(P)/B}$ "$\mathring{\mathcal{U}}$ is a non-principal $V[G]-\kappa$-complete $V[G]$-ultrafilter over λ". Define $\mathcal{I}$ in $V[G]$ by:
if $X\subset\lambda$, then $X\in\mathcal{I}$ iff for no $p\in j(P)/B$, $p \Vdash^{V[G]}_{j(P)/B}$ "$X\in\mathring{\mathcal{U}}$".

(1) Let $X\subset Y\subset\lambda$, and $Y\in\mathcal{I}$.
By the way of contradiction assume that $X\notin\mathcal{I}$. Hence for some $p\in j(P)/B$, $p \Vdash^{V[G]}_{j(P)/B}$

"$X\in\mathring{\mathcal{U}}$". Also $p \Vdash_{j(P)/B}^{V[G]}$ "$X\subset Y$". Then $p \Vdash_{j(P)/B}^{V[G]}$ "$Y\in\mathring{\mathcal{U}}$", and so $Y\notin\mathcal{I}$, a contradiction.

(2) Let $\{X_\alpha : \alpha < \xi\}\subset\mathcal{I}$, $\xi < \kappa$.
By the way of contradiction assume that $\bigcup\{X_\alpha : \alpha < \xi\}\in\mathcal{I}$. Then for $p\in j(P)/B$, $p \Vdash_{j(P)/B}^{V[G]}$ "$\bigcup\{X_\alpha : \alpha < \xi\}\in\mathring{\mathcal{U}}$". But then $p \Vdash_{j(P)/B}^{V[G]}$ "$(\exists\alpha<\xi)(X_\alpha\in\mathring{\mathcal{U}})$". Hence there are $q\in j(P)/B$ and $\alpha < \xi$ so that $q \le p$ and $q \Vdash_{j(P)/B}^{V[G]}$ "$X_\alpha\in\mathring{\mathcal{U}})$", hence $X_\alpha\notin\mathcal{I}$, a contradiction.

(3) $\emptyset\in\mathcal{I}$, for no p can force "$\emptyset\in\mathring{\mathcal{U}}$".

(4) $\lambda\notin\mathcal{I}$, for $\Vdash_{j(P)/B}^{V[G]}$ "$\lambda\in\mathring{\mathcal{U}}$".

(5) if $\alpha\in\lambda$, then $\{\alpha\}\in\mathcal{I}$, for no p can force "$\{\alpha\}\in\mathring{\mathcal{U}}$".

To embed $\wp(\lambda)/\mathcal{I}$ into $D = Comp(j(P)/B)$, notice that

(i) $X\in\mathcal{I}$ iff $\|X\in\mathring{\mathcal{U}}\|_D = O_D$, and

(ii) $X, Y\notin\mathcal{I}$ and $X = Y \ (mod\ \mathcal{I})$, then $\|X\in\mathring{\mathcal{U}}\|_D = \|Y\in\mathring{\mathcal{U}}\|_D$. For $(X-Y),(Y-X)\in\mathcal{I}$ and so then $\|(X-Y)\in\mathring{\mathcal{U}}\|_D = \|(Y-X)\in\mathring{\mathcal{U}}\|_D$, thus $\|X\in\mathring{\mathcal{U}}\|_D = \|(X\cap Y)\in\mathring{\mathcal{U}}\|_D = \|Y\in\mathring{\mathcal{U}}\|_D$.

For $[X]\in\wp(\lambda)/\mathcal{I}$ define $h([X]) = \|X\in\mathring{\mathcal{U}}\|_D$. By (i) and (ii), this is a well-defined mapping from $\wp(\lambda)/\mathcal{I}$ into D. Let $[X] \le [Y]$. Then $(X-Y)\in\mathcal{I}$ and so $\|X\in\mathring{\mathcal{U}}\|_D \le \|Y\in\mathring{\mathcal{U}}\|_D$, hence $h([X]) \le h([Y])$. Also, if $[X] \ne [Y]$, then $\|X\in\mathring{\mathcal{U}}\|_D \ne \|Y\in\mathring{\mathcal{U}}\|_D$, so $h([X]) \ne h([Y])$. Thus h is an embedding. □

Chapter 2.

Model I.

A model with an $\aleph_1$-complete $\aleph_2$-saturated ideal over ω_1, and which satisfies Chang's conjecture.

(Kunen's model, see [K₂].)

We shall start with a huge embedding $j:V\rightarrow M$ with critical point κ. We shall do a (finite support,κ)-iteration of Silver's collapse S.

Def. 39: Let γ, δ be regular cardinals, $\gamma < \delta$. Silver's collapse of δ to γ^+ is a poset $S(\gamma,\delta)$ defined by: $s\in S(\gamma,\delta)$ iff

(39.1) $s\subset\delta\times\wp(\gamma\times\delta)$ is a function with $dom(s)\subset\delta$;

(39.2) $|s| \le \gamma$;

(39.3) there is $\beta\in\gamma$ so that for every $\alpha\in dom(s)$, $s(\alpha)\subset\beta\times\alpha$ is a function with $dom(s(\alpha))\subset\beta$;

(39.4) if $s, t\in S(\gamma,\delta)$, then $s \le t$ iff $dom(t)\subset dom(s)$ and for every $\alpha\in dom(t)$, $t(\alpha)\subset s(\alpha)$.

Note: If δ is inacc., then $S(\gamma,\delta)$ is a $<\gamma$-closed δ-c.c. poset and $\Vdash_{S(\gamma,\delta)}$ "$2^\gamma = \gamma^+ = \delta$" (see [J],[K₁]).

Lemma 39: Silver's collapse satisfies (28.1) - (28.6), and also (32.1).

Proof: Left to the reader. □

Lemma 40: Let $I_\alpha = [\alpha]^{<\omega}$ for every limit $\alpha \le \kappa$. Then $I = \langle I_\alpha : \alpha \le \kappa\rangle$ satisfies (29.1) - (29.3), and (33.1).

Proof: Easy, and hence left to the reader. □

Let $\mathcal{P} = \langle P_\alpha : \alpha \le \kappa\rangle$ be the (I,κ)-iteration of S in V (see Lemma 30). Then $\hat{\mathcal{P}} = j(\mathcal{P}) = \langle \hat{P}_\alpha : \alpha \le j(\kappa)\rangle$ is the $(j(I), j(\kappa))$-iteration of S in V by Lemma 34.

Lemma 41: P_κ satisfies the κ-c.c. in V. (and so $j(P_\kappa)$ satisfies $j(\kappa)$-c.c. in V.)

Proof: A sketch:

It is carried by induction in the usual way. The limit case is standard, for we are using finite support iteration (e.g. see [K$_1$], [B]). Thus we shall prove that each P_α for α a successor satisfies the κ-c.c. For P_0 it is known. Consider $P_{\alpha+1}$. If α is not inacc., then $P_{\alpha+1} = P_\alpha * \{\emptyset\}$, and so satisfies the κ-c.c. as P_α does. For α inacc. it is a bit harder.

Let $\{\langle p_\xi, \mathring{q}_\xi\rangle : \xi \in \kappa\}$ be an antichain in $P_{\alpha+1}$. Let $\mathring{C} \in V^{P_\alpha \upharpoonright V_\alpha}$ be a name for $S(\alpha, \kappa)$ as defined in $V^{P_\alpha \upharpoonright V_\alpha}$. Since $|P_\alpha \upharpoonright V_\alpha| \leq \alpha < \kappa$, by pigeon-hole argument there exists $X \in [\kappa]^\kappa$ and $p \in P_\alpha \upharpoonright V_\alpha$ so that $p_\xi \upharpoonright V_\alpha = p$ for every $\xi \in X$. Then $p_\xi \Vdash_{P_\alpha}$ "$\mathring{q}_\xi \in \mathring{C}$" for any $\xi \in X$. By Lemma 19, using 30.10, $p_\xi \upharpoonright V_\alpha \Vdash_{P_\alpha \upharpoonright V_\alpha}$ "$\mathring{q}_\xi \in \mathring{C}$" for any $\xi \in X$, and so $p \Vdash_{P_\alpha \upharpoonright V_\alpha}$ "$\mathring{q}_\xi \in \mathring{C}$" for any $\xi \in X$, and hence $p \Vdash_{P_\alpha \upharpoonright V_\alpha}$ "$|\mathring{q}_\xi| \leq \alpha$". By Lemma 12, we can assume WLOG that $|\mathring{q}_\xi| \leq \gamma$ for some $\alpha < \gamma < \kappa$, as $|P_\alpha \upharpoonright V_\alpha| \leq \alpha < \gamma$ and hence satisfies the γ-c.c. By the Δ-system lemma (see e.g. [K$_1$]), there must be $Y \in [X]^\kappa$, $\mathring{q} \in V^{P_\alpha \upharpoonright V_\alpha}$, $|\mathring{q}| \leq \gamma$, so that $dom(\mathring{q}_\xi) \cap dom(\mathring{q}_\rho) = dom(\mathring{q})$ whenever $\xi, \rho \in Y$. Hence for every $\xi, \rho \in Y$, $p \Vdash_{P_\alpha \upharpoonright V_\alpha}$ "$\mathring{q}_\xi \cap \mathring{q}_\rho = \mathring{q}$", and so $p \Vdash_{P_\alpha \upharpoonright V_\alpha}$ "$\mathring{q}_\xi \mathbin{\mathcal{I}} \mathring{q}_\rho$". Then $p_\xi \upharpoonright V_\alpha \Vdash_{P_\alpha \upharpoonright V_\alpha}$ "$\mathring{q}_\xi \mathbin{\mathcal{I}} \mathring{q}_\rho$", and by Lemma 17, using (30.11), $p_\xi \Vdash_{P_\alpha}$ "$\mathring{q}_\xi \mathbin{\mathcal{I}} \mathring{q}_\rho$". Since $\langle p_\xi, \mathring{q}_\xi\rangle \supset\subset \langle p_\rho, \mathring{q}_\rho\rangle$, it follows that $p_\xi \supset\subset p_\rho$. Therefero $\{p_\xi : \xi \in Y\}$ is an antichain of size κ in P_α, which contradicts the induction hypothesis.□

Let G_1 be P-generic over V. Let Q denote $S(\kappa, j(\kappa))$ as defined in $V[G_1]$. Let $\mathring{Q}$ be a V^P-term so that $(\mathring{Q})^{G_1} = Q$. Let $B = P * \mathring{Q}$. Let G_2 be Q-generic over $V[G_1]$. Let $G = G_1 * G_2$. By Lemma 34, B can be regularly embedded into $j(P)$. By Lemma 26 there is an elementary embedding $\hat{\jmath}: V[G_1] \to M[H_1]$ extending j and definable in $V[H_1]$. Since $\hat{\jmath}''S(\kappa, j(\kappa))^{V[G_1]} \subset S(j(\kappa), j(j(\kappa)))^{M[H_1]}$, for every $A \subset \hat{\jmath}''S(\kappa, j(\kappa))^{V[G_1]}$, A directed, $|A| \leq j(\kappa)$, and $A \in M[H_1]$, there is $s \in S(j(\kappa), j(j(\kappa)))^{M[H_1]}$ so that $s << A$ (s is the set union of A; note that for every $s \in S(\kappa, j(\kappa))$, the β [see (39.3)] is not moved by j, hence $j(s)$ has the same β as s, and that 's why the union of A is a condition from $S(j(\kappa), jj(\kappa))$). Therefore, by Lemma 37, there exists a non-principal $V[G]$-κ-complete $V[G]$-ultrafilter over κ in $V[H_1]$. By Lemma 38 there is a κ-complete ideal $\mathcal{I}$ over κ so that $\wp(\kappa)/\mathcal{I}$ can be embedded into $Comp(j(P)/B)$. Since $j(P)$ satisfies the $j(\kappa)$-c.c. in V, $j(P)/B$ satisfies the $j(\kappa)$-c.c. in $V[G]$. Hence $Comp(j(P)/B)$ satisfies the $j(\kappa)$-c.c. in $V[G]$, and so $\wp(\kappa)/\mathcal{I}$ satisfies the $j(\kappa)$-c.c. in $V[G]$ as well; and so, $V[G] \models$ "$\mathcal{I}$ is $j(\kappa)$-saturated". Since $V[G] \models$ "$\aleph_1 = \kappa$ and $\aleph_2 = j(\kappa)$", $V[G] \models$ "$\mathcal{I}$ is an $\aleph_1$-complete, $\aleph_2$-saturated ideal over ω_1".

Now to show that Chang's conjecture holds in $V[G]$: by Lemma 35 there are H $j(B)$-generic over V and an elementary embedding $i: V[G] \to M[H]$ definable in $V[H]$ and extending $\hat{\jmath}$ so that if $X \in V$, $Y \in V[H]$, $Y \subset X$, $|Y| \leq j(\kappa)$, and $Y \subset M[H]$, then $Y \in M[H]$. Let $\mathcal{A}$ be a structure of type $(\aleph_1, \aleph_2)$ (i.e. of type $(\kappa, j(\kappa))$) in $V[G]$. WLOG assume that its universe is $j(\kappa)$. Then $i(\mathcal{A})$ is a structure of type $(\aleph_1, \aleph_2)$ in $M[H]$. Since $i''\mathcal{A} \subset M[H]$, $i''\mathcal{A} \in V[H]$ and has size $\leq j(\kappa)$, $i''\mathcal{A} \in M[H]$ and it is not hard to prove that $M[H] \models$ "$i''\mathcal{A}$ is a structure of type $(\kappa, j(\kappa))$, it is an elementary substructure of $i(\mathcal{A})$, $|\kappa| = \aleph_0$ and $|j(\kappa)| = \aleph_1$". Hence $M[H] \models$ "$i(\mathcal{A})$ has an elementary substructure of type $(\aleph_0, \aleph_1)$". By the elementarity of i, $V[G] \models$ "$\mathcal{A}$ has an elementary substructure of type $(\aleph_0, \aleph_1)$". □

Note: If GCH holds in V, then GCH also holds in $V[G]$.

Model II.

A model with an $\aleph_1$-complete $\aleph_3$-saturated ideal over ω_3.

(Magidor's model - see [M].)

We shall start with a huge embedding $j: V \to M$ with critical point κ. We shall do a (finite support,κ)-iteration of Magidor's collapse D.

Def. 42: Let γ, δ be regular cardinals, $\gamma < \delta$. Magidor's γ, δ collapse is a poset $D(\gamma, \delta)$ defined by:

$$D(\gamma, \delta) = \begin{cases} S(\omega_0, \delta), & \text{if } \gamma = \omega_0; \\ S(\gamma^+, \delta), & \text{otherwise,} \end{cases}$$

where S is Silver's collapse (see Def. 39).

Note: if δ is inacc. and γ regular so that $\omega < \gamma < \delta$, then $D(\gamma, \delta)$ is a γ-closed, δ-c.c. poset.

Lemma 43: Magidor's collapse satisfies (28.1)-(28.6), (32.1).

Proof: See Lemma 39. □

For every α limit so that $\omega < \alpha \leq \kappa$ define $I_\alpha = [\alpha]^{<\omega}$. Then $I = \langle I_\alpha : \text{limit } \alpha \leq \kappa\rangle$ satisfies (29.1) - (29.3), (33.1) (see Lemma 40). Let $\mathcal{P} = \langle P_\alpha : \alpha \leq \kappa\rangle$ be the (I, κ)-iteration of D in V. Then $\hat{\mathcal{P}} = j(\mathcal{P}) = \langle \hat{P}_\alpha : \alpha \leq j(\kappa)\rangle$ is the $(j(I), j(\kappa))$-iteration of M in V by Lemma 34. By induction in the usual way (see Lemma 41) it is easy to show that P_κ satisfies the κ-c.c. in V, and so $j(P_\kappa) = \hat{P}_{j(\kappa)}$ satisfies the $j(\kappa)$-c.c. in V. Let P denote P_κ. Let G_1 be P-generic over V. Let Q denote $D(\kappa, j(\kappa)) = S(\kappa^+, j(\kappa))$ as defined in $V[G_1]$. Let $\mathring{Q}$ be a V^P-term so that $(\mathring{Q})^{G_1} = Q$. Let $B = P * \mathring{Q}$. Let G_2 be Q-generic over $V[G_1]$. Let $G = G_1 * G_2$. By Lemma 34, B can be regularly embedded into $j(P)$. Since $j(P)$ satisfies the $j(\kappa)$-c.c. in V, by Lemma 26 there is an elementary embedding $\hat{\jmath}: V[G_1] \to M[H_1]$ extending j, definable in $V[H_1]$. Since $\hat{\jmath}''D(\kappa, j(\kappa))^{V[G_1]} \subset D(j(\kappa), j(j(\kappa)))^{M[H_1]}$, for every $A \subset \hat{\jmath}''D(\kappa, j(\kappa))^{V[G_1]}$, A directed, $|A| \leq j(\kappa)$, and $A \in M[H_1]$, there is $s \in D(j(\kappa), j(j(\kappa)))^{M[H_1]}$ so that $s << A$; s is the set union of A. Since Q is κ-closed, by Lemma 36 there is a non-principal $V[G]$-κ-complete $V[G]$-ultrafilter over $j(\kappa)$ in $V[H_1]$. Therefore by Lemma 38 there is a κ-complete ideal $\mathcal{I}$ over $j(\kappa)$ so that $\wp(j(\kappa))/\mathcal{I}$ can be embedded into $Comp(j(P)/B)$. Since $j(P)$ satisfies the $j(\kappa)$-c.c. in V, $j(P)/B$ satisfies the $j(\kappa)$-c.c. in $V[G]$. Hence $Comp(j(P)/B)$ satisfies the $j(\kappa)$-c.c. in $V[G]$, and so $\wp(j(\kappa))/\mathcal{I}$ satisfies the $j(\kappa)$-c.c. in $V[G]$. In other words, $V[G] \models$ "$\mathcal{I}$ is $j(\kappa)$-saturated". Since $V[G] \models$ "$\aleph_1 = \kappa$ and $\aleph_3 = j(\kappa)$", $V[G] \models$ "$\mathcal{I}$ is an $\aleph_1$-complete, $\aleph_3$-saturated ideal over ω_3". □

Note: If GCH holds in V, then GCH also holds in $V[G]$.

Model III.

A model with an $\aleph_1$-complete $(\aleph_2, \aleph_2, \aleph_0)$-saturated ideal over ω_1, and which satisfies Chang's conjecture.

(Laver's model, see [L].)

We shall start with a huge embedding $j: V \to M$ with critical point κ. We shall do an $(\omega$-Easton,$\kappa)$-iteration of Easton's collapse E.

Def. 44: Let γ, δ be regular cardinals, $\gamma < \delta$. Easton's collapse of δ to γ^+ is a poset $E(\gamma, \delta)$ defined by: $s \in E(\gamma, \delta)$ iff

(44.1) $s \subset \delta \times \wp(\gamma \times \delta)$ is a function with $dom(s) \subset \delta$;

(44.2) $dom(s)$ is a γ-Easton subset of δ;

(44.3) there is $\beta \in \gamma$ so that for every $\alpha \in dom(s)$, $s(\alpha) \subset \beta \times \alpha$ is a function with $dom(s(\alpha)) \subset \beta$;

(44.4) if $s, t \in s(\gamma, \delta)$, then $s \leq t$ iff $dom(t) \subset dom(s)$ and for every $\alpha \in dom(t)$, $t(\alpha) \subset s(\alpha)$.

Note: If δ is Mahlo, then $E(\gamma, \delta)$ is a $<\gamma$-closed δ-c.c. poset and $\Vdash_{E(\gamma,\delta)}$ "$2^\gamma = \gamma^+ = \delta$" (see [L]).

Lemma 45: Let δ be Mahlo and let γ be regular so that $\gamma < \delta$. Let $\mathcal{A}$ be a family of γ-Easton subsets of δ so that $|\mathcal{A}| \geq \delta$. Then there is a family $\mathcal{B} \subset \mathcal{A}$, $|\mathcal{B}| \geq \delta$ so that $\mathcal{B}$ forms a Δ-system with root $\Delta \subset \sigma$ for some $\sigma < \delta$.

Proof: WLOG assume that $|\mathcal{A}| = \delta$. Let $A = \{\beta \in \delta : \beta \text{ regular}\}$, and let $\mathcal{A} = \{X_\beta : \beta \in A\}$. By Lemma 4, for each X_β there is some σ_β so that $\gamma \leq \sigma_\beta < \delta$ and $X_\beta \subset \sigma_\beta$. Let $B = \{\beta \in A : \sigma_\beta \leq \beta\}$.

(a) Assume that B is stationary in δ.

For every $\beta \in B-\gamma$ define $f(\beta) =$ "the least τ so that $X_\beta \cap \beta \subset \tau$". Since $|X_\beta \cap \beta| < \beta$ and β is regular, f is regressive. By Fodor's theorem there are stationary $C \subset B-\gamma$ and $\sigma < \delta$ so that $f''C = \{\sigma\}$. Thus, if $\beta \in C$, $X_\beta \cap \beta \subset \sigma$ and $X_\beta \subset \sigma_\beta \subset \beta$, so $X_\beta \subset \sigma$.
Thus $\mathcal{D} = \{X \in \mathcal{A} : X \subset \sigma\}$ has size δ. Now apply the Δ-system lemma to $\mathcal{D}$ to obtain a Δ-system $\mathcal{B} \subset \mathcal{D}$ of size δ. Then Δ, the root of $\mathcal{B}$, is a subset of σ.

(b) Assume that B is not stationary in δ.
Then there is a cub C in δ so that $B \cap C = \emptyset$. $D = A \cap C$ is stationary and if $\beta \in D$, then $\beta \notin B$ and so $\beta < \sigma_\beta$. Define a regressive function f on $D-\gamma$ by $f(\beta) =$ "the least τ so that $X_\beta \cap \beta \subset \tau$". By Fodor's theorem there are a stationary $E \subset D-\gamma$ and $\sigma < \delta$ so that $f''E = \{\sigma\}$. So for all $\beta \in E$, $X_\beta \cap \beta \subset \sigma$ and $\beta < \sigma_\beta$. By induction choose a sequence $\langle \beta_\alpha : \alpha < \delta \rangle \subset E$ so that $\sigma_{\beta_\alpha} < \beta_{\alpha+1}$ for all $\alpha < \delta$. Let $\mu < \nu < \delta$ and let $\xi \in X_{\beta_\mu} \cap X_{\beta_\nu}$. Then $\xi \in X_{\beta_\mu} \subset \sigma_{\beta_\mu} \subset \beta_\nu$, so $\xi \in X_{\beta_\nu} \cap \beta_\nu \subset \sigma$. Thus $X_{\beta_\mu} \cap X_{\beta_\nu} \subset \sigma$ whenever $\mu, \nu < \sigma$. Now apply Δ-system lemma to $\{X_{\beta_\alpha} \cap \sigma : \alpha < \delta\}$. So there is $F \in [\delta]^\delta$ so that $\{X_{\beta_\alpha} \cap \sigma : \alpha \in F\}$ is a Δ-system with root $\Delta \subset \sigma$. Then $\mathcal{B} = \{X_{\beta_\alpha} : \alpha \in F\}$ is also a Δ-system with the same root Δ. □

Lemma 46: Easton's collapse satisfies (28.1) - (28.6), (32.1), if one replaces "inacc." by "Mahlo".

Proof: Left to the reader. □

Lemma 47: Let I_α is the ideal of ω-Easton subsets of α, for every limit $\alpha \leq \kappa$. Then $I = \langle I_\alpha : \text{limit } \alpha \leq \kappa \rangle$ satisfies (29.1) - (29.3), and (33.1) with respect to j, if one replaces "inacc." by "Mahlo".

Proof: Left to the reader. □

Let $\mathcal{P} = \langle P_\alpha : \alpha \leq \kappa \rangle$ be the (I, κ)-iteration of E in V as described in Lemma 30 with "inacc." replaced by "Mahlo". One can check that (30.7) - (30.11) still hold true with this replacement. Using the fact that κ is huge (in fact for this measurability suffices), the set of Mahlo cardinals bellow κ is stationary in κ. Since all Mahlo cardinals $\leq j(\kappa)$ in M are the same as in V (see Lemma 4), conclusions of Lemma 34 still hold. Hence $j(\mathcal{P}) = \langle \hat{P}_\alpha : \alpha \leq j(\kappa) \rangle$ is the $(j(I), j(\kappa))$-iteration of E in V, and $P_\kappa * E(\kappa, j(\kappa))^{P_\kappa}$ can be regularly embedded into $j(P_\kappa)$ in V.

Lemma 48: P_κ satisfies the $(\kappa, \kappa, <\kappa)$-c.c..

Proof: By Induction.

(a) Let α be the least Mahlo. We shall show that P_α satisfies the (κ, κ, σ)-c.c.. WLOG assume that $\alpha < \sigma$.
Since α is the least Mahlo, P_α is isomorphic to $P_0 = E(\omega, \kappa)$. Let $X \in [E(\omega, \kappa)]^\kappa$. By Lemma 45 there is $X_1 \in [X]^\kappa$ so that $\{dom(p) : p \in X_1\}$ is a Δ-system with root $\Delta \subset \nu < \kappa$. WLOG assume $\sigma < \nu$. Since κ is inacc., there are less than κ possibilities for $p|\nu$. Thus, by pigeon-hole argument, there is $Y \in [X_1]^\kappa$ so that if $p_1 \neq p_2 \in Y$, then $p_1|\nu = p_2|\nu$. Let $Z \in [Y]^\nu$. Define $q \in E(\omega, \kappa)$ by $dom(q) = \bigcup\{dom(p) : p \in Z\}$ and $q(\alpha) = p(\alpha)$ for any $p \in Z$ so that $\alpha \in dom(p)$. Then $q \leq p$ for all $p \in Z$, since a union of ν ω-Easton sets is ν-Easton, hence $dom(q)$ is ν-Easton and so $dom(q)-\nu$ is ω-Easton. But $dom(q) \cap \nu = dom(p) \cap \nu$ for any $p \in Z$ and hence ω-Easton. So $dom(q)$ is ω-Easton. Therefore $q \in E(\omega, \kappa)$ and $q << Z$.

(b) Assume that α has an immediate Mahlo predecessor γ. We are going to show that P_α satisfies the (κ, κ, σ)-c.c. for $\sigma < \kappa$. WLOG assume that $\alpha < \sigma$.
Let $\mathring{E} \in V^{P_\gamma \upharpoonright V_\gamma}$ be a name for $E(\gamma, \kappa)$ as defined in $V^{P_\gamma \upharpoonright V_\gamma}$. $p \in P_\alpha$ iff $supp(p) \subset \gamma+1$, $p|\gamma \in P_\gamma$, $p(\gamma) \in V^{P_\gamma \upharpoonright V_\gamma}$, and $p|\gamma \Vdash_{P_\gamma}$ "$p(\gamma) \in \mathring{E}$". Since $|P_\gamma \upharpoonright V_\gamma| \leq \gamma < \kappa$, there are $p \in P_\gamma \upharpoonright V_\gamma$, and $X_1 \in [\kappa]^\kappa$ so that $p_\xi \upharpoonright V_\gamma = p$ for every $\xi \in X_1$. Since each $p_\xi|\gamma \Vdash_{P_\gamma}$ "$p_\xi(\gamma) \in \mathring{E}$", be Lemma

19, using (30.10), $(p_\xi|\gamma)\upharpoonright V_\gamma \Vdash_{P_\gamma\upharpoonright V_\gamma}$ "$p_\xi(\gamma)\in\mathring{E}$", and so $p \Vdash_{P_\gamma\upharpoonright V_\gamma}$ "$p_\xi(\gamma)\in\mathring{E}$" for all $\xi\in X_1$. Define $A_\xi = \{\delta\in\kappa : p \Vdash_{P_\gamma\upharpoonright V_\gamma}$ "$\delta\in dom(p_\xi(\gamma))$"$\}$, for $\xi\in X_1$. Let $B_\xi = A_\xi - \sigma$. Then each B_ξ is an ω-Easton subset of κ:

If not, then for some regular $\tau \geq \omega$, $|B_\xi \cap \tau| = \tau$ (so $\tau > \sigma > \alpha > \gamma$). Let $B_\xi \cap \tau = \{\delta_\eta : \eta < \tau\}$. Then each $\delta_\eta \in \tau$. For every $\eta < \tau$, $p \Vdash_{P_\gamma\upharpoonright V_\gamma}$ "$\delta_\eta\in dom(p_\xi(\gamma))$". Since $P_\gamma\upharpoonright V_\gamma$ preserves τ, $p \Vdash_{P_\gamma\upharpoonright V_\gamma}$ "$|dom(p_\xi(\gamma))\cap\tau| = \tau$", which is a contradiction.

By Lemma 45 there is $X_2\in[X]^\kappa$ so that $\{B_\xi : \xi\in X_2\}$ form a Δ-system with the root $\Delta\subset\nu$, for some $\nu < \kappa$. WLOG assume that $\sigma < \nu$. By smallness of V_ν, there is $X_3\in[X_2]^\kappa$ so that $p_\xi\cap V_\nu = p_\rho\cap V_\nu$ whenever $\xi, \rho\in X_3$.

Let $Y\in[X_3]^\sigma$. Since P_γ satisfies the (κ,κ,σ)-c.c., as by the induction hypothesis it satisfies the $(\kappa,\kappa,<\kappa)$-c.c., there is $r\in P_\gamma$ so that $r \leq p_\xi|\gamma$ for every $\xi\in Y$. Since the root of $\{B_\xi : \xi\in X_2\}$ is a subset of ν, for every $\delta > \nu$, at most one $\xi\in Y$ satisfies that $p \Vdash_{P_\gamma\upharpoonright V_\gamma}$ "$\delta\in dom(p_\xi(\gamma))$". So, $p \Vdash_{P_\gamma\upharpoonright V_\gamma}$ "$\bigcup\{dom(p_\xi(\gamma)) : \xi\in Y\}$ *is a σ-Easton subset of* κ". It follows that (for $\sigma < \nu$) $p \Vdash_{P_\gamma\upharpoonright V_\gamma}$ "$\bigcup\{dom(p_\xi(\gamma)) : \xi\in Y\}$ *is a γ-Easton subset of* κ", and so $p \Vdash_{P_\gamma\upharpoonright V_\gamma}$ "$\bigcup\{p_\xi(\gamma) : \xi\in Y\}\in\mathring{E}$". Let $\mathring{t}\in V^{P_\gamma\upharpoonright V_\gamma}$ be so that $p \Vdash_{P_\gamma\upharpoonright V_\gamma}$ "$\mathring{t} = \bigcup\{p_\xi(\gamma) : \xi\in Y\}$". Then for every $\xi\in Y$, $p \Vdash_{P_\gamma\upharpoonright V_\gamma}$ "$\mathring{t} \leq p_\xi$ *in* $\mathring{E}$". By (5*) from the proof of Lemma 30, $r\upharpoonright V_\gamma \leq (p_\xi|\gamma)\upharpoonright V_\gamma = p$, so $r\upharpoonright V_\gamma \Vdash_{P_\gamma\upharpoonright V_\gamma}$ "$\mathring{t} \leq p_\xi$ *in* $\mathring{E}$", for every $\xi\in Y$. By Lemma 17, using (30.11), $r \Vdash_{P_\gamma}$ "$\mathring{t} \leq p_\xi$ *in* $\mathring{E}$", for every $\xi\in Y$. Now define t so that $t|\gamma = r$, $t(\gamma) = \mathring{t}$, and $t(\xi) = \emptyset$ for all $\gamma < \xi < \alpha$. $t\in P_\alpha$, and $t \leq p_\xi$ for every $\xi\in Y$.

(c) Assume that α has a cofinal sequence of smaller Mahlo cardinals. Then the support is (by Lemma 3) is of size smaller than α, and in fact direct limits are taken. The proof now continues along standard lines, using Lemma 45 to obtain the required Δ-system of supports (see e.q. [K1], [B]).□

Since $j(\mathcal{P})$ is the same kind of iteration in V, and since $j(\kappa)$ is Mahlo in V, we also proved that $j(P_\kappa)$ satisfies the $(j(\kappa), j(\kappa), <j(\kappa))$-c.c. in V.

As before, let P denote P_κ, let Q denote $E(\kappa, j(\kappa))$ as defined in V^P, and let B denote $P * Q$.

Lemma 49: $\Vdash_B$ "$j(P)/B$ satisfies the $(j(\kappa), j(\kappa), <\kappa)$-c.c.".

Proof: Let $\Vdash_B$ "$j(P)/B = \{s_\alpha : \alpha < j(\kappa)\}$". Let $\mathring{X}\in V^B$ and $b_0\in B$ so that $b_0 \Vdash_B$ "$\mathring{X}\in [j(\kappa)]^{j(\kappa)}$". There is $Y_0\in[j(\kappa)]^{j(\kappa)}$ so that for any $\alpha\in Y_0$ there is $b\in B$, $b\leq b_0$ and $b \Vdash_B$ "$\alpha\in\mathring{X}$". For each $\alpha\in Y_0$ choose one such b and denote it b_α. Since $B = P * Q$, for each $\alpha\in Y_0$ there are $p_\alpha\in P$ and $q_\alpha\in Q$ so that $b_\alpha = \langle p_\alpha, q_\alpha\rangle$. Hence, for every $\alpha\in Y_0$, $\langle p_\alpha, q_\alpha\rangle \Vdash_B$ "$\alpha\in\mathring{X}$". Since $|P| < j(\kappa)$, there are $Y_1\in[Y_0]^{j(\kappa)}$ and $p\in P$ so that $\langle p_\alpha, q_\alpha\rangle = \langle p, q_\alpha\rangle$ whenever $\alpha\in Y_1$. Thus, for every $\alpha\in Y_1$, $\langle p, q_\alpha\rangle \Vdash_B$ "$\alpha\in\mathring{X}$". For any $\alpha\in Y_1$, $\langle p, q_\alpha, s_\alpha\rangle\in j(P)$. By the proof of Lemma 48,

(*) there is $Y_2\in[Y_1]^{j(\kappa)}$ so that the coordinatewise union of $\{\langle p, q_\alpha, s_\alpha\rangle : \alpha\in C\}$ is a condition from $j(P)$ whenever $C\in[Y_2]^{<j(\kappa)}$.

Let G_1 be P-generic over V so that $p\in G_1$. Now switch to $V[G_1]$. Since $j(\kappa)$ still is Mahlo here (as $|P| < j(\kappa)$), there are $Y_3\in[Y_2]^{j(\kappa)}$ and $\sigma < j(\kappa)$ so that $\{dom(q_\alpha) : \alpha\in Y_3\}$ form a Δ-system with root $\Delta\subset\sigma$. By pigeon-hole argument there are $Y_4\in[Y_3]^{j(\kappa)}$ and $q\in Q$ so that $q_\alpha|\sigma = q$ for every $\alpha\in Y_4$. Hence

(**) $\{dom(q_\alpha) : \alpha\in Y_4\}$ form a Δ-system with root $\Delta\subset\sigma$, and $q_\alpha|\sigma = q$ for every $\alpha\in Y_4$.

Let $\mathring{Y}_5 \in V[G_1]^Q$ so that $\Vdash^{V[G_1]}_Q$ "$\alpha \in \mathring{Y}_5$ iff $\alpha \in Y_4$ & $q_\alpha \in \underline{G_2}$", where $\underline{G_2}$ is the canonical name for the Q-generic filter over $V[G_1]$. Let $\mathring{Y} \in V[G_1]^Q$ so that $\Vdash^{V[G_1]}_Q$ "$\alpha \in \mathring{Y}$ iff $\alpha \in \mathring{X} \cap \mathring{Y}_5$". Let G_2 be Q-generic over $V[G_1]$ so that $q \in G_2$. Let $G = G_1 * G_2$. Then G is B-generic over V.

(***) $V[G] \models$ "$|Y| = j(\kappa)$", where $Y = (\mathring{Y})^{G_2}$.

It will suffice to prove that $q \Vdash^{V[G_1]}_Q$ "$|\mathring{Y}| = j(\kappa)$", since $q \in G_2$. Let us assume by the way of contradiction that $q \not\Vdash^{V[G_1]}_Q$ "$|\mathring{Y}| = j(\kappa)$". There are $\bar{q} \leq \dot{q}$ and $\nu < j(\kappa)$ (WLOG assume that $\sigma \leq \nu$) so that $\bar{q} \Vdash^{V[G_1]}_Q$ "$\mathring{Y} \subset \nu$". Since $\bar{q} \leq q$, by (**) there is $\tau \in Y_4 - \nu$ so that $(dom(\bar{q}) \cap dom(q_\tau)) - \sigma = \emptyset$. Thus $\bar{q}$ and q_τ are compatible, i.e. there is $q' \leq \bar{q}, q_\tau$. Since $q_\tau \Vdash^{V[G_1]}_Q$ "$q_\tau \in \underline{G_2}$", $q_\tau \Vdash^{V[G_1]}_Q$ "$\tau \in \mathring{Y}_5$". Hence $q' \Vdash^{V[G_1]}_Q$ "$\tau \in \mathring{Y}_5$". Since $q_\tau \Vdash^{V[G_1]}_Q$ "$\tau \in \mathring{X}$", $q' \Vdash^{V[G_1]}_Q$ "$\tau \in \mathring{X}$". Therefore $q' \Vdash^{V[G_1]}_Q$ "$\tau \in \mathring{Y}$". On the other hand, $\bar{q} \Vdash^{V[G_1]}_Q$ "$\mathring{Y} \subset \nu$", and so $q' \Vdash^{V[G_1]}_Q$ "$\mathring{Y} \subset \nu$", a contradiction as $\tau \geq \nu$.

(****) $V[G_1] \models$ "$(\forall D \in [Y_3]^{<\kappa})(\exists t \in Q * j(P)/B)(t = \langle \bigcup\{q_\alpha : \alpha \in D\}, s\rangle << \{\langle q_\alpha, s_\alpha\rangle : \alpha \in D\})$"

Let $V[G_1] \models$ "$D \in [Y_3]^{<\kappa}$". Then $V[G_1] \models$ "$D \in [Y_2]^{<\kappa}$". Since P satisfies the κ-c.c. in V, and since $Y_2 \in V$, there is $C \in [Y_3]^{<\kappa}$ so that $V[G_1] \models$ "$D \subset C$". By (*), the coordinatewise union of $\{\langle p, q_\alpha, s_\alpha\rangle : \alpha \in C\}$ is a condition from $j(P)$. Since $p \in G_1$, in $V[G_1]$, the coordinatewise union of $\{\langle q_\alpha, s_\alpha\rangle : \alpha \in C\}$ is a condition from $Q * j(P)/B$. Since $D \subset C$, the coordinatewise union of of $\{\langle q_\alpha, s_\alpha\rangle : \alpha \in D\}$ is a condition t from $Q * j(P)/B$. The condition t has the form $\langle \bigcup\{q_\alpha : \alpha \in D\}, s\rangle$ for some V^B-term s so that $\Vdash_B$ "$s \in j(P)/B$", and clearly $t << \{\langle q_\alpha, s_\alpha\rangle : \alpha \in D\}$.

(*****) $V[G] \models$ "$(\forall Z \in [Y]^{<\kappa})(\exists s \in j(P)/B)(s << \{s_\alpha : \alpha \in Z\})$.

If $V[G] \models$ "$Z \in [Y]^{<\kappa}$", then $V[G] \models$ "$Z \in [Y_3]^{<\kappa}$". Since $Y_3 \in V[G_1]$, and since Q is κ-closed, $Z \in V[G_1]$. By (****), in $V[G_1]$ there is a condition $t = \langle \bigcup\{q_\alpha : \alpha \in Z\}, s\rangle \in Q * j(P)/B$ so that $t << \{\langle q_\alpha, s_\alpha\rangle : \alpha \in Z\}$. Since $Z \subset Y$, and so $Z \subset (\mathring{Y}_5)^{G_2}$, each q_α, $\alpha \in Z$, is in G_2. Since $\{q_\alpha : \alpha \in Z\} \in V[G_1]$, and G_2 is Q-generic over $V[G_1]$, $\bigcup\{q_\alpha : \alpha \in Z\} \in G_2$. Thus $V[G] \models$ "$s << \{s_\alpha : \alpha \in Z\}$".

If $\alpha \in Y$, then $b_0 \geq \langle p, q_\alpha\rangle$ and $\langle p, q_\alpha\rangle \in G$, hence $b_0 \in G$. Thus there is $b_1 \leq b_0$ so that $b_1 \Vdash_B$ "$(\forall Z \in [\mathring{Y}]^{<\kappa})(\exists s \in j(P)/B)(s << \{s_\alpha : \alpha \in Z\})$. □

Let G_1 be P-generic over V, let G_2 be Q-generic over $V[G_1]$. Then $G = G_1 * G_2$ is B-generic over V. Let G_3 be $j(P)/B$-generic over $V[G]$ (possible by Lemma 34). Then $H_1 = G_1 * G_2 * G_3$ is $j(P)$-generic over V. By Lemma 26 there is an elementary embedding $\hat{j}: V[G_1] \to M[H_1]$ definable in $V[H_1]$ and extending j. Similarly as in Model I, for every directed $A \subset \hat{j}''E(\kappa, j(\kappa))^{V[G_1]}$, $|A| < j(\kappa)$, and $A \in V[H_1]$, there is $q \in E(j(\kappa), j(j(\kappa)))^{V[H_1]}$ so that $q << A$; the set union of A. Thus, by Lemma 37 there is a non-principal $V[G]$-κ-complete $V[G]$-ultrafilter over κ in $V[H_1]$. By Lemma 38 there is a κ-complete ideal $\mathcal{I}$ over κ in $V[G]$, so that $\wp(\kappa)/\mathcal{I}$ can be embedded into $Comp(j(P)/B)$. By Lemma 49, $j(P)/B$ satisfies the $(j(\kappa), j(\kappa), <\kappa)$-c.c. in $V[G]$. Since $\kappa = \aleph_1$ and $j(\kappa) = \aleph_2$ in $V[G]$, $\mathcal{I}$ is ω_1-complete $(\aleph_2, \aleph_2, \aleph_0)$-saturated ideal over ω_1.

Note:

(1) If GCH holds in V, then it also holds in $V[G]$.

(2) $V[G]$ also satisfies Chang's conjecture (see Model I).

(3) Laver showed (see [L]) that from the existence of an ω_1-complete $(\aleph_2, \aleph_2, \aleph_0)$-saturated ideal over ω_1 follows that $\binom{\aleph_2}{\aleph_1} \to \binom{\aleph_0}{\aleph_1}^{1,1}_2$. Juhazs and Hajnal (private communication to Laver) showed that adding $\aleph_1$ Cohen reals destroys the partition relation, hence destroys all ω_1-complete $(\aleph_2, \aleph_2, \aleph_0)$-saturated ideals over ω_1. Since adding $\aleph_1$ Cohen reals is a σ,finite-c.c. forcing, it preserves (see [BT]) ω_1-complete $\aleph_2$-saturated ideals over ω_1. Hence if N is a model with an ω_1-complete $\aleph_2$-saturated ideal over ω_1, and if P is a forcing notion for adding $\aleph_1$ Cohen reals,

than any generic extension of N via P is a model with an ω_1-complete $\aleph_2$-saturated ideal over ω_1, and with no ω_1-complete $(\aleph_2, \aleph_2, \aleph_0)$-saturated ideal over ω_1.

References:

[B] Baumgartner, J.E., Iterated forcing, Surveys in Set Theory, London Math. Soc. Lecture Note Series, 87, Cambridge University Press, 1983.

[BT] Baumgartner, J.E., Taylor A., Saturation properties of ideals in generic extensions I, II, Trans. Amer. Math. Soc. 270 (1982), no.2, 557-574, and 271 (1982), no.2, 587-609.

[F_1] Franek, F., Some results about saturated ideals and about isomorphisms of κ-trees, Ph.D. thesis, University of Toronto, 1983.

[F_2] Franek, F., Certain values of completeness and saturatedness of a uniform ideal rule out certain sizes of the underlying index set, Canadian Mathematical Bulletin, 28 (1985), pp. 501-505

[FL] Forman, M., Laver,R., There is a model in which every $\aleph_2$-chromatic graph of size $\aleph_2$ has an $\aleph_1$-chromatic subgraph of size $\aleph_1$, a handwritten note.

[FMS] Forman, M., Magidor, M., Shelah, S. Marten's Maximum, Saturated Ideals and non-regular ultrafilters I,II, preprint, to apper.

[J] Jech, T.T., Set theory, Academic Press, 1980.

[K_1] Kunen, K., Set theory, North-Holland, 1980.

[K_2] Kunen, K., Saturated ideals, J. Symbolic Logic 43 (1978), no.1, 65-76.

[L] Laver, R., An $(\aleph_2, \aleph_2, \aleph_0)$-saturated ideal on $\aleph_1$, Logic Colloquium 10, 173-180, North-Holland, 1982.

[M] Magidor, M., On the existence of non-regular ultrafilters and the cardinality of ultrapowers, Trans. Amer. Math. Soc. 249 (1979), no.1., 97-111.

[MK] Magidor, M., Kanamori, A., The evolution of large cardinal axioms in set theory, Higher Set Theory, 99-275, Lecture Notes in Math., 669, Springer, Berlin, 1978.

[Mi] Mitchell, W.J., Hypermeasurable cardinals, Logic Colloquium '78 (Mons, 1978), pp.303-316, Stud. Logic Foundations Math., 97, North-Holland, 1979.

[S] Solovay, R.M., Real-valued measurable cardinals, Axiomatic Set Theory, (Proc. Sympos. Pure Math., Vol XIII, Part I, Univ. California, Los Angeles, California, 1967), 397-428.

[SRK] Solovay, R.M., Reinhardt, W.N., Kanamori, A., Strong axioms of infinity and elementary embeddings, Ann. Math. Logic 13 (1978), no.1, 73-116.

[U] Ulam, S., Zur Masstheorie in der algemainem Mengenlehre, Fund. Math. 16 (1930).

[W] Woodin, H., An $\aleph_1$-dense ideal on $\aleph_1$, handwritten notes.

McMaster University, Dept. of Comp. Sci. and Systems, Hamilton, Ontario, Canada L8S 4K1

ULTRAFILTERS AND RAMSEY THEORY -- AN UPDATE

Neil Hindman[1]
Howard University
Washington, D.C. 20059, USA

1. Introduction.

The "update" in the title refers to the earlier survey [37] which I wrote in 1979. That survey dealt with results in Ramsey Theory and related results in the theory of ultrafilters, including some results on the algebraic structure of the compact left-topological semigroups $(\beta N,+)$ and $(\beta N,\bullet)$.

I will write here about things in the same general areas which I have learned since 1979. (Although many of these things were known then to others.)

Probably the most significant fact which I was not aware of when [37] was written is that the structure theory of compact left topological semigroups had already been fairly well developed. We present some of this background material in Section 2, focusing especially on the extension of the operation of an arbitrary semigroup S to its Stone-Čech compactification βS. In Section 3 we deal with questions of continuity, featuring some unpublished results of Eric van Douwen. Section 4 has results about the semigroups $(\beta N,+)$ and $(\beta N,\bullet)$ while in Section 5 we present results about the algebraic structure of βS for more or less arbitrary semigroups S.

Section 6 consists of some applications of ultrafilters to areas of mathematics besides Ramsey Theory.

For the remainder of this paper we restrict our attention to Ramsey Theory, broadly interpreted. In Section 7 we deal with density problems in Ramsey Theory. Section 8 presents some recent powerful combinatorial applications of algebraic properties of ultrafilters.

The editors have informed me that they want to have room to publish other papers. Accordingly, most results will be presented but not proved. The exceptions will be those results which have not previously been published and results which can be easily explained in a line or two. Our set theoretic notation is more or less standard. We mention only that if $n \in \omega$, then $n = \{m \in \omega: m < n\}$ and that we distinguish between ω and $N = \omega\backslash\{0\}$. (Algebraically this distinction can be important.)

1. The author gratefully acknowledges support received from the National Science Foundation via grant DMS-8520873.

We use some special algebraic notation. If S is a commutative semigroup with operation "+" we write for $A \subseteq S$, $FS(A) = \{\Sigma F: F \in [A]^{<\omega}\}$ and if $\langle x_n \rangle_{n<\lambda}$ is a λ-sequence in S we write $FS(\langle x_n \rangle_{n<\lambda}) = \{\Sigma_{n \in F} x_n: F \in [\lambda]^{<\omega}\}$. If the operation is "$\bullet$" we define FP(A) and $FP(\langle x_n \rangle_{n<\lambda})$ analogously.

2. Basic Algebraic Structure of βS.

Given a semigroup $(S,\bullet)$ and $x \in S$ one defines the functions $\lambda_x: S \to S$ and $\rho_x: S \to S$ by $\lambda_x(y) = x \bullet y$ and $\rho_x(y) = y \bullet x$. If S is also a topological space and, for each $x \in S$, λ_x is continuous we say S is a left topological semigroup. (Likewise, if for each $x \in S$, ρ_x is continuous we say S is a right topological semigroup.) If S is both a left topological and right topological semigroup, it is called a semitopological semigroup. If the operation $\bullet : S \times S \to S$ is continuous, S is called a topological semigroup.

The left-right terminology has not been standardized. As we use it, an operation is left-continuous provided multiplication on the left by any x is continuous. Some other people say the operation is left-continuous provided the operation is continuous in the left variable -- what we call right continuity.

Given a discrete space X, we take the Stone-Čech compactification βX of X to be {p: p is an ultrafilter on X}. That is $p \in \beta X$ if and only if (1) $p \subseteq P(X)$, (2) $X \in p$, (3) $\emptyset \notin p$, (4) $A \in p$ and $B \in p$ implies $A \cap B \in p$, (5) $A \in p$ and $A \subseteq B \subseteq X$ implies $B \in p$, and (6) $A \cup B \in p$ implies $A \in p$ or $B \in p$. Given $A \subseteq X$, $\bar{A} = \{p \in \beta X: A \in p\}$. Then $\{\bar{A}: A \subseteq X\}$ define a basis for the topology of βX (and a basis for the closed sets as well). Given $x \in X$ we identify x with $e(x) = \{A \subseteq X: x \in A\}$, the principal ultrafilter generated by x. In this way we pretend that $X \subseteq \beta X$.

In 1951 Arens [1] showed how to define a multiplication on the second dual of a Banach algebra. Using essentially these methods Day [19, Section 6] showed how to extend the operation of a discrete semigroup S to the dual $C^*(S)^*$ of the space $C^*(S)$ of bounded real valued functions on S. Since βS may be embedded in $C^*(S)^*$ (via $\gamma(p)(f) = f^\beta(p)$), Day had thus obtained implicitly an extension of the operation to βS.

It is easy to describe the extension directly. One starts with $\bullet$ defined on $S \times S$ and first extends it to $\beta S \times S$. Given $y \in S$, $\rho_y: S \to S \subseteq \beta S$ so ρ_y has a continuous extension $\rho_y^\beta: \beta S \to \beta S$. For $p \in \beta S \backslash S$ we define $p \bullet y = \rho_y^\beta(p)$. Now one extends the operation to $\beta S \times \beta S$. Given $p \in \beta S$, $\lambda_p: S \to \beta S$ so λ_p has a continuous extension $\lambda_p^\beta: \beta S \to \beta S$. For $q \in \beta S \backslash S$, we define $p \bullet q = \lambda_p^\beta(q)$. From the construction one has trivially that λ_ρ is continuous and that, for each $x \in S$, ρ_x is continuous.

To see that the operation is associative, let $p,q,r \in \beta S$. Observe that $p \bullet (q \bullet r) = (\lambda_p \circ \lambda_q)(r)$ and $(p \bullet q) \bullet r = \lambda_{p\bullet q}(r)$ so, since $\lambda_p \circ \lambda_q$ and $\lambda_{p\bullet q}$ are continuous it suffices to show that for each $s \in S$, $(\lambda_\rho \circ \lambda_q)(s) = (\lambda_{p\bullet q})(s)$, i.e. $p \bullet (q \bullet s) = (p \bullet q) \bullet s$. Now $p \bullet (q \bullet s) = (\lambda_\rho \circ \rho_s)(q)$ and $(p \bullet q) \bullet s = (\rho_s \circ \lambda_p)(q)$ so, again by continuity, it suffices to show that for $t \in S$, $(\lambda_\rho \circ \rho_s)(t) = (\rho_s \circ \lambda_p)(t)$, i.e. $p \bullet (t \bullet s) = (p \bullet t) \bullet s$. But $p \bullet (t \bullet s) = \rho_{t\bullet s}(p)$ and $(p \bullet t) \bullet s = (\rho_s \circ \rho_t)(p)$ and for each $x \in S$, $\rho_{t\bullet s}(x) = (\rho_s \circ \rho_t)(x)$ by the associativity of $\bullet$ on S.

A crucial observation (due originally to Glazer) about the operation is that, given $p,q \in \beta S$ and $A \subseteq S$, $A \in p \bullet q$ if and only if $\{x \in S: A/x \in p\} \in q$, where $A/x = \{y \in S: y \bullet x \in A\}$. (If the operation on S is denoted by +, we write $A - x = \{y \in S: y + x \in S\}$.)

There are alternative ways to view the extension of $\bullet$ to βS which also only involve the notion of ultrafilters. We look first at the method used by van Douwen in [21].

2.1 __Definition__. Let X be a topological space, let B be a set, let A be a filter on B and let $\langle a_x \rangle_{x\in B}$ be an indexed set in X. Given $y \in X$, $A\text{-}\lim\langle a_x \rangle_{x\in B} = y$ if and only if for each neighborhood U of y, $\{x \in B: a_x \in U\} \in A$.

The notion of $A\text{-}\lim\langle a_x \rangle_{x\in B}$ is due (in the case A is an ultrafilter) to Frolík in [23] (using different notation). It is presented in essentially the same generality as here by Furstenberg [25, p. 180].

The basic facts about this notion are easy to derive, including the fact that if $p \in \beta B$ and X is a compact Hausdorff space, then there is a unique $y \in X$ with $p\text{-}\lim\langle a_x \rangle_{x\in B} = y$.

The method used by van Douwen is to define, for $p \in \beta S$ and $s \in S$, $p \bullet s = p\text{-}\lim \langle t \bullet s \rangle_{t\in S}$. Then, for p and q in βS, one defines $p \bullet q = q\text{-}\lim \langle p \bullet s \rangle_{s\in S}$. To see that the definitions agree with the earlier construction, first let $p \in \beta S$, $s \in S$ and $A \in p\text{-}\lim \langle t \bullet s \rangle_{t\in S}$. Then $\{t \in S: t \bullet s \in A\} \in p$. That is $A/s \in p$. Hence $A \in p \bullet s$ as earlier defined. Thus $p\text{-}\lim \langle t \bullet s \rangle_{t\in S} \subseteq p \bullet s$ and since both are ultrafilters equality holds. Similarly, one sees that $A \in q\text{-}\lim \langle p \bullet s \rangle_{s\in S}$ if and only if $\{s \in S: A/s \in p\} \in q$.

Another characterization is from the folklore, told to me by Blass. (The ultrafilter $p \otimes q$ defined below is the same as $t[q \bullet p]$ as defined in [16, p. 157].)

2.2 __Definition__. Given sets X and Y, p an ultrafilter on X, and q an ultrafilter on Y, define

$$p \otimes q = \{A \subseteq X \times Y: \{y \in Y: \{x \in X: (x,y) \in A\} \in p\} \in q\}.$$

Now, given our discrete semigroup S, we have $\bullet : S \times S \to S \subseteq \beta S$ so that it has a continuous extension $\bullet^{\beta}: \beta(S \times S) \to \beta S$. It is then a fact that, for $p,q \in \beta S$, $p \bullet q = \bullet^{\beta}(p \circledast q)$. To see this, let $B \in \bullet^{\beta}(p \circledast q)$ and pick $A \in p \circledast q$ such that $\bullet^{\beta}[A] \subseteq B$. Let $C = \{y \in S: \{x \in S: (x,y) \in A\} \in p\}$. Then $C \in q$ and, given $y \in C$, $\{x \in S: (x,y) \in A\} \subseteq B/y$ so that $C \subseteq \{y \in S: B/y \in p\}$. That is, $B \in p \bullet q$ as required.

However it is constructed, we have that $(\beta S,\bullet)$ is a compact (Hausdorff) left topological semigroup. There is a fundamental structure theorem for compact left topological semigroups due to Ruppert [48, Satz 2]. (Or see [9], where the theory is derived in reasonably elementary detail.)

2.3 Theorem (Ruppert). Let T be a compact left topological semigroup. Then

(i) T has a unique smallest two-sided ideal K.

(ii) K has idempotents and, given an idempotent $e \in T$, the following are equivalent:
- (a) $e \in K$;
- (b) $K = TeT$;
- (c) eT is a minimal right ideal of T;
- (d) Te is a minimal left ideal of T;
- (e) eTe (with the inherited operation) is a group (and a maximal subgroup of T).

(iii) Each minimal right (respectively left) ideal of T is of the form eT (respectively Te) for some idempotent $e \in K$.

(iv) $K = \cup\{eTe: e$ is an idempotent in $K\} = \cup\{eT: e$ is an idempotent in $K\}$ $= \cup\{Te: e$ is an idempotent in $K\}$.

(v) All maximal subgroups of K are algebraically isomorphic and maximal subgroups of K which are contained in the same minimal left ideal are isomorphic via a homeomorphism.

(A right ideal of T is a subset $A \neq \emptyset$ such that $A \bullet T \subseteq A$. Left and two-sided ideals are defined analogously.)

Since βS is the Stone-Čech compactification of S, we have that, if T is any compact left topological semigroup and φ is a (necessarily continuous since S is discrete) homomorphism from S to T then there is a continuous extension $\varphi^{\beta}: \beta S \to T$. It is further quite easy to show that if, for each $s \in S$, $\rho_{\varphi(s)}: T \to T$ is continuous, then φ^{β} is a homorphism. It was shown in [9, Theorem 4.5], that a similar universal object exists for any Hausdorff semitopological semigroup S. This result was significantly improved [38, Theorem 2.10] by dropping any separation assumptions and any assumed relationship between the topological and algebraic structure of S.

2.4 Theorem (Hindman and Milnes). Let S be a semigroup which is also a topological space. There exist a compact Hausdorff left topological semigroup δS and a continuous homomorphism $e: S \to \delta S$ such that $e[S]$ is dense in δS and $\rho_{e(s)}$ is continuous for each $s \in S$. Further if T is a compact Hausdorff left topological semigroup, $\varphi: S \to T$ is a continuous homomorphism, and $\rho_{\varphi(s)}$ is continuous for each $s \in S$, then there is a continuous homomorphism $\eta: \delta S \to T$ such that $\eta \circ e = \varphi$.

A recent observation of Ruppert [47] is of considerable interest in the context of βS because its hypotheses hold whenever S is commutative, as can be routinely verified.

2.5 Theorem (Ruppert). Let T be a left topological semigroup with dense center. If L is a left ideal in T then cl(L) is an ideal.

Thus, in particular, if T is compact and L is any minimal left ideal, then cl(L) is the closure of the minimal ideal of T. (See Section 5.)

3. Right Continuity at Points of $\beta S\backslash S$ -- Some Contributions of Eric van Douwen.

In 1978 and 1979, van Douwen wrote a manuscript [21] dealing with βS where S was a discrete cancellative semigroup. Several of the results in this paper were published, with his permission, in [37]. At that time he was promising to publish the results soon. However, by 1985 when I spoke to him about this matter he had decided not to publish these results; he felt that most of them were already known or had been improved on by others in the meantime. This judgement was largely true. We present in this section the most significant exceptions.

These results respond to a question asked by Raimi in correspondence. Raimi, concerned with $(\beta N,+)$, knew that if $p \in \beta N\backslash N$, then ρ_p is not continuous. (See for example [37, Theorem 10.12].) He asked whether one could show that the restriction of ρ_p to $\beta N\backslash N$ is not (ever or always) continuous. Eric answered this question rather nicely in a fairly general context.

Unfortunately I only have his answers in the form of a letter which included no proofs. Accordingly the sequence of lemmas which I present here are my own and I was unable to prove one of his results in the full generality which he had.

Recall that $p \in \beta S$ is a uniform ultrafilter if and only if for each $A \in p$, $|A| = |S|$. Recall also that an ultrafilter is λ- complete provided whenever $\mathcal{A} \subseteq p$ and $|\mathcal{A}| < \lambda$ one has $\cap\mathcal{A} \in p$.

3.1 <u>Definition</u>. Let S be an infinite discrete semigroup.

(a) $U(S) = \{p \in \beta S: p \text{ is uniform}\}$.

(b) For $p \in U(S)$, $R(p) = \{q \in U(S)$: $\rho_q|_{U(S)}$ is continuous at $p\}$.

(c) For $p \in U(S)$, $J(p) = \{q \in U(S)$: the restriction of $\cdot$ to $U(S) \times U(S)$ is continuous at $(p,q)\}$.

3.2 <u>Lemma</u>. Let S be a right cancellative semigroup with $|S| = \lambda \geq \omega$ and let $\langle a_\sigma \rangle_{\sigma<\lambda}$ enumerate S. Let $D \in [S]^\lambda$. There is a one-to-one λ-sequence $\langle y_\sigma \rangle_{\sigma<\lambda}$ in D such that, if for $\sigma < \lambda$ one has $B_\sigma = \{y_\tau a_\delta$: $\tau < \sigma$ and $\delta \leq \sigma\}$, then given $\eta \leq \sigma < \lambda$, $y_\sigma a_\eta \notin B_\sigma$.

Proof. Construct $\langle y_\sigma \rangle_{\sigma<\lambda}$ inductively. Given B_σ, one has $|B_\sigma| \leq |\sigma| \cdot |\sigma|$ so for each $\eta \leq \sigma$, $|\rho_{a_\eta}^{-1}[B_\sigma]| \leq |\sigma| \cdot |\sigma|$ (by right cancellation) so that $|\bigcup_{\eta\leq\sigma} \rho_{a_\eta}^{-1}[B_\sigma]| < \lambda$. Pick $y_\sigma \in D \backslash \bigcup_{\eta\leq\sigma} \rho_{a_\eta}^{-1}[B_\sigma]$. □

3.3 <u>Lemma</u>. Let S be a with $|S| = \lambda \geq \omega$ and let $\langle a_\sigma \rangle_{\sigma<\lambda}$ enumerate S. Let $q \in \beta S$ such that q is not ω^+-complete. There exists $g: \lambda \to \omega$ such that for each $n < \omega$, $\{a_\sigma: g(\sigma) > n\} \in q$.

Proof. Pick $\{A_n: n < \omega\} \subseteq q$ such that $\bigcap_{n<\omega} A_n \notin q$. Define $g(\sigma) = 0$ if $a_\sigma \in \bigcap_{n<\omega} A_n$ and otherwise $g(\sigma) = n$ where n is the first index with $a_\sigma \notin A_n$. Let $n < \omega$ and suppose $\{a_\sigma: g(\sigma) > n\} \notin q$. Then $\bigcup_{m\leq n}\{a_\sigma: g(\sigma) = m\} \in q$ so one may pick m with $\{a_\sigma: g(\sigma) = m\} \in q$. If $m = 0$ one has $(\bigcap_{n<\omega} A_n) \cup (S\backslash A_0) \in q$. Otherwise one has $S\backslash A_m \in q$. In either case one has a contradiction. □

3.4 <u>Lemma</u>. Let S be a cancellative semigroup with $|S| = \lambda \geq \omega$ and let $\langle a_\sigma \rangle_{\sigma<\lambda}$ enumerate S. Let $D \in [S]^\lambda$ and let $\langle y_\sigma \rangle_{\sigma<\lambda}$ be as guaranteed by Lemma 3.2. Let $p \in U(S)$ with $\{y_\sigma: \sigma < \lambda\} \in p$. If there is some $q \in R(p)$ which is not ω^+-complete, then p is a P-point of U(S).

Proof. Let such q be given. Pick $g: \lambda \to \omega$ as guaranteed by Lemma 3.3. To see that p is a P-point of U(S), let $\{C_n: n < \omega\} \subseteq p$. We need to produce $E \in p$ such that $|E\backslash C_n| < \lambda$ for each $n < \omega$.

For each $\sigma < \lambda$ let B_σ be as in Lemma 3.2 and let $H_\sigma = \{y_\tau: \tau \geq \sigma\} \cap \bigcap_{m\leq g(\sigma)} C_m$. Then each $H_\sigma \in p$. Let $A = \bigcup_{\sigma<\lambda} H_\sigma \cdot a_\sigma$. Then $S = \{x: A/x \in p\}$ so $A \in p \cdot q$. Since $q \in R(p)$, pick $E \in p$ such that $\rho_q[\hat{E} \cap U(S)] \subseteq \hat{A}$. We may presume that $E \subseteq \{y_\sigma: \sigma < \lambda\}$. We claim that for each $\sigma < \lambda$, $|E\backslash H_\sigma| < \lambda$.

Suppose not and pick $\sigma < \lambda$ with $|E\backslash H_\sigma| = \lambda$ and pick $r \in U(S)$ with $E\backslash H_\sigma \in r$. Let $n = g(\sigma)$. Now $A \in r \cdot q$, since $r \in \hat{E}$, so $\{x \in S: A/x \in r\} \in q$. Also $\{a_\tau: \tau > \sigma\} \in q$ since q is uniform. And $\{a : g(\tau) \geq n\} \in q$ by the choice of g. We may thus pick a point in $\{x \in S\colon A/x \in r\} \cap \{a_\tau\colon \tau > \sigma\} \cap \{a_\tau\colon g(\tau) \geq n\}$. That is, we have $\tau > \sigma$ with $g(\tau) \geq n$ such that $A/a_\tau \in r$. Since $\tau > \sigma$ and $g(\tau) \geq n = g(\sigma)$ we have $H_\tau \subseteq H_\sigma$. Since $E\backslash H_\sigma \in r$ we have $E\backslash H_\tau \in r$. We show that $(A/a_\tau) \cap (E\backslash H_\tau) \subseteq \{y_\gamma\colon \gamma \leq \tau\}$, contradicting the choice of r as a uniform ultrafilter. Suppose instead we have $\gamma > \tau$ with $y_\gamma \in (A/a_\tau) \cap (E\backslash H_\tau)$. Now $y_\gamma a_\tau \in A$ so pick μ with $y_\gamma \bullet a_\tau \in H_\mu \bullet a_\mu$. Pick $\eta \geq \mu$ such that $y_\eta \in H_\mu$ and $y_\gamma \bullet a_\tau = y_\eta \bullet a_\mu$.

Now, if $\gamma > \eta$ then $y_\gamma \bullet a_\tau \in B_\gamma$ while if $\eta > \gamma$ we have $y_\eta \bullet a_\mu \in B_\eta$. In either case we contradict Lemma 3.2. Thus we must have $\gamma = \eta$. But then by left cancellation we have $a_\tau = a_\mu$ and hence $\tau = \mu$. But then $y_\gamma = y_\eta \in H_\mu = H_\tau$ while $y_\gamma \in E\backslash H_\tau$, a contradiction.

Thus we have that for each $\sigma < \tau$, $|E\backslash H_\sigma| < \lambda$. Now given n, pick σ such that $g(\sigma) > n$. Then $H_\sigma \subseteq C_n$ so $E\backslash C_n \subseteq E\backslash H_\sigma$ so $|E\backslash C_n| < \lambda$ as required. □

Recall that a cardinal λ is Ulam-measurable if and only if there is an ω^+-complete non-principal ultrafilter on λ. (See [16, Chapter 8] for discussion of this notion.)

The assumption that λ is not Ulam-measurable in the following theorem is mine. It was not in van Douwen's letter and is presumably not needed. (At any rate it is consistent that there are no Ulam-measurable cardinals.)

3.5 <u>Theorem</u> (van Douwen). Let S be a cancellative semigroup with $|S| = \lambda \geq \omega$. If λ is not Ulam-measurable then $\{p \in U(S)\colon R(p) = \emptyset\}$ is dense in U(S).

Proof. Let V be open in U(S) and pick $D \in [S]^\lambda$ with $\bar{D} \cap U(S) \subseteq V$. Pick a one-to-one λ-sequence $\langle y_\sigma \rangle_{\sigma<\lambda}$ in D as guaranteed by Lemma 3.2. Let $E = \{y_\sigma\colon \sigma < \lambda\}$. Now $\bar{E} \cap U(S)$ is compact (and infinite) so is not a P-space [26, Problem 4K]. Pick $p \in \bar{E} \cap U(S)$ such that p is not a P-point. Suppose $R(p) \neq \emptyset$ and pick $q \in R(p)$. Since λ is not Ulam-measurable, q is not ω^+-complete. But this contradicts Lemma 3.4. □

Notice in particular that since (N,+) is cancellative and ω is not Ulam-measurable, Raimi's question is certainly answered by Theorem 3.5.

I quote now (with substitution of notation) from Erik's letter announcing these results. "One would perhaps think one must be able to get all of U(S) in Theorem (3.5), at least in a special case like S = (N,+) or at least get the result that $J(p) = \emptyset$ for all p. Not so:"

3.6 <u>Theorem</u> (van Douwen). Let S be a cancellative semigroup with $|S| = \omega$. The following statements are equivalent.

(a) $\{p \in U(S)\colon\ J(p) = U(S)\}$ is dense in U(S).

(b) $\{p \in U(S)\colon R(p) \neq \emptyset\}$ is dense in U(S).

(c) $\beta N\backslash N$ has a P-point.

Therefore (a) and (b) are consistent with but independent from ZFC.

Proof. That (a) implies (b) is trivial.

To see that (b) implies (c), enumerate S as $\langle a_n \rangle_{n<\omega}$ and pick a one-to-one sequence $\langle y_n \rangle_{n<\omega}$ in S as guaranteed by Lemma 3.2. Let $E = \{y_n : n < \omega\}$ and pick $p \in \overline{E} \cap U(S)$ such that $R(p) \neq \phi$. By Lemma 3.4 and the fact that ω is not Ulam-measurable p is a P-point of U(S). Since U(S) and $\beta N \backslash N$ are homeomorphic, $\beta N \backslash N$ has a P-point.

To see that (c) implies (a), we show that if p is a P-point of U(S), then $J(p) = U(S)$. (It is easy to see that if $\beta N \backslash N$ has a P-point, then the set of P-points of U(S) is dense in U(S).) Let p be a P-point of U(S) and let $q \in U(S)$. To see that $\bullet|_{U(S)\times U(S)}$ is continuous at (p,q), let $A \in p \bullet q$. Let $B = \{x \in S: A/x \in p\}$. Then $B \in q$. Since p is a P-point of U(S), pick $C \in p$ such that, for each $x \in B$, $|C\backslash(A/x)| < \omega$. We claim $\bullet[(\overline{C} \times \overline{B}) \cap (U(S) \times U(S))] \subseteq \overline{A}$. To this end let $(r,s) \in (\overline{C} \times \overline{B}) \cap (U(S) \times U(S))$. We show $B \subseteq \{x \in S: A/x \in r\}$ so that $A \in r \bullet s$. To this end, let $x \in B$. Then $|C\backslash(A/x)| < \omega$. Since $C \in r$ and $r \in U(S)$, $A/x \in r$ as required.

The last remark about consistence and independence refers of course to the famous results of W. Rudin [46] and Shelah [50]. □

In closing this section we turn to another contribution of van Douwen's -- in this case not a result but a question. In [34], I had presented a proof that, if the Continuum Hypothesis holds and the (then unproved) Finite Sum Theorem was valid then there existed $p \in \beta N \backslash N$ such that, for each $A \in p$, $\{x \in A: A - x \in p\} \in p$. Since we now know that this simply says $p + p = p$, and since idempotents are known to exist in any compact left topological semigroup [22], the existence of such p is known with no special assumptions. (See [4] for a discussion of the origins of the idempotent theorem.) I felt there was no longer anything of interest in [22]. However, in conversation in July of 1985, Eric noted that the proof in [22] actually established the existence of an ultrafilter with a basis consisting of sets of the form FS(N) for $B \in [N]^{\omega}$. He asked whether one could prove the existence of such an ultrafilter in ZFC. This question directly inspired three papers [32] (by me), [11] (by Blass) and [12] (by Blass and me). The last of these papers included the answer to Erik's question.

We found out, after obtaining this result, that the question had been independently asked -- and answered -- by Matet in 1985. (Matet's results will appear in [42].)

3.7 __Theorem__. (Matet and, independently but later, Blass and Hindman). If there is $p \in \beta N$ with a base of sets of the form FS(B) for $B \in [N]^\omega$, then there exists a P-point in $\beta N \backslash N$.

4. The Semigroups $(\beta N,+)$ and $(\beta N,\cdot)$.

For some time after the Glazer-Galvin proof of the Finite Sum Theorem (see [37, Section 8]) it was an open question as to whether there was some $p \in \beta N \backslash N$ with $p = p + p = p \bullet p$. The existence of such an ultrafilter was known to imply that, given $r \in N$, if $N = \cup_{i<r} A_i$ then there would exist $i < r$ and infinite B with $FS(B) \cup FP(B) \subset A_i$. That conclusion is now known to be false. In fact, letting $PS(B) = \{x + y: x,y \in B \text{ and } x \neq y\}$ and $PP(B) = \{x \bullet y: x,y \in B \text{ and } x \neq y\}$ we have:

4.1 __Theorem__ ([30]). There exist $\langle A_i \rangle_{i<7}$ such that $N = \cup_{i<7} A_i$ and for no i is there an infinite B with $PS(B) \cup PP(B) \subseteq A_i$. (It is not assumed that $B \subseteq A_i$.)

Likewise the possibility that $p + p = p = p \bullet p$ is known to fail badly:

4.2 __Theorem__ ([33]). There do not exist p and q in $\beta N \backslash N$ with $p + q = p \bullet q$.

This leaves a very annoying question (first asked by van Douwen in correspondence): Do there exist p,q,r, and s in $\beta N \backslash N$ with $p + q = r \bullet s$?

As we have already seen, combinatorial statements frequently have a corresponding algebraic statement in βN. For example, Theorem 4.1 may be rephrased as:

4.3 __Theorem__ ([30]). There is a partition of $\beta N \backslash N$ into seven open-and-closed subsets so that, for no $p \in \beta N$ do $p + p$ and $p \bullet p$ lie in the same cell of the partition.

By comparison, I know of no nice combinatorial statement equivalent to the solution of $p + q = r \bullet s$. It is simply one of those questions which is annoying because we can't answer it.

Several results of van Douwen exhibiting "bad" behavior of $(\beta N,+)$ and $(\beta N,\bullet)$ were presented in [37]. In [28], we showed that left cancellation is much better behaved.

4.4 __Theorem__. Let $C = \{p \in \beta N \backslash N$: left cancellation holds at p in $(\beta N,+)$ and in $(\beta N,\bullet)\}$. Then C has dense interior in $\beta N \backslash N$.

As we have seen in Section 2, the minimal ideal of a compact left topological semigroup is the smallest two-sided ideal. Let us denote by M and K the minimal ideals of $(\beta N,+)$ and $(\beta N,\cdot)$ respectively. It is easy to see that if $p \in M$, then left cancellation in $(\beta N,+)$ fails at p.

4.5 **Theorem** ([28]). There exists $p \in clM$ such that left cancellation holds at p.

4.6 **Theorem** ([28] and [33]). $M \cap K = \phi$ but $clM \cap K \neq \phi$.

Recall from Section 2 that M is the union of pairwise isomorphic groups. It is an obvious question, first raised in correspondence by Karl Hofmann, as to what these groups look like. In [39] we obtained the following result. For later reference we note that the subgroups produced are all contained in $\cap_{n\in N} \overline{Nn}$.

4.7 **Theorem** (Hindman and Pym). Let p be an idempotent in $(\beta N,+)$. Then $p + \beta N + p$ contains a copy of the free semigroup on 2^c generators. If $p \in M$, then $p + \beta N + p$ contains a copy of the free group on 2^c generators.

This result has recently been extended [41].

4.8 **Theorem** (Lisan). There exist c pairwise disjoint topological and algebraic copies of $\cap_{n\in N} \overline{N2^n}$ which miss clM. In particular there exist copies of the free group on 2^c generators which miss clM.

We remark that Pym [44] has recently shown that the entire structure of $\cap_{n\in N} \overline{N2^n}$ arises in a natural way in βS for many different S, where S is assumed to have much less structure than a semigroup. (We are prevented from being more precise by the amount of terminology which would have to be introduced.)

5. The Ideal Structure of βS.

In [35] we were able to characterize the minimal ideal of βS and its closure, with no special assumptions on S.

5.1 **Theorem**. Let $p \in \beta S$. Then p is in the minimal ideal of βS if and only if for each $A \in p$, there exists $F \in [S]^{<\omega}$ such that for all $y \in S$, $(\cup_{t\in F} A/t)/y \in p$. Also p is in the closure of the minimal ideal if and only if for each $A \in p$ there exists $F \in [S]^{<\omega}$ such that, for all $G \in [S]^{<\omega}$ there exists $x \in A$ with

$x \bullet G \subseteq \bigcup_{t \in F}(A/t)$.

If S is commutative, it is an immediate consequence of Theorem 2.5 that the closure of the minimal ideal of βS is again an ideal of βS. In fact this assumption is not needed.

5.2 **Theorem** ([35]). The closure of the minimal ideal of βS (for any discrete semigroup S) is again an ideal of βS.

However, we showed [36] that $T = \bigcap_{n \in N} \overline{Nn}$ is a compact subsemigroup, the closure of whose minimal ideal is not an ideal.

$(\beta S \backslash S) \bullet (\beta S \backslash S)$ is usually an ideal of βS, hence must contain the minimal ideal. In [53] using the notion "inflatable" (or [54] using the weaker notion of "α-inflatable for some α") Umoh showed that $(\beta S \backslash S) \bullet (\beta S \backslash S)$ does not have to be much bigger than the minimal ideal. (The notion of "inflatable" is too complicated to list here. It lies strictly between "cancellative" and "right cancellative".)

5.3 **Theorem** (Umoh). Let S be a countable inflatable semigroup with identity. There exist points in the closure of the minimal ideal which are not equal to $q \bullet r$ for any $q, r \in \beta S \backslash S$.

If $\mathcal{A}$ is a filter on S then $\hat{\mathcal{A}}$ is a closed subset of βS and all closed subsets of βS are of this form. In [17], Davenport obtained the following simple characterization of when $\hat{\mathcal{A}}$ is a subsemigroup.

5.4 **Theorem** (Davenport). Let $\mathcal{A}$ be a filter on S. Then $\hat{\mathcal{A}}$ is a subsemigroup of βS if and only if for each $A \in \mathcal{A}$ and each $B \in P(S) \backslash \mathcal{A}$ there exists $F \in [S \backslash B]^{<\omega}$ such that $\bigcup_{t \in F} A/t \in \mathcal{A}$.

The virtue of a characterization such as this is that it refers only to properties of S and $\mathcal{A}$ which are usually easy to check. Davenport also obtained a reasonably simple characterization of the minimal ideal of $\hat{\mathcal{A}}$ given that $\mathcal{A}$ satisfied certain reasonably common conditions.

6. Some Useful Notations Involving Ultrafilters.

In [10], Blass introduced the "generalized quantifier" $(p\ x)\varphi(x)$, where p is an ultrafilter on a set S, to mean $\varphi(x)$ is true for p-almost all x. That is $(p\ x)\ \varphi(x)$

if and only if $\{x \in S: \varphi(x)\} \in p$. This quantifier has the nice property that it "commutes with negation and conjunction and therefore with all propositional connectives."

I will not be concerned here with the applications in [10], but rather with some other applications which Blass told me about. These applications depend on the simple fact that, given $p,q \in \beta S$ and a formula φ, if (under all interpretations of the unlisted free variables in $\varphi(x)$) one has $(p\ x)\ \varphi(x) \leftrightarrow (q\ x)\ \varphi(x)$, then $p = q$. (To see this let $\varphi(x)$ be "$x \in A$". Then, if $A \subseteq S$, the statement $(p\ x)\varphi(x)$ is the statement that $A \in p$.)

Now as usual assume S is a semigroup and $p,q \in \beta S$. Observe that $(p \bullet q\ x)\ \varphi(x) \leftrightarrow (q\ y)(p\ z)\varphi(z \bullet y)$. Indeed, $\{x \in S: \varphi(x)\}/y = \{z \in S: \varphi(z \bullet y)\}$. Thus:

$$\begin{aligned}(p \bullet q\ x)\varphi(x) &\leftrightarrow \{y \in S: \{x \in S: \varphi(x)\}/y \in p\} \in q \\ &\leftrightarrow \{y \in S: \{z \in S: \varphi(z \bullet y)\} \in p\} \in q \\ &\leftrightarrow \{y \in S: (p\ z)\ \varphi(z \bullet y)\} \in q \\ &\leftrightarrow (q\ y)(p\ z)\ \varphi(z \bullet y)\end{aligned}$$

Consequently we have a simple proof of associativity. Let $p,q,r \in \beta S$.

$$\begin{aligned}((p \bullet q) \bullet r\ x)\ \varphi(x) &\leftrightarrow (r\ u)(p \bullet q\ z)\ \varphi(z \bullet u) \\ &\leftrightarrow (r\ u)(q\ v)(p\ w)\ \varphi((w \bullet v) \bullet u) \\ &\leftrightarrow (r\ u)(q\ v)(p\ w)\ \varphi(w \bullet (v \bullet u)) \\ &\leftrightarrow (q \bullet r\ y)(p\ w)\ \varphi(w \bullet y) \\ &\leftrightarrow (p \bullet (q \bullet r)\ x)\ \varphi(x).\end{aligned}$$

Another application is a proof of the fact from Section 2 that $\bullet^{\beta}(p \otimes q) = p \bullet q$. To see this observe first that for $f: S \to T$ and $p \in \beta S$, one has $(f^{\beta}(p)\ x)\ \varphi(x) \leftrightarrow (p\ y)\ \varphi(f(y))$. (If $B = \{x \in T: \varphi(x)\}$ and $B \in f^{\beta}(p)$, pick $A \in p$ with $f^{\beta}[\bar{A}] \subseteq \bar{B}$. Then $A \subseteq \{y \in S: \varphi(f(y))\}$. Thus $(f^{\beta}(p)\ x)\ \varphi(x) \to (p\ y)\ \varphi(f(y))$. Since this quantifier commutes with negation, the result follows.) Also, directly from Definition 2.2 we have $(p \otimes q\ x)\ \varphi(x) \leftrightarrow (q\ y)(p\ z)\ \varphi((z,y))$. We thus have

$$\begin{aligned}(\bullet^{\beta}(p \otimes q)\ x)\ \varphi(x) &\leftrightarrow (p \otimes q\ u)\ \varphi(\bullet(u)) \\ &\leftrightarrow (q\ y)(p\ z)\ \varphi(z \bullet y) \\ &\leftrightarrow (p \bullet q\ x)\ \varphi(x).\end{aligned}$$

The other notation and its applications were told to me by Scott Williams. He did not know the origin, but said it is "well known in Prague".

Given a compact Hausdorff space X, and $f: X \to X$ let f^n be n-fold composition. (That is $f^1 = f$, $f^{n+1} = f^n \circ f$.) Given $p \in \beta N$ and $x \in X$, define $f^p(x) = p\text{-}\lim \langle f^n \rangle_{n \in N}$. (See Definition 2.1.)

6.1 Lemma. Let X be a compact Hausdorff space and let $f\colon X \to X$ be continuous. If $p,q \in \beta N$, then $f^q \circ f^p = f^{p+q}$.

Proof. Let $x \in X$ and let $y = f^q(f^p(x))$. To see that $y = f^{p+q}(x)$, let U be an open neighborhood of y and let $A = \{n \in N\colon f^n(x) \in U\}$. Let $B = \{n \in N\colon f^n(f^p(x)) \in U\}$. Then $B \in q$. We show that $B \subseteq \{n \in N\colon A - n \in p\}$ so that $A \in p + q$ as required. Let $n \in B$. Since f is continuous and $f^n(f^p(x)) \in U$, pick an open neighborhood V of $f^p(x)$ with $f^n[V] \subseteq U$. Let $C = \{m \in N\colon f^m(x) \in V\}$. Then $C \in p$ and given $m \in C$, $f^n(f^m(x)) \in U$ so that $n + m \in A$. Thus $C \subseteq A - n$ so $A - n \in p$. □

The assumption of continuity in Lemma 6.1 is needed. To see this let $X = Z \cup \{-\infty, +\infty\}$, the usual two point compactification of the integers and define $f\colon X \to X$ by $f(n) = -n - 1$ if $n \geq 0$, $f(n) = -n$ if $n < 0$, and $f(+\infty) = f(-\infty) = 0$. Observe that if $n \in N$ then $f^{2n}(0) = n$ and $f^{2n+1}(0) = -n - 1$ and hence $f^{2n+1}(+\infty) = n$. Now let $p,q \in \beta N$ with $N2 \in p$ and $N2 + 1 \in q$ (so that $N2 + 1 \in p + q$). Then $f^q(f^p(0)) = f^q(+\infty) = +\infty \neq -\infty = f^{p+q}(0)$.

Recall that $x \in X$ is a recurrent point of f provided x is in the forward orbit closure of f, i.e. $x \in cl\{f^n(x)\colon n \in N\}$.

6.2 Theorem. Let X be a compact Hausdorff space, let $x \in X$ and let $f\colon X \to X$ be continuous. The following are equivalent.

(a) x is a recurrent point of f.

(b) There exists $p \in \beta N$ with $f^p(x) = x$.

(c) There exists $p \in \beta N$ with $p + p = p$ such that $f^p(x) = x$.

Proof. To see that (a) implies (b), let $\mathcal{A} = \{\{n \in N\colon f^n(x) \in U\}\colon U$ is a neighborhood of $x\}$. Then $\mathcal{A}$ has the finite intersection property so pick $p \in \beta N$ with $\mathcal{A} \subseteq p$. Then $f^p(x) = x$.

To see that (b) implies (c), let $T = \{p \in \beta N\colon f^p(x) = x\}$. Then T is closed. (If $p \in \beta N \backslash T$, then for some neighborhood U of x, $\{n \in N\colon f^n(x) \in U\} \notin p$. For this U, if $B = \{n \in N\colon f^n(x) \notin U\}$, then $\bar{B}$ is a neighborhood of p missing T.) Also T is a subsemigroup of $(\beta N,+)$. Indeed, if $p,q \in T$, then $f^{p+q}(x) = f^q(f^p(x)) = f^q(x) = x$. Thus by [22], there is an idempotent $p \in T$.

To see that (c) implies (a), pick p with $f^p(x) = x$. Let U be an open neighborhood of x. Then $\{n \in N\colon f^n(x) \in U\} \in p$ so is non-empty. □

7. Density Results in Ramsey Theory.

A new area of Ramsey Theory was opened in 1974 with the proof [51] of Szemerédi's Theorem. A new powerful tool in this area was provided by Furtenberg's proof of Szemerédi's Theorem using ergodic theory [24]. (See [25] and [3] for extensive descriptions and additional references.)

Recall that the ordinary upper density of a set $A \subset N$ is $\bar{d}(A) = \lim_{n\to\infty} \sup \{|A \cap \{1,2,\ldots,n\}|/n\colon n \in N\}$. A more natural notion of density from the point of view of ergodic theory is what, following [25], we have agreed to call the Banach density of A. That is, $d^*(A) = \sup\{\alpha\colon$ there exist increasing sequences $\langle x_n\rangle_{n\in N}$ and $\langle t_n\rangle_{n\in N}$ with, for each n, $|A \cap \{x_n + 1,\ldots, x_n + t_n\}|/t_n \geq \alpha\}$. It is easy to see that if $d^*(A) = \alpha$ then there exist increasing sequences $\langle x_n\rangle_{n\in N}$ and $\langle t_n\rangle_{n\in N}$ with $\lim_{n\to\infty} |A \cap \{x_n + 1, x_n + 2,\ldots, x_n + t_n\}|/t_n = \alpha$.

Several subsets of βN are definable in terms of density. As with many other combinatorially defined subsets, these are definable in the form $\{p \in \beta N\colon$ for all $A \in p, \varphi(A)\}$. Any such set is automatically closed. (If $A \in p$ such that $\neg\varphi(A)$, then $\hat{A}$ is a neighborhood of p missing the defined set.) Another useful notion is that of a "divisible" statement introduced by Glasner [27]. That is a statement φ about subsets of N is divisible if and only if (i) $\varphi(N)$ and $\neg\varphi(\phi)$; (ii) if $A \subset B \subset N$ and $\varphi(A)$, then $\varphi(B)$; and (iii) if $A,B \subset N$ and $\varphi(A \cup B)$, then $\varphi(A)$ or $\varphi(B)$. It is easy to see that if φ is a divisible statement, then $\{p \in \beta N\colon$ for all $A \in p, \varphi(A)\} \neq \phi$. (Or see [37, Theorem 6.7] or [36, Lemma 2.5].)

7.1 <u>Definition</u>. (a) $\Delta = \{p \in \beta N\colon$ for all $A \in p$, $\bar{d}(A) > 0\}$.
(b) $\Delta^* = \{p \in \beta N\colon$ for all $A \in p$, $d^*(A) > 0\}$.
(c) $\Delta_1 = \{p \in \beta N\colon$ for all $A \in p$, there exists $k \in N$ with $d^*(\bigcup_{t=1}^{k} A - t) = 1\}$.

By the above remarks, Δ and Δ^* are non-empty. By Theorem 5.1 we have $\Delta_1 = \mathrm{cl}M$, where M is the minimal ideal of $(\beta N,+)$. We also have Δ is a right ideal of $(\beta N,+)$ and of $(\beta N,\cdot)$ [37, Theorem 10.8] (due to van Douwen), Δ_1 is an ideal of $(\beta N,+)$ and a right ideal of $(\beta N,\cdot)$ [28, Lemma 3.5, Theorem 3.9], and Δ^* is an ideal of $(\beta N,+)$ and a right ideal of $(\beta N,\cdot)$ [33, Theorem 7.12].

In [2], Bergelson introduced the following notion.

7.2 <u>Definition</u>. A set $B \subseteq N$ is a set of nice combinatorial recurrence if and only if for all $\epsilon > 0$ and all $A \subseteq N$, if $\bar{d}(A) > 0$ then there exists $n \in B$ with

$\bar{d}(A \cap A - n) \geq \bar{d}(A)^2 - \epsilon$.

Bergelson showed that if C is any infinite subset of N, then $D(C) = \{x - y: x,y \in C \text{ and } x > y\}$ is a set of nice combinatorial recurrence. He then established [2] the following generalization of Schur's Theorem.

7.3 Theorem (Bergelson). Let $m \in N$ and let $N = \cup_{i<m} A_i$. Then there exists $i < m$ such that $\bar{d}(A_i) > 0$ and for every $\epsilon > 0$ $\bar{d}(\{n \in A_i: \bar{d}(A_i \cap A_i - n) \geq \bar{d}(A_i)^2 - \epsilon\}) > 0$.

Bergelson has recently told me in conversation of the following theorem, whose proof we are presenting with his permission.

7.4 Theorem (Bergelson). Assume that whenever $B \in [N]^\omega$, one has that $\{x^2: x \in FS(B)\}$ is a set of nice combinatorial recurrence. Let $p \in \Delta$ such that $p + p = p$. Then for all $A \in p$ and all $\epsilon > 0$, $\{x \in A: A - x \in p \text{ and } \bar{d}(A \cap A - x^2) \geq \bar{d}(A)^2 - \epsilon \text{ and } \bar{d}(A \cap A - x) \geq \bar{d}(A)^2 - \epsilon\} \in p$.

Proof. By [5, Lemma 2.1] we have $\{x \in A: A - x \in p \text{ and } \bar{d}(A \cap A - x) \geq \bar{d}(A)^2 - \epsilon\} \in p$. (We essentially duplicate the above cited proof in what follows.) Let $B = \{x \in N: \bar{d}(A \cap A - x^2) \geq \bar{d}(A)^2 - \epsilon\}$ and suppose that $B \notin p$. Since $p + p = p$, pick (see for example [37, Theorem 8.6]) $C \in [N]^\omega$ with $FS(C) \subseteq N\backslash B$. Pick by assumption $x \in FS(C)$ such that $\bar{d}(A \cap A - x^2) \geq \bar{d}(A)^2 - \epsilon$. But then $x \in B$, a contradiction. □

The interest in Theorem 7.3 is strengthened by Bergelson's announcement (in conversation) that he and Furstenberg have proved that the hypothesis is true. As a consequence, they easily obtain a non linear Ramsey Theory result: If $N = \cup_{i<m} A_i$ then there are some $i < m$ and $x,y,z \in A_i$ with $x + y^2 = z$. (To see this let $p \in \Delta$ with $p + p = p$ and pick $i < m$ with $A_i \in p$. Let $\epsilon = \bar{d}(A_i)^2/2$ and let $B = \{x \in A_i: A_i - x \in p \text{ and } \bar{d}(A_i \cap A_i - x^2) \geq \bar{d}(A_i)^2 - \epsilon \text{ and } \bar{d}(A_i \cap A_i - x) \geq \bar{d}(A_i)^2 - \epsilon\}$. Pick $y \in B$. Pick $x \in A_i \cap A_i - y^2$ and let $z = x + y^2$.)

In [37] we presented the following Theorem of Raimi [45]: There exists $E \subseteq N$ such that whenever $m \in N$ and $N = \cup_{i<m} A_i$ there exist $i < m$ and $k \in N$ with $|(A_i + k) \cap E| = \omega$ and $|(A_i + k)\backslash E| = \omega$. Using properties of a probability space, Bergelson and Weiss [7] have generalized this result.

7.5 <u>Theorem</u>. (Bergelson and Weiss). There exists $E \subseteq N$ such that whenever $A \subseteq N$ and $\bar{d}(A) > 0$, there is some $k \in N$ with $\bar{d}((A + k) \cap E) > 0$ and $\bar{d}((A + k)\backslash E) > 0$.

Call a family F of subsets of the set Z of integers translation invariant provided, whenever $F \in F$ and $k \in Z$ one has $F + k \in Z$. Call such a family partition regular if, whenever $m \in N$ and $N = \cup_{i<m} A_i$ there exist $i < m$ and $F \in F$ with $F \subseteq A_i$. In [3], Bergelson made the following conjecture: If F is a translation invariant partition regular family of <u>finite</u> subsets of Z and if $A \subseteq N$ with $d^*(A) > 0$, then there is $F \in F$ with $F \subseteq A$. (The most famous instance of the validity of Bergelson's conjecture has F consisting of all length k arithmetic progressions.)

Davenport and I made the following simple observation: If F is a partition regular translation invariant set of finite subsets of Z and $A = \{p \in \beta N$: for all $A \in p$ there exist $F \in F$ with $F \subseteq A\}$, then A is a closed ideal of $(\beta N,+)$. (Partition regularity yields that $A \neq \phi$. Translation invariance yields that A is a right ideal. To see that A is a left ideal, let $p \in A$, $q \in \beta N$, and $A \in q + p$. Pick $F \in F$ with $F \subseteq \{x \in N\colon A - x \in q\}$. Since $|F| < \omega$, $\cap_{x\in F} A - x \in q$. If $t \in \cap_{x\in F} A - x$, then $t + F \subseteq A$.) Bergelson's conjecture is easily seen to be equivalent to the assertion that for any such A, $\Delta^* \subseteq A$. Since Δ_1 is the smallest closed ideal of $(\beta N,+)$ one does always get $\Delta_1 \subseteq A$ and, fairly easily, that $\Delta_1 \neq A$. Further by definition, $p \in \Delta_1$ if and only if for each $A \in p$, there exists k with $d^*(\cup_{t=1}^{k} A - t) = 1$. Also, by [29, Theorem 3.8], $p \in \Delta^*$ if and only if for each $A \in p$ and each $\epsilon > 0$ there exists k with $d^*(\cup_{t=1}^{k} A - t) > 1 - \epsilon$. The similarity between these descriptions led us to believe that perhaps no closed ideals of $(\beta N,+)$ could be found strictly between Δ_1 and Δ^*, (so one would have a proof of Bergelson's conjecture).

Observe that, by Theorem 2.5, if $p \in \beta N\backslash N$, then $cl((\beta N\backslash N) + p)$ is a closed ideal of $(\beta N,+)$. Call such an ideal "subprincipal". The answer which we obtained [18] is vastly different than the one we wanted:

7.6 <u>Theorem</u>. (Davenport and Hindman). Δ_1 is the intersection of subprincipal closed ideals lying strictly between it and Δ^*.

Presumably one of the main reasons we were unable to obtain our desired result is that Bergelson's conjecture is false. We are grateful to Imre Ruzsa for permission to present his unpublished proof of this fact. (It is inspired by his [49, Theorem 1].)

Recall that a set $B \subseteq N$ is syndetic if and only if B has bounded gaps; that is, there exists $k \in N$ with $N = \cup_{t=1}^{k} B - t$. Also B is piecewise syndetic if and only if there exists k with $d^*(\cup_{t=1}^{k} B - t) = 1$.

7.7 Lemma. Let A be piecewise syndetic. Then there exist a syndetic set B and an increasing sequence $\langle y_n \rangle_{n \in N}$ such that $\{y_n + x: n \in N, x \in B, \text{ and } x \leq n\} \subseteq A$.

Proof. Pick k such that $d^*(\cup_{t=1}^{k} A - t) = 1$. For each n pick $x_n \in N$ with $\{x_n + 1, x_n + 2, \ldots, x_n + n\} \subseteq \cup_{t=1}^{k} A - t$ and $x_{n+1} > x_n$. Choose a subsequence $\langle y_n \rangle_{n \in N}$ of $\langle x_n \rangle_{n \in N}$ so that

(1) for each $n \in N$, $\{y_n + 1, y_n + 2, \ldots, y_n + n\} \subseteq \cup_{t=1}^{k} A - t$ and

(2) For n,m, and s in N and $t \in \{1,2,\ldots,k\}$, if $s \leq n \leq m$, then $y_n + s + t \in A$ if and only if $y_m + s + t \in A$.

(See the proof of [28, Lemma 3.4] for a detailed description of how to do this.)

Let $B = \{n \in N: y_n + n \in A\}$. Then by (2) we have immediately that $\{y_n + x: n \in N, x \in B, \text{ and } x \leq n\} \subseteq A$. To see that B is syndetic we show $N = \cup_{t=1}^{k} B - t$. Let $m \in N$ and pick $t \in \{1,2,\ldots,k\}$ with $y_m + m + t \in A$. Let $n = m + t$. Then $y_n + n \in A$ so $n \in B$. Thus $m \in \cup_{t=1}^{k} B - t$. □

7.8 Theorem (Ruzsa). Bergelson's conjecture is false. That is there exist A with $d(A) > 0$ and a partition regular translation invariant family F of finite subsets of Z such that no member F of F is contained in A.

Proof. Pick any A with $d(A) > 0$ such that A is not piecewise syndetic. (The sets constructed in Section 11 of [37] are such sets. For a simpler example consider $\{n \in N: \text{for all } k > 3 \text{ and all } m, \text{ if } 2^{k-1} < m < 2^{k-1} + k, \text{ then } n \not\equiv m \pmod{2^k}\}$.)

Since A is not piecewise syndetic we have (by simply negating the definition) that there exists $b: N \to N$ such that for all $g,x \in N$ there exists $y \in \{x + 1, x + 2, \ldots, x + b(g)\}$ with $\{y + 1, y + 2, \ldots, y + g\} \cap A = \emptyset$. (That is a gap of length g in A can be found within $b(g)$ of any point.) We may presume b is an increasing function.

Let $F = \{\{a_1, a_2, \ldots, a_k\}: k \in N \setminus \{1\}, \text{ each } a_i \in Z, a_1 < a_2 < \ldots < a_k, \text{ and } b(\max\{a_{i+1} - a_i: 1 \leq i < k\}) < k\}$. F is clearly translation invariant. To see that F is partition regular, let $m \in N$ and let $N = \cup_{i<m} C_i$. Pick $j < m$ such that C_j is piecewise syndetic. (For example, let $p \in \Delta_1$ and pick j such that $C_j \in p$.) By Lemma 7.7 pick a syndetic set B and an increasing sequence $\langle y_n \rangle_{n \in N}$ such that $\{y_n + x: n \in N, x \in B, \text{ and } x \leq n\} \subseteq C_j$.

Since B is syndetic, pick $g \in N$ such that $N = \cup_{t=1}^{g} B - t$. Let $k = b(g) + 1$. Pick $a_1 \in B$ and inductively for $1 \leq i < k$, pick $a_{i+1} \in \{a_i + 1, a_i + 2, \ldots, a_i + g\} \cap B$. Let $d = \max\{a_{i+1} - a_i: 1 \leq i < k\}$. Then $d \leq g$ so $b(d) \leq b(g) < k$ so $\{a_1, a_2, \ldots, a_k\} \in F$. Let $n = a_k$. Then $\{y_n + a_1, y_n + a_2, \ldots, y_n + a_k\} \subseteq C_j$ and $\{y + a, y + a, \ldots, y + a\} \in F$.

Now suppose we have some $F \in \mathcal{F}$ with $F \subseteq A$. Pick k such that $F = \{a_1, a_2, \ldots, a_k\}$ with $a_1 < a_2 < \ldots < a_k$. Let $x = a_1 - 1$ and let $g = \max\{a_{i+1} - a_i : 1 \le i < k\}$. Note $b(g) < k$ by the definition of $\mathcal{F}$. Pick $y \in \{x + 1, x + 2, \ldots, x + b(g)\}$ with $\{y + 1, y + 2, \ldots, y + g\} \cap A = \emptyset$. Now $y \le x + b(g) \le x + k - 1 = a_1 - 1 + k - 1 < a_k$. Pick the least i such that $y < a_i$ and note $i \ge 2$. Now $a_i \in A$ and $\{y + 1, y + 2, \ldots, y + g\} \cap A = \emptyset$ so $a_i \ge y + g + 1$. Since $a_{i-1} \le y$ we have $a_i - a_{i-1} > g$, a contradiction. □

Since the proof of Theorem 7.8 works on any A which is not piecewise syndetic, one obtains counterexamples with density arbitrarily close to 1. However, the size of the finite sets involved is always unbounded.

The following result of Kříž [40] is much stronger since only pair sets are used. Its proof is also much more complicated.

7.9 Theorem (Kříž). Let $\epsilon > 0$. There exist a set A with $d(A) > 1/2 - \epsilon$ and a partition regular translation invariant family $\mathcal{F}$ of two element subsets of $\mathbb{Z}$ such that no $F \in \mathcal{F}$ is contained in A.

8. New Combinatorial Applications of Ultrafilters.

In 1982 Tim Carlson proved a remarkable theorem, whose proof utilizes ultrafilters, and which has as corollaries numerous earlier results in Ramsey Theory. This theorem initially circulated in notes by Prikry. It now can be found as Theorem 3 of [13]. Unfortunately, and perhaps unavoidably, one must develop a large amount of terminology to state Carlson's Theorem and we will not do this here.

A recent result [4] addresses the issue of whether one can find solutions to different Ramsey type problems all lying in the same cell of a partition. (For example, if $m \in \mathbb{N}$ and $\mathbb{N} = \cup_{i<m} A_i$ one can certainly find $i < m$ and $j < m$ so that $\bar{d}(A_i) > 0$ and $|\{x \in \mathbb{N}: x^2 \in A_j\}| = \omega$. On the other hand, it is easy to prevent $i = j$.) The result extends earlier work of mine [31] and joint work with Deuber [20].

The proof of this result is very simple, producing an ultrafilter every member of which has the listed properties. It utilizes the simple fact, using alternatively $(\beta\mathbb{N},+)$ and $(\beta\mathbb{N},\cdot)$, that if L is a left ideal of a semigroup and R is a right ideal, then $L \cap R \ne \emptyset$. Given $A \subseteq \mathbb{N}$, $D(A) = \{x - y: x, y \in A \text{ and } y < x\}$.)

8.1 Theorem. (Bergelson and Hindman) Let $m \in \mathbb{N}$ and let $\mathbb{N} = \cup_{i<m} A_i$. There exists $i < m$ such that

(a) A_i contains solutions to all partition regular systems of homogeneous linear equations with integer coefficients.

(b) One can inductively choose a sequence $\langle x_n \rangle_{n<\omega}$ in A_i so that $FS(\langle x_n \rangle_{n<\omega}) \subseteq A_i$ and for each n, given $\langle x_j \rangle_{j<n}$, the set of choices for x_n

has positive upper density.

(c) A_i is piecewise syndetic.

(d) there is some k such that for each n there exists x with $x \bullet \{1,2,\ldots,n\} \subseteq \bigcup_{t=1}^{k} A_i/t$

(e) for each partition regular translation invariant family F of finite subsets of Z there exists $F \in F$ with $F \subseteq A_i$.

(f) there is a syndetic set B with $D(B) \subseteq D(A_i)$.

(g) for each $\epsilon > 0$, $\bar{d}(\{n \in A_i: \bar{d}(A_i \cap A_i - n) > \bar{d}(A_i)^2 - \epsilon\}) > 0$.

(h) For each $k \in N$, $\bar{d}(\{m \in A_i: \bar{d}(\cap_{t=0}^{k} A - tm) > 0\}) > 0$.

(i) There exists $B \in [N]^\omega$ with $FP(B) \subseteq A_i$.

Our final application utilizes an old method of proof of Ramsey's Theorem using ultrafilters. (See [14, page 39].) When I first saw this proof over ten years ago I was quite unimpressed. It essentially takes a standard proof and replaces appeals to the pigeon hole principle (if $m \in N$ and $N = \bigcup_{i<m} A_i$, then some A_i is infinite) with references to a non-principal ultrafilter p (if $m \in N$ and $N = \bigcup_{i<m} A_i$, then some $A_i \in p$). However, Bergelson pointed out that we could probably get stronger results if we used special ultrafilters. Indeed, this is so. For example utilizing p such that $p + p = p$, one obtains the Milliken-Taylor Theorem ([43], [52]). In [6] we display the results when we utilize a "combinatorially large ultrafilter". (That is, an ultrafilter used to produce Theorem 8.1.)

I will illustrate the method here with a simple result utilizing an ultrafilter every member of which contains arbitrarily long arithmetic progressions. (These exist by van der Waerden's Theorem.) The method of proof differs somewhat from [6] because we utilize here the product $\circledast$ introduced in Definition 2.2.

8.2 <u>Theorem</u>. (Bergelson and Hindman.) Let $m \in N$ and let $[N]^2 = \bigcup_{i<m} A_i$. Then there exist $i < m$ and $\langle B_n \rangle_{n \in N}$ such that each B_n is an arithmetic progression of length n and $\{\{x,y\}$: there exist $n,t \in N$ with $t \neq n$ and $x \in B_t$ and $y \in B_n\} \subseteq A_i$.

Proof. Pick $p \in \beta N$ such that each member of p contains arbitrarily long arithmetic progressions. Observe that $L = \{(x,y): x,y \in N \text{ and } x > y\} \in p \circledast p$. (For, given y, $\{x \in N: (x,y) \in L\}$ is cofinite, hence in p. Thus $\{y \in N: \{x \in N: (x,y) \in L\} \in p\} = N \in p$.) Now given i, let $C_i = \{(x,y) \in L: \{x,y\} \in A_i\}$ and pick i such that $C_i \in p \circledast p$.

It thus suffices to show that whenever $C \in p \circledast p$, there exists a sequence $\langle B_n \rangle_{n \in N}$ with each B_n an arithmetic progression of length n and $\{(x,y): x > y$ and there exist $n,t \in N$ with $t \neq n$ and $x \in B_t$ and $y \in B_n\} \subseteq C$.

Let $D = \{y \in N: \{x \in N: (x,y) \in C\} \in p\}$. We choose $\langle B_n \rangle_{n \in N}$ inductively with each $B_n \subseteq D$. Given $n > 1$ and $\langle B_t \rangle_{t=1}^{n-1}$ we let $a = \max B_{n-1}$ and let $E_n = D \cap \bigcap_{t=1}^{n-1} \bigcap_{z \in B_t} \{x \in N: (x,z) \in C\} \cap \{x \in N: x > a\}$. Then since each $B_t \subseteq D$ we have $E_n \in p$. Pick a length n arithmetic progression $B_n \subseteq E_n$. Then $\langle B_n \rangle_{n \in N}$ is as required. □

REFERENCES

1. R. Arens, The adjoint of a bilinear operator, Proc. Amer. Math. Soc. 2 (1951), 839-848.

2. V. Bergelson, A density statement generalizing Schur's Theorem, J. Comb. Theory (Series A) 43 (1986), 338-343.

3. V. Bergelson, Ergodic Ramsey Theory, in Logic and Combinatorics ed., S. Simpson, Contemporary Mathematics 69 (1987), 63-87.

4. V. Bergelson and N. Hindman, A combinatorially large cell of a partition of N, J. Comb. Theory (Series A), to appear.

5. V. Bergelson and N. Hindman, Density versions of two generalizations of Schur's Theorem, J. Comb. Theory (Series A), to appear.

6. V. Bergelson and N. Hindman, Ultrafilters and multidimensional Ramsey theorems, manuscript.

7. V. Bergelson and B. Weiss, Translation properties of sets of positive upper density, Proc. Amer. Math. Soc. 94 (1985), 371-376.

8. J. Berglund and N. Hindman, Filters and the weak almost periodic compactification of a discrete semigroup, Trans. Amer. Math. Soc. 284 (1984), 1-38.

9. J. Berglund, H. Junghenn, and P. Milnes, Compact right topological semigroups and generalizations of almost periodicity, Lecture Notes in Math. 663, Springer-Verlag, Berlin (1978).

10. A. Blass, Selective ultrafilters and homogeneity, Annals of Pure and Applied Logic, to appear.

11. A. Blass, Ultrafilters related to Hindman's finite-unions theorem and its extensions, in Logic and Combinatorics, ed. S. Simpson, Contemporary Mathematics 65 (1987), 89-124.

12. A. Blass and N. Hindman, On strongly summable ultrafilters and union ultrafilters, Trans. Amer. Math. Soc., 304 (1987), 93-99.

13. T. Carlson, Some unifying principles in Ramsey Theory, Discrete Math., to appear.

14. W. Comfort, Some recent applications of ultrafilters to topology, in Proceedings of the Fourth Prague topological Symposium, 1976, ed. J. Novak, Lecture Notes in Math. 609 (1977), 34-42.

15. W. Comfort, Ultrafilters: an interim report, Surveys in General Topology, Academic Press, New York, 1980, 33-54.

16. V. Comfort and S. Negrepontis, The theory of ultrafilters, Springer-Verlag, Berlin, 1974.

17. D. Davenport, The algebraic properties of closed subsemigroups of ultrafilters on a discrete semigroup, Dissertation, Howard University, 1987.

18. D. Davenport and N. Hindman, Subprincipal closed ideals in βN, Semigroup Forum, to appear.

19. M. Day, Amenable semigroups, Illinois J. Math. 1 (1957), 509-544.

20. V. Deuber and N. Hindman, Partitions and sums of (m,p,c)-sets, J. Comb. Theory (Series A) 45 (1987), 300-302.

21. E. van Douwen, The Čech-Stone compactification of a discrete cancellative groupoid, manuscript.

22. R. Ellis, Lectures on topological dynamics, Benjamin, New York, 1969.

23. Z. Frolík, Sums of ultrafilters, Bull. Amer. Math. Soc. 73 (1967), 87-91.

24. H. Furstenberg, Ergodic behavior of diagonal measures and a theorem of Szemeredi on arithmetic progressions, J. d'Analyse Math. 31 (1977), 204-256.

25. H. Furstenberg, Recurrencee in ergodic theory and combinatorial number theory, Princeton University Press, Princeton, 1981.

26. L. Gillman and M. Jerison, Rings of continuous functions, van Nostrand, Princeeton, 1960.

27. S. Glasner, Divisibility properties and the Stone-Cech compactification, Canad. J. Math. 32 (1980), 993-1007.

28. N. Hindman, Minimal ideals and cancellation in βN, Semigroup Forum 25 (1982), 291-310.

29. N. Hindman, On density, translates, and pairwise sums of integers, J. Comb. Theory (Series A) 33 (1982), 147-157.

30. N. Hindman, Partitions and pairwise sums and products, J. Comb. Theory (Series A) 37 (1984), 46-60.

31. N. Hindman, Ramsey's Theorem for sums, products and arithmetic progressions, J. Comb. Theory (Series A) 38 (1985), 82-83.

32. N. Hindman, Summable ultrafilters and finite sums, in Logic and Combinatorics, ed. S. Simpson, contemporary Mathematics 65 (1987), 263-274.

33. N. Hindman, Sums equal to products in βN, Semigroup Forum 21 (1980), 221-255.

34. N. Hindman, The existence of certain ultrafilters on N and a conjecture of Graham and Rothschild, Proc. Amer. Math. Soc. 36 (1972), 341-346.

35. N. Hindman, The ideal structure of the space of κ-uniform ultrafilters on a discrete semigroup, Rocky Mountain J. Math. 16 (1986), 685-701.

36. N. Hindman, The minimal ideals of a multiplicative and additive subsemigroup of βN, Semigroup Forum 32 (1985), 283-292.

37. N. Hindman, Ultrafilters and combinatorial number theory, in Number Theory Carbondale 1979, ed. M. Nathanson, Lecture Notes in Math. 751 (1979), 119-184.

38. N. Hindman and P. Milnes, The LMC compactification of a topologized semigroup, Czech. Math. J., to appear.

39. N. Hindman and J. Pym, Free groups and semigroups in βN, Semigroup Forum 30 (1984), 177-193.

40. I. Kříž, Large independent sets in shift-invariant graphs. Solution of Bergelson's problem, Graphs and Combinatorics 3 (1987), 145-158.

41. A. Lisan, Free groups in βN which miss the minimal ideal, Semigroup Forum, to appear.

42. P. Matet, Some filters of partitions, manuscript.

43. K. Milliken, Ramsey's Theorem with sums or unions, J. Comb. Theory (Series A) 18 (1975), 276-290.

44. J. Pym, Semigroup structure in Stone-Cech compactifications, manuscript.

45. R. Raimi, Translation properties of finite partitions of the positive integers, Fund. Math. 61 (1968), 253-256.

46. W. Rudin, Homogeneity problems in the theory of Cech compactifications, Duke Math. J. 23 (1956), 409-419.

47. W. Ruppert, In a left topological semigroup with dense center the closure of any left ideal is an ideal, Semigroup Forum, to appear.

48. W. Ruppert, Rechstopologische Halbgruppen, J. Reine Angew. Math. 261 (1973), 123-133.

49. I. Ruzsa, Difference sets and the Bohr topology I., manuscript.

50. S. Shelah, Proper forcing, Lecture Notes in Math. 940 (1982).

51. E. Szemerédi, On sets of integers containing no k elements in arithmetic progression, Acta. Arith. 27 (1975), 199-245.

52. A. Taylor, A canonical partition relation for finite subsets of ω, J. Comb. Theory (Series A) 21 (1976), 137-146.

53. H. Umoh, Ideals of the Stone-Cech compactification of semigroups, Semigroup Forum 32 (1985), 201-214.

54. H. Umoh, The ideal of products in $\beta S\backslash S$, Dissertation, Howard University, 1987.

CONCERNING STATIONARY SUBSETS OF $[\lambda]^{<\kappa}$

Pierre MATET*

Freie Universität Berlin, Institut für Mathematik II,
Arnimallee 3, 1000 Berlin 33, West Germany

0. Introduction

We shall start by presenting, in Section 1, some simple characterizations of those subsets of $[\lambda]^{<\kappa}$ that contain a closed unbounded set. Then, in Section 2 we use one of those characterizations to obtain a two-cardinal version of the well-known stationary reflection property of weakly compact cardinals. Finally, in Sections 3 and 4 we investigate the properties of some normal ideals over $[\lambda]^{<\kappa}$ which can be associated with certain games.

The author would like to thank Hans-Dieter Donder and Jean-Pierre Levinski for helpful discussions.

Throughout this paper κ will denote a fixed uncountable regular cardinal and λ a fixed limit ordinal $\geq\kappa$.

Let us now recall some definitions.

Let A be a set of size $\geq\kappa$. We set $[A]^{<\kappa} = \{a \subset A : |a| < \kappa\}$. Let $C \subseteq [A]^{<\kappa}$ be given. C is said to be unbounded in case for every $a \in [A]^{<\kappa}$, there is a $b \in C$ with $a \subseteq b$. C is closed if for every sequence $a_\alpha \in C$, $\alpha < \gamma < \kappa$, such that $a_\beta \subseteq a_\alpha$ for $\beta < \alpha$, we have $\bigcup_{\alpha<\gamma} a_\alpha \in C$. It is well-known that C is closed iff C is closed under directed unions of size $<\kappa$. C is said to be strongly closed if $\cup d \in C$ for every nonempty $d \in [C]^{<\kappa}$. We let $NS_{\kappa,A}$ (respectively $SNS_{\kappa,A}$) denote the collection of those subsets T of $[A]^{<\kappa}$ with the property that there exists an unbounded $C \subseteq [A]^{<\kappa}$ such that C is closed (resp. strongly closed) and $C \cap T = 0$.

Given an ideal I over a set X, we set $I^+ = \{B \subseteq X : B \notin I\}$, and $I^* = \{B \subseteq X : X - B \in I\}$. For each $D \in I^+$, we let $I|D = \{B \subseteq X : B \cap D \in I\}$.

---

*The author gratefully acknowledges the support of the Deutsche Forschungsgemeinschaft.

An ideal I over $[\lambda]^{<\kappa}$ is said to be normal if I^* is closed under diagonal intersections.

Finally we set $\alpha^* = \alpha - \{0\}$ for every ordinal α.

1. Closed unbounded subsets of $[\lambda]^{<\kappa}$

Fix a one-to-one function $j : \lambda \times \lambda \to \lambda$ with $j(0,0) = 0$, and let E_j^κ denote the set of all $a \in [\lambda]^{<\kappa}$ such that $j(\alpha,\beta) \in a$ for all $\alpha,\beta \in a$ with $\alpha < \beta$.

Let us first make this easy remark.

LEMMA 1.1. Given $f : [\lambda]^{n+1} \to \lambda$, $n \in \omega^*$, there exists $g : [\lambda]^n \to \lambda$ such that $f[[a]^{n+1}] \subseteq a$ for all infinite $a \in E_j^\kappa$ with $g[[a]^n] \subseteq a$.

PROOF. Define g so that $f(d) = g(\{j(d_k,d_{k+1}) : k < n\})$ whenever d_p, $p \le n$, is the increasing enumeration of $d \in [\lambda]^{n+1}$.

The following result is essentially due to Baumgartner (see [7], page 115).

PROPOSITION 1.2. Let D be a closed unbounded subset of $[\lambda]^{<\kappa}$. Then there exists $g : \lambda \to \lambda$ with the property that $a \in D$ for all $a \in E_j^\kappa$ such that $a \cap \kappa \in \kappa^*$ and $g[a] \subseteq a$.

PROOF. We shall first define a sequence g_n, $n \in \omega^*$, of functions from λ to λ. By induction on the size of d, define $a_d \in D$, $d \in [\lambda]^{<\omega}$, such that $d \subseteq a_d$, and $a_c \subseteq a_d$ for $c \subseteq d$. Define $g_1 : \lambda \to \lambda$ by letting $g_1(\alpha) = \alpha + 1$. Given $n \in \omega^*$, let $f_{n+1} : [\lambda]^{n+1} \to \lambda$ satisfy the following conditions. Suppose $d \in [\lambda]^n$, and let d_p, $p < n$, be the increasing enumeration of d. Then $f_{n+1}(d \cup \{d_{n-1} + 1\})$ equals the order type of a_d ; $f_{n+1}(d \cup \{d_{n-1} + 2k + 2\}) = \alpha$ whenever $k < n$, $\alpha \in a_d$, and $\alpha \cap a_d$ has order type d_k; and $f_{n+1}(d \cup \{d_{n-1} + 2k + 3\}) = \beta$ whenever $k < n$, $\beta \in a_{d-\{d_k\}}$, and $\beta \cap a_{d-\{d_k\}}$ has order type d_k. Now use Lemma 1.1 to find $g_{n+1} : \lambda \to \lambda$ such that $f_{n+1}[[a]^{n+1}] \subseteq a$ for all infinite $a \in E_j^\kappa$ with $g_{n+1}[a] \subseteq a$. Suppose $a \in E_j^\kappa$ is such that $a \neq 0$, $a \cap \kappa \in \kappa$, and $g_n[a] \subseteq a$ for all $n \in \omega^*$. Then, clearly, $a = \bigcup\{a_d : d \in [a]^{<\omega} - \{0\}\}$, and since D is closed under directed unions of size $<\kappa$, we have $a \in D$. Finally, define $g : \lambda \to \lambda$ so that for all $\alpha < \lambda$, $g(j(0,\alpha)) = \alpha + 1$, and $g(j(n,n+1+\alpha)) = g_n(\alpha)$ whenever $n \in \omega^*$.

We mention this immediate corollary.

COROLLARY 1.3. $NS_{\kappa,\lambda} = SNS_{\kappa,\lambda} \mid E_j^\kappa$.

The following is especially useful because of its simplicity.

PROPOSITION 1.4. Let D be a closed unbounded subset of $[\lambda]^{<\kappa}$. Then there exists $h : \lambda \times \lambda \to \lambda$ such that $a \in D$ whenever $a \neq 0$, $a \cap \kappa \in \kappa$, and $h[a \times a] \subseteq a$.

PROOF. Let $g : \lambda \to \lambda$ be as in the statement of Proposition 1.2. Now define h so that for every $\alpha < \lambda$, $h(\alpha,\alpha) = \alpha + 1$, $h(\alpha,\alpha + 1) = 0$, $h(\alpha,\alpha + 2) = g(\alpha)$, and $h(\alpha,\beta + 2) = j(\alpha,\beta)$ whenever $\beta > \alpha$.

We shall need the following easy generalization of Proposition 1.2.

PROPOSITION 1.5. Suppose A is a set of ordinals such that $\kappa \subseteq A$ and $\cup A \notin A$. Let $j : A \times A \to A$ be a one-to-one function with $j(0,0) = 0$, and let $D \in NS^*_{\kappa,A}$. Then there exists $g : A \to A$ such that $a \in D$ for all $a \in [A]^{<\kappa}$ such that $a \cap \kappa \in \kappa^*$, $j[a \times a] \subseteq a$, and $g[a] \subseteq a$.

2. Stationary reflection

For the remainder of this paper λ will be assumed to be a cardinal.

We start by recalling some definitions.

NWC_κ denotes the set of all $A \subseteq \kappa$ such that there is a tree $(T, \leq_T)$ with the following properties: (1) $T = A$; (2) $\alpha \leq_T \beta$ implies $\alpha \leq \beta$; (3) the set of immediate successors of each $\alpha \in T$ is nonstationary; (4) if δ is a limit ordinal and α,β are both at level δ, and if $\{\gamma : \gamma <_T \alpha\} = \{\gamma : \gamma <_T \beta\}$, then $\alpha = \beta$; and (5) T has no branch of length κ. κ is said to be weakly compact in case κ is strongly inaccessible and $\kappa \notin NWC_\kappa$. Note that our definition of the weakly compact ideal NWC_κ differs from that given on page 91 of [1], which is incomplete.

We let NSh_κ denote the set of all $A \subseteq \kappa$ such that there are $f_\alpha : \alpha \to \alpha$, $\alpha \in A$, with the property that for every $g : \kappa \to \kappa$, there exists $\beta < \kappa$ such that $g \restriction \beta \neq f_\alpha \restriction \beta$ for all $\alpha \in A - \beta$.

$NSh_{\kappa,\lambda}$ denotes the set of all $B \subseteq [\lambda]^{<\kappa}$ such that there are $f_a : a \to a$, $a \in B$, with the property that for every $g : \lambda \to \lambda$, there exists $b \in [\lambda]^{<\kappa}$ such that $g \restriction b \neq f_a \restriction b$ for all $a \in B$ with $b \subseteq a$. κ is said to be λ-Shelah in case $[\lambda]^{<\kappa} \notin NSh_{\kappa,\lambda}$.

The following collects some well-known results (see [1], [4], [5] and [9]).

PROPOSITION 2.1. (i) Assume κ is λ-Shelah. Then κ is weakly compact, and $NSh_{\kappa,\lambda}$ is a normal ideal over $[\lambda]^{<\kappa}$ that extends $NS_{\kappa,\lambda}$.

(ii) $NSh_{\kappa,\kappa} = \{B \subseteq [\kappa]^{<\kappa} : B \cap \kappa \in NSh\}$.

(iii) If κ is strongly inaccessible, then $NWC_\kappa = NSh_\kappa$.

The following, which is a two-cardinal version of Theorem 2.9. of [1], strengthens the theorem of [6].

PROPOSITION 2.2. Assume κ is λ-Shelah. Let $T \in NS^+_{\kappa,\lambda}$, and let $R \in NWC^*_\kappa$ be a set of regular uncountable cardinals. Then the set of all $a \in [\lambda]^{<\kappa}$ such that $a \cap \kappa \in R$ and $T \cap [a]^{<a\cap\kappa} \in NS^+_{a\cap\kappa,a}$ lies in $NSh^*_{\kappa,\lambda}$.

PROOF. Let H denote the set of all $a \in [\lambda]^{<\kappa}$ with $a \cap \kappa \in R$. We claim that $H \in NSh^*_{\kappa,\lambda}$. Suppose otherwise. Then the set $K = \{a \in [\lambda]^{<\kappa} : a \cap \kappa \in \kappa - R\}$ lies in $NSh^+_{\kappa,\lambda}$. Pick $f_\alpha : \alpha \to \alpha$, $\alpha \in \kappa - R$, with the property that for every $g : \kappa \to \kappa$, there exists $\beta < \kappa$ such that $g \restriction \beta \neq f_\alpha \restriction \beta$ whenever $\alpha \in \kappa - R$ with $\alpha \geq \beta$. Now select $h_a : a \to a$, $a \in K$, so that $h_a \restriction a \cap \kappa = f_{a\cap\kappa}$. Then one can find $k : \lambda \to \lambda$ with the property that for every $\beta < \kappa$, there exists $a \in K$ with $\beta \subseteq a$ and $k \restriction \beta = h_a \restriction \beta$. This easily leads to a contradiction.

Now let D denote the set of all $a \in H$ such that $T \cap [a]^{<a\cap\kappa} \in NS_{a\cap\kappa,a}$. Suppose $D \in NSh^+_{\kappa,\lambda}$. Fix a bijection $j : \lambda \times \lambda \to \lambda$ with $j(0,0) = 0$, and let E denote the set of all $a \in [\lambda]^{<\kappa}$ such that $j[a \times a] \subseteq a$. Denote by B the set of all $a \in D \cap E$ with $\bigcup a \notin a$. We clearly have $B \in NSh^+_{\kappa,\lambda}$. Now using Proposition 1.5, pick for each $a \in B$, $h_a : a \to a$ such that $b \notin T$ whenever $b \in E \cap [a]^{<a\cap\kappa}$, $b \cap \kappa \in \kappa^*$ and $h_a[b] \subseteq b$. Select $g : \lambda \to \lambda$ such that for all $b \in [\lambda]^{<\kappa}$, there exists $a \in B$ with $b \subseteq a$ and $g \restriction b = h_a \restriction b$. Now choose $b \in T \cap E$ with $b \cap \kappa \in \kappa^*$ and $g[b] \subseteq b$. Then one can find $a \in B$ such that $b \in [a]^{<a\cap\kappa}$ and $g \restriction b = h_a \restriction b$, a contradiction.

3. Game ideals

For the duration of this section μ will denote a fixed infinite regular cardinal less than κ.

A subset D of $[\lambda]^{<\kappa}$ is said to be μ-closed if for every sequence $a_\alpha \in D$, $\alpha < \mu$, such that $a_\beta \subset a_\alpha$ for $\beta < \alpha$, we have $\bigcup_{\alpha<\mu} a_\alpha \in D$. We let $NS^\mu_{\kappa,\lambda}$ denote the collection of all those $T \subseteq [\lambda]^{<\kappa}$ such that $T \cap D = 0$ for some μ-closed unbounded set $D \subseteq [\lambda]^{<\kappa}$.

The easy proof of the following is left to the reader.

PROPOSITION 3.1. $NS^{\mu}_{\kappa,\lambda}$ is a normal ideal over $[\lambda]^{<\kappa}$ that extends $NS_{\kappa,\lambda}$.

For each $T \subseteq [\lambda]^{<\kappa}$, we define a two-person game $G^{\mu}_{\kappa,\lambda}(T)$ consisting of μ moves, with player I moving first at limit stages. First I chooses an element a_o of $[\lambda]^{<\kappa}$; II answers by playing $b_o \in [\lambda]^{<\kappa}$ with $a_o \subseteq b_o$; then I selects $a_1 \in [\lambda]^{<\kappa}$ with $b_o \subseteq a_1$; II answers by playing $b_1 \supseteq a_1$, etc. I thus produces a sequence a_α, $\alpha < \mu$, such that $a_\beta \subseteq a_\alpha$ for $\beta < \alpha$. II wins iff $\bigcup_{\alpha<\mu} a_\alpha \in T$.

We remark that the games $G^{\mu}_{\kappa,\kappa}(T)$, $T \subseteq \kappa$, have already been considered in the literature, in particular (in the case $\kappa = \omega_1$) in connection with the axiom of determinacy (e.g. see [10], pp. 26-32).

Since, for $\mu > \omega_1$, the games $G^{\mu}_{\kappa,\lambda}(T)$ have uncountable length, we include a definition of strategy to avoid misunderstandings.

A strategy for such a game $G^{\mu}_{\kappa,\lambda}(T)$ is simply a function, with values on $[\lambda]^{<\kappa}$, and defined on sequences of length $<\mu$ of members of $[\lambda]^{<\kappa}$. A strategy σ is a winning strategy for player I (respectively for player II) if $\bigcup_{\alpha<\mu} c_\alpha$ does not belong to T (resp. does belong to T) whenever $c_\alpha \in [\lambda]^{<\kappa}$, $\alpha < \mu$, is a sequence satisfying the following three conditions: (1) $c_\beta \subseteq c_\alpha$ for $\beta < \alpha$; (2) $c_o = \sigma(0)$ (resp. $c_1 = \sigma(c_o)$); and (3) Given $n \in \omega$ and a limit ordinal $\beta \in \mu$ with $\beta + n \neq 0$, we have $c_\gamma = \sigma(c_\alpha : \alpha < \gamma)$ for $\gamma = \beta + 2n$ (resp. $\gamma = \beta + 2n + 1$).

We shall need the following piece of notation. Given a strategy τ for II in the game $G^{\mu}_{\kappa,\lambda}(T)$, we define a function $\hat{\tau}$ with the following property. Suppose that I plays a_α, $\alpha < \mu$, and that II plays by τ, thus producing b_α, $\alpha < \mu$. Then we let $\hat{\tau}(a_\alpha : \alpha \leq \beta) = b_\beta$ for every $\beta < \mu$.

Let $NGS^{\mu}_{\kappa,\lambda}$ denote the collection of all those $T \subseteq [\lambda]^{<\kappa}$ such that I has a winning strategy in $G^{\mu}_{\kappa,\lambda}(T)$.

PROPOSITION 3.2. (i) Given $T \subseteq [\lambda]^{<\kappa}$, $T \in NGS^{\mu}_{\kappa,\lambda}$ iff II has a winning strategy in the game $G^{\mu}_{\kappa,\lambda}([\lambda]^{<\kappa} - T)$.

(ii) $NS^{\mu}_{\kappa,\lambda} \subseteq NGS^{\mu}_{\kappa,\lambda}$.

(iii) $NGS^{\mu}_{\kappa,\lambda}$ is a normal ideal over $[\lambda]^{<\kappa}$.

PROOF. (i): Let $T \subseteq [\lambda]^{<\kappa}$ be fixed. Let us first assume that I has a winning strategy σ in the game $G^{\mu}_{\kappa,\lambda}(T)$. We need to find a winning strategy τ for player II in $G^{\mu}_{\kappa,\lambda}([\lambda]^{<\kappa} - T)$. Suppose n and β are ordinals such that $n \in \omega$, $\beta \in \mu$ and $\beta = \cup\beta$. We let $\hat{\tau}(a_\alpha : \alpha \leq \beta + n) = \sigma(c_\gamma : \gamma \leq \beta + 2n + 1)$, where (1) $c_o = \sigma(0)$; (2) $c_1 = a_o \cup c_o$; and (3) if δ and p are ordinals such that $\delta = \bigcup\delta$, $p \in \omega$ and $0 < \delta + p \leq \beta + n$, then $c_{\delta+2p} = \sigma(c_\gamma : \gamma < \delta + 2p)$, and $c_{\delta+2p+1} = a_{\delta+p} \cup c_{\delta+2p}$.

Now for the reverse direction: Suppose τ is a winning strategy for II in $G^\mu_{\kappa,\lambda}([\lambda]^{<\kappa} - T)$. We are looking for a winning strategy σ for player I in $G^\mu_{\kappa,\lambda}(T)$. First put $\sigma(0) = \hat{\tau}(0)$. Now assume $n \in \omega$ and $\beta \in \mu$ are such that $\beta = \bigcup\beta$ and $\beta + n \neq 0$. We set $\sigma(c_\alpha : \alpha < \beta + 2n) = \hat{\tau}(a_\gamma : \gamma \leq \beta + n)$, where (1) $a_0 = 0$; (2) if $\delta \leq \beta + n$ is a nonzero limit ordinal, then $a_\delta = \bigcup_{\gamma<\delta} a_\gamma$; and (3) if δ and p are ordinals such that $\delta = \bigcup\delta$, $p \in \omega$ and $\delta + p < \beta + n$, then $a_{\delta+p+1} = c_{\delta+2p+1}$.

(ii) is straightforward.

(iii): $NGS^\mu_{\kappa,\lambda}$ is easily seen to be an ideal over $[\lambda]^{<\kappa}$. We only show normality. Thus suppose T_γ and τ_γ, $\gamma < \lambda$, are such that $T_\gamma \subseteq [\lambda]^{<\kappa}$ and τ_γ is a winning strategy for II in $G^\mu_{\kappa,\lambda}(T_\gamma)$. We define a winning strategy σ for II in the game $G^\mu_{\kappa,\lambda}(\underset{\gamma<\lambda}{\triangle} T_\gamma)$ by letting

$$\hat{\sigma}(a_\alpha : \alpha \leq \beta) = \bigcup_{\delta\leq\beta} \bigcup_{\gamma\in a_\delta - \bigcup_{\alpha<\delta} a_\alpha} \hat{\tau}_\gamma(a_\xi : \delta \leq \xi \leq \beta).$$

Donder and Levinski have investigated the relationship between the three ideals $NS_{\kappa,\lambda}$, $NS^\mu_{\kappa,\lambda}$ and $NGS^\mu_{\kappa,\lambda}$. Their results will appear elsewhere.

Given an ideal I over $[\lambda]^{<\kappa}$, the principle $\diamondsuit_{\kappa,\lambda}[I]$ asserts the existence of a sequence $t_a \subseteq a$, $a \in [\lambda]^{<\kappa}$, such that $\{a \in [\lambda]^{<\kappa} : t_a = B \cap a\} \in I^+$ for all $B \subseteq \lambda$. Such a sequence is said to be a $\diamondsuit_{\kappa,\lambda}[I]$-sequence.

In [8] Jech observed that $\diamondsuit_{\kappa,\lambda}[NS_{\kappa,\lambda}]$ can be forced by adding $\lambda^{<\kappa}$ Cohen subsets of κ to the ground model. We now show that $\diamondsuit_{\kappa,\lambda}[NGS^\mu_{\kappa,\lambda}]$ holds as well in the generic extension.

Thus let M be a transitive model of ZFC. Assume $2^{<\kappa} = \kappa$ holds in M. Let P consist, in M, of all those functions p such that $|dom(p)| < \kappa$, $dom(p) \subset \{(\alpha,a) \in \lambda \times [\lambda]^{<\kappa} : \alpha \in a\}$, and $ran(p) \subseteq 2$, ordered by reverse inclusion. Define $u : P \to [\lambda]^{<\kappa}$ by letting u(p) be the union of all those a such that $(\alpha,a) \in dom(p)$ for some $\alpha \in a$.

PROPOSITION 3.3. Let G be P-generic over M, and set $t_a = \{\alpha \in a : (\bigcup G)(\alpha,a) = 1\}$ for all $a \in [\lambda]^{<\kappa}$. Then the following hold:

(i) In M[G], t_a, $a \in [\lambda]^{<\kappa}$, is a $\diamondsuit_{\kappa,\lambda}[NGS^\mu_{\kappa,\lambda}]$-sequence.

(ii) Assume that $\nu^{<\mu} < \kappa$ holds in M for all infinite cardinals $\nu < \kappa$, and that Y is, in M, a stationary subset of κ consisting of infinite cardinals of cofinality μ. Set $W = \{a \in [\lambda]^{<\kappa} : |a| \in Y\}$. Then in $M[G]$, t_a, $a \in [\lambda]^{<\kappa}$, is a $\diamondsuit_{\kappa,\lambda}[NGS^{\mu}_{\kappa,\lambda}|W]$-sequence.

PROOF. We only show (ii), as the proof of (i) is similar. For each $\alpha < \kappa$, let E_α denote the set of all increasing functions h such that $dom(h) \in \mu^*$ and $\alpha \in ran(h) \subseteq \alpha + 1$. Let $p \in G$ and f, T, τ in $M[G]$ be such that p forces that $f \in 2^\lambda$, that $T \subseteq [\lambda]^{<\kappa}$ and that τ is a winning strategy for II in the game $G^{\mu}_{\kappa,\lambda}(T)$. By induction on $\alpha < \kappa$, define, in M, $p_\alpha \in P$, a_α, g_α so that:

(1) $p_0 = p$, and $a_0 = g_0 = 0$;

(2) $p_{\alpha+1} \leq p_\alpha$, and $a_{\alpha+1} \subset u(p_{\alpha+1})$;

(3) $g_{\alpha+1} : a_{\alpha+1} \to 2$;

(4) $p_{\alpha+1}$ forces that $f \restriction a_{\alpha+1} = g_{\alpha+1}$ and that $a_{\alpha+1} = \bigcup_{h \in E_\alpha} \hat{\tau}(u(p_{h(\beta)}) : \beta \in dom(h))$;

(5) if α is an infinite limit ordinal, then $p_\alpha = \bigcup_{\beta<\alpha} p_\beta$, $a_\alpha = \bigcup_{\beta<\alpha} a_\beta$ and $g_\alpha = \bigcup_{\beta<\alpha} g_\beta$.

Set $A = \bigcup_{\alpha<\kappa} a_\alpha$, and define a bijection $k : A \to \kappa$ so that $k[a_\alpha] \in \kappa$ for all $\alpha < \kappa$. Put $C = \{\alpha<\kappa : k[a_\alpha] = \alpha\}$. As C is a closed unbounded subset of κ, it is possible to select $\rho \in Y \cap C$. Now define a function $q : a_\rho \times \{a_\rho\} \to 2$ by letting $q(\alpha, a_\rho) = g_\rho(\alpha)$. Clearly, $p_\rho \cup q$ forces that $a_\rho \in W \cap T$ and that $(\cup G)(\alpha, a_\rho) = f(\alpha)$ for all $\alpha \in a_\rho$.

4. The nonstationary ideal over $[\lambda]^{<\omega_1}$

This last section is devoted to the special case $\kappa = \omega_1$. We shall use games to obtain several characterizations of the nonstationary subsets of $[\lambda]^{<\omega_1}$.

Let us first make the following observation.

PROPOSITION 4.1. Let A be a set of size $\geq\omega_1$, and let D be a closed unbounded subset of $[A]^{<\omega_1}$. Then there exists $h : A \times A \to A$ such that $a \in D$ whenever $a \in [A]^{<\omega_1}$ is such that $a \neq 0$ and $h[a \times a] \subseteq a$.

PROOF. By Theorem 1.4 of [2] and the proofs of Proposition 1.2 and Proposition 1.4.

Given $T \subseteq [\lambda]^{<\omega_1}$, we define two more two-person games $H^{\omega}_{\omega_1,\lambda}(T)$ and

$K^{\omega}_{\omega_1,\lambda}(T)$. Either game lasts ω moves, player I making the first move. In $H^{\omega}_{\omega_1,\lambda}(T)$, I and II alternately pick members of $[\lambda]^{<\omega}$, thus building a sequence c_n, $n \in \omega$. II wins $H^{\omega}_{\omega_1,\lambda}(T)$ just in case $\bigcup_{n\in\omega} c_n \in T$. The game $K^{\omega}_{\omega_1,\lambda}(T)$ is defined similarly, where now the choices are made from λ, and to win II must insure that the set of chosen ordinals lies in T.

PROPOSITION 4.2. Let $T \subseteq [\lambda]^{<\omega_1}$ be given. Then the following are equivalent:

(i) $T \in NS_{\omega_1,\lambda}$.

(ii) $T \in NGS^{\omega}_{\omega_1,\lambda}$.

(iii) I has a winning strategy in $H^{\omega}_{\omega_1,\lambda}(T)$.

(iv) I has a winning strategy in $K^{\omega}_{\omega_1,\lambda}(T)$.

(v) II has a winning strategy in $H^{\omega}_{\omega_1,\lambda}([\lambda]^{<\omega_1} - T)$.

(vi) II has a winning strategy in $K^{\omega}_{\omega_1,\lambda}([\lambda]^{<\omega_1} - T)$.

PROOF. (i) → (ii) holds by Proposition 3.2.

(ii) → (i) : Let τ be a winning strategy for II in the game $G^{\omega}_{\omega_1,\lambda}([\lambda]^{<\omega_1} - T)$. For each $n \in \omega$, we define a function $f_n : \lambda^{n+1} \to [\lambda]^{<\omega_1}$ by letting $f_n(\alpha_o,\ldots,\alpha_n) = \tau(a_o,\ldots,a_{2n})$, where $a_o = \{\alpha_o\}$ and for every $p<n$, $a_{2p+1} = \tau(a_o,\ldots,a_{2p})$ and $a_{2p+2} = a_{2p+1} \cup \{\alpha_{p+1}\}$. Let C denote the collection of all nonempty $a \in [\lambda]^{<\omega_1}$ such that $f_n[a^{n+1}] \subseteq a$ whenever $n \in \omega$. Clearly C is closed and unbounded. Given $a \in C$, let α_n, $n \in \omega$, be an enumeration of a. Then $a = \bigcup_{n\in\omega} f_n(\alpha_o,\ldots,\alpha_n)$, and consequently $a \notin T$.

(v) → (i), (vi) → (i) and (iv) → (i) are proved as (ii) → (i).

(i) → (vi): Assume $T \in NS_{\omega_1,\lambda}$. Then by Proposition 4.1, there exists $g : \lambda \times \lambda \to \lambda$ such that $a \notin T$ for every nonempty $a \in [\lambda]^{<\omega_1}$ with $g[a \times a] \subseteq a$. We shall define a winning strategy τ for player II in $K^{\omega}_{\omega_1,\lambda}([\lambda]^{<\omega_1} - T)$. Suppose $n \in \omega$ and $\alpha_p \in \lambda$, $p \le 2n$, are given such that $\alpha_{2k+1} = \tau(\alpha_o,\ldots,\alpha_{2k})$ for all $k < n$. Let a denote the closure of the set $\{\alpha_p : p \le 2n\}$ under g. We first define a function h from a to the power set of $\bigcup_{m\in\omega} \omega^{m+1}$, as follows. Given $x \in a$, one can represent x (possibly in more than one way) as a sequence of g's, parentheses and α_k's. For each such sequence t, suppress all parentheses of t, and substitute in t 0 for each g and k + 1 for each α_k. The resulting sequences of natural numbers form the image of x under h. For instance, if $x = g(g(g(\alpha_3,\alpha_5),\alpha_2),g(\alpha_o,\alpha_5))$, then $(0,0,0,4,6,3,0,1,6) \in h(x)$. Then define $r : \bigcup_{m\in\omega} \omega^{m+1} \to \omega$ by letting

$r(q_o,\dots,q_p) = \prod_{i \le p} w_i^{q_i}$, where w_j, $j < \omega$, is the increasing enumeration of all prime numbers >1. Finally, if there exist $x \in a$ and $y \in h(x)$ with $2n+1 = r(y)$, then set $\tau(\alpha_o,\dots,\alpha_{2n}) = x$; otherwise put $\tau(\alpha_o,\dots,\alpha_{2n}) = 0$.

(i) → (v) and (i) → (iv) are proved similarly to (i) → (vi).

Finally, the proof of (iii) ↔ (v) is left to the reader.

REFERENCES

[1] J.E. BAUMGARTNER, Ineffability properties of cardinals II, "Logic, Foundations of Mathematics and Computability Theory", Butts and Hintikka (eds), Reidel, Dordrecht (Holland), 1977, 87 - 106.

[2] J.E. BAUMGARTNER, Applications of the proper forcing axiom, "Handbook of Set-Theoretic Topology", Kunen and Vaughan (eds), North-Holland, Amsterdam, 1984, 913 - 959.

[3] D.M. CARR, The minimal normal filter on $P_\kappa\lambda$, Proc. Amer. Math. Soc. 86 (1982), 316 - 320.

[4] D.M. CARR, $P_\kappa\lambda$-generalizations of weak compactness, Z. Math. Logik Grundlag. Math. 31 (1985), 393 - 401.

[5] D.M. CARR, The structure of ineffability properties of $P_\kappa\lambda$, Acta Math. Hungar. 47 (1986), 325 - 332.

[6] D.M. CARR, A note on the λ-Shelah property, Fund. Math. 128 (1987), 197-198.

[7] C.A. DI PRISCO and W. MAREK, Some aspects of the theory of large cardinals, "Mathematical Logic and Formal Systems", Alcantara (ed), Lecture Notes Pure Appl. Math. 94, Dekker, New York, 1985, 87 - 139.

[8] T.J. JECH, Some combinatorial problems concerning uncountable cardinals, Ann. Math. Logic 5 (1973), 165 - 198.

[9] C.A. JOHNSON, Some partition relations for ideals on $P_\kappa\lambda$, preprint.

[10] E.M. KLEINBERG, "Infinitary Combinatorics and the Axiom of Determinateness", Lecture Notes in Math. 612, Springer, Berlin, 1977.

[11] T.K. MENAS, On strong compactness and supercompactness, Ann. Math. Logic 7 (1974), 327 - 359.

When Hereditarily Collectionwise Hausdorffness Implies Regularity

Zs. Nagy
Mathematical Institute of Hungarian Academy of Sciences
Budapest, V., Realtanoda U. 13-15, H-1364, Pf. 127 Hungary

and

S. Purisch*
Department of Mathematics, Tennessee Technological University
Cookeville, Tennessee 38505, USA

ABSTRACT: Every collectionwise Hausdorff (CWT_2) first countable space is regular. Generalizations of first countability are considered. If X is hereditarily CWT_2($HCWT_2$), blobular, of character $\aleph_1$, and contains no first countable S space, then X is regular. The existence of a first countable S space implies the existence of a nonregular, $HCWT_2$, lob space of character $\aleph_1$. If it is consistent that there are no $HCWT_2$ analogues of S spaces for all regular ordinals τ, then the following is consistent. Every $HCWT_2$ globular space is regular.

A space is <u>collectionwise Hausdorff</u> (CWT_2) if the points in each discrete closed subset can be separated by a pairwise disjoint collection of open sets. So a space is hereditarily CWT_2 ($HCWT_2$) if the above holds for each discrete subset.

In [P_1] and [P_2] $HCWT_2$ was used in characterizing monotone normality or total orderability for scattered spaces. Often regularity

---

<u>AMS (MOS) subject classification (1980).</u> Primary 54A25, 54D10, 54D15, Secondary 54A35, 54E99.
<u>Key words and phases.</u> Hereditarily collectionwise Hausdorff, regularity, lob, blobular, globular, S space, right separated, scattered.
* The second author is pleased to thank the International Research and Exchanges Board and the Hungarian Academy of Science for Support during the preparation of this paper.

was obtained for free. This lead to the present investigation. In particular we are interested in lob spaces ([N]), that is a space in which every point has a linearly ordered (with respect to inclusion) neighborhood base. We also are interested in the more general class of globular spaces (named by B. Scott). A space is globular if each point p has a neighborhood subbase $\cup\mathcal{C}_p$ where $|\mathcal{C}_p| < \aleph_0$ and each member of $\mathcal{C}_p$ is a decreasing (perhaps transfinite) sequence of open sets.

EXAMPLE 1. There exists a nonregular $HCWT_2$ space.

A simple example is $\omega_1\cup\{p\}$ where ω_1 has its usual order topology, and a basic neighborhood of p is of the form $\{p\}\cup\{\alpha+1:\alpha\in C\}$ where C is a closed unbounded subset of ω_1.

In a space X an element p can be separated from a subspace C if there are disjoint open sets V and W such that $p\in V$ and $C\subset W$.

THEOREM 1. Every CWT_2 first countable space is regular.

Proof. Let X be first countable, T_2, and not regular. Let C be closed in X and $p\in X-C$ such that p cannot be separated from C. Let $\{O_n\}_{n<\omega}$ be a decreasing neighborhood base of p. For each $n<\omega$ pick $x_n\in\bar{O}_n\cap C$. Since X is T_2, $D = \{x_n:n<\omega\}\cup\{p\}$ is closed and discrete. Since the points of D cannot be separated by a pairwise disjoint collection of open sets, X is not CWT_2.

EXAMPLE 2. ([W]). There exists a $HCWT_2$ first countable space which is not normal.

A space is scattered if each of its nonempty sets has an isolated point. The length of a scattered space X is the least ordinal α such that the αth derived set $X^{(\alpha)}$ is empty.

It is not difficult to show every CWT_2 scattered space of finite length is regular (in fact paracompact). The next example shows a limitation of this result and another of theorem 1. The construction is a modification of a technique used in [D].

EXAMPLE 3. There exist a $HCWT_2$, countable, scattered space which is not regular. Trivially, each point is a G_δ.

In the ordinal space $\omega^\omega + 1$ we will redefine the topology at the last point $\omega^\omega = p$.

For each $n<\omega$ define $\mathcal{A}_n \subset \mathcal{P}(\omega^{n+1} - (\omega^n \cup (\omega^{n+1})^{(1)}))$ as follows. $\mathcal{A}_0 = \{\{1\}\}$. Assume $\mathcal{A}_{n-1}$ has been defined. For each $m<\omega$ let $\omega^n m + \mathcal{A}_{n-1}$ denote the translate of $\mathcal{A}_{n-1}$ to $\omega^n(m+1) - \omega^n m$. That is, $B \in \omega^n m + \mathcal{A}_{n-1}$ iff there exists $B' \in \mathcal{A}_{n-1}$ such that $B = \{\omega^n m + \alpha : \alpha \in B'\}$. Define $A \in \mathcal{A}_n$ iff either

1) for some nonzero $m<\omega$

$A = (\omega^n(m+1) - (\omega^n m \cup (\omega^{n+1})^{(1)})) \cup \bigcup\{$one member of $\omega^n i + \mathcal{A}_{n-1} : 0<i<m\}$, or

2) $A = \bigcup\{$one member of $\omega_n m + \mathcal{A}_{n-1} : 0<m<\omega\}$.

A subbasic neighborhood of p is $\{p\} \cup \bigcup\{$one member of $\mathcal{A}_n : n \in \omega - k\}$, k fixed.

Note by induction it is easy to show for all $n<\omega$ $\bigcap \mathcal{A}_n \neq \emptyset$ and for all $\mathcal{B} \subset \mathcal{A}_n$ if $1 \le |\mathcal{B}| = k \le n+1$, then type $(\bigcap \mathcal{B}) \ge \omega^{n+1-k}$.

Hence, with this topology $\omega^\omega + 1$ is not regular since every basic neighborhood of p has order type $\omega^\omega + 1$ and is disjoint from $(\omega^\omega)^{(1)}$.

To see with this topology $\omega^\omega + 1$ is $HCWT_2$ note that ω^ω is $HCWT_2$, $\overline{(\bigcup \mathcal{A}_n)} \subset (\omega^{n+1}+1) - (\omega^n+1)$ = the interval $[\omega^n+1, \omega^{n+1}]$ for all $n<\omega$, and by induction it is easy to show for all $n<\omega$ and all discrete $D \subset \omega^{n+1}+1$ that there exist $A \in \mathcal{A}_n$ such that $\overline{A} \cap D = \emptyset$.

Obviously this space is countable and scattered.

An <u>S space</u> is regular, hereditarily separable, and not Lindelöf. The existence of an S space is independent of ZFC (see [R]). A space is <u>right separated</u> if there is a well order on it under which each initial segment is open.

THEOREM 2. If X is lob, $HCWT_2$, of character $\aleph_1$, and contains no first countable S space, then X is regular.

Proof. Let X be lob, T_2, of character $\aleph_1$, containing no first countable S space, and not regular. Let C and p be as in the proof of theorem 1, and let $\mathcal{O} = \{O_\alpha\}_{\alpha<\tau}$ be a neighborhood base of p which is a decreasing sequence of open sets and τ is ω or ω_1. If $\tau=\omega$, then as in the proof of theorem 1, X is not CWT_2. So let $\tau=\omega_1$.

For each $\alpha<\omega_1$ choose $x_\alpha \in \overline{O}_{\alpha'} \cap C$ where α' is the first ordinal $<\omega_1$ such that for all $\beta<\alpha$ $x_\beta \notin \overline{O}_{\alpha'}$. Then $Y = \{x_\alpha : \alpha<\omega_1\}$ is right separated. So Y is first countable since it is lob. If Y is not CWT_2, then X is not $HCWT_2$ and we are done. So let Y be CWT_2. By theorem 1 Y is regular. Since X contains no first countable S space, Y contains a discrete (perhaps not closed) subset Y' of cardinality $\aleph_1$. Since $Y' \cup \{p\}$ is discrete and its points cannot be separated by a pairwise disjoint collection of open sets, X is not $HCWT_2$.

EXAMPLE 4. The existence of a first countable S space implies the existence of a nonregular, $HCWT_2$, lob space of character $\aleph_1$.

Let $X = \{x_\alpha : \alpha<\omega_1\}$ be a first countable S space. Our example is the set $(X\times(\omega+1)) \cup \{p\}$ with the following topology. $X\times\omega$ is discrete. For $\alpha<\omega_1$ a basic neighborhood of $\langle x_\alpha, \omega\rangle$ is $O\times\{n : m<n\le\omega\}$, where O is a neighborhood of x_α in X and $m<\omega$. A basic neighborhood of p is $(\{x_\alpha : \tau<\alpha<\omega_1\}\times\omega) \cup \{p\}$ for $\tau<\omega_1$.

NOTE. In theorem 2 and example 4 if we only require the points where regularity fails to have linearly ordered neighborhood bases, then any S space will do.

We want to generalize theorem 2 to globular spaces. Surprisingly the proof is difficult. For that reason theorem 2 was proven separately.

In the definition of a globular space if for each point p we have $|\mathcal{C}_p| \le 2$, then the space is called <u>blobular</u>. Note if a globular space has character $\aleph_1$, then the space is blobular.

THEOREM 3. If X is blobular, $HCWT_2$, of character $\aleph_1$, and contains no first countable S space, then X is regular.

Proof. Let X be blobular, T_2, of character $\aleph_1$, containing no first countable S space, and not regular. Let p and C be as in theorem 1. If p has a linearly ordered neighborhood base, then by the proofs of theorems 1 and 2, X is not $HWCT_2$.

So let p not have a linearly ordered neighborhood base. Then p has a neighborhood subbase $\mathcal{O} \cup \mathcal{U}$ where $\mathcal{O} = \{O_n\}_{n<\omega}$ and $\mathcal{U} = \{U_\alpha\}_{\alpha<\omega_1}$ are strictly decreasing sequences of open sets.

For each $n<\omega$ let $C_n = C \cap \bigcap_{\alpha<\omega_1} \overline{O_n \cap U_\alpha}$. Then $\{C_n\}_{n<\omega}$ is decreasing, and for each $n<\omega$ C_n is closed. Since X is T_2, $\bigcap_{n<\omega} C_n = \emptyset$.

Case 1. For each $n<\omega$ $C_n \neq \emptyset$. Choose $x_n \in C_n$ for all $n<\omega$. Then $\{x_n\}_{n<\omega}$ is closed, discrete, and cannot be separated from p. So X is not CWT_2.

Note if C contains a countable subset D which cannot be separated from p, then for all $n<\omega$ $C_n \cap D$ is infinite. So case 1 holds.

Case 2. There exists $m<\omega$ such that $C_m = \emptyset$. If $X-\{p\}$ is not $HCWT_2$, then neither is X and we are done. So let $X-\{p\}$ be $HCWT_2$.

The closure of any countable subset of C can be separated from p. For suppose not. Then there exists a countable $N \subset C$ such that $\overline{N}$ cannot be separated from p. Since $C_m = \emptyset$, there exists a cofinal $W \subset \omega_1$ such that $\{\overline{O_m \cap U_\alpha} \cap \overline{N} : \alpha \in W\}$ is strictly decreasing. For each $\alpha \in W$ choose $x_\alpha \in (\overline{O_m \cap U_\alpha} - \overline{O_m \cap U_{\alpha'}}) \cap \overline{N}$ where $\alpha' = \min\{\beta : \alpha < \beta \in W\}$. Let $Y = \{x_\alpha : \alpha \in W\}$. Then Y is right separated. So Y is first countable since it is globular. As in the proof of theorem 2 since X contains no first countable S space and $X-\{p\}$ is $HCWT_2$, Y contains a discrete subset Y' of cardinality $\aleph_1$. Since $Y' \cup N$ is separable, it is not $HCWT_2$. This contradicts $X-\{p\}$ being $HCWT_2$. Hence the closure of any countable subset of C can be separated from p.

We define $Y = \{x_\alpha : \alpha<\omega_1\}$ by induction on α. Let $\alpha = \beta + n$, where β is a limit ordinal and $n<\omega$, and for $\gamma<\alpha$ assume x_γ has been defined. Let $\delta(\alpha) = \min\{\delta : \text{for all } \gamma<\alpha\ x_\gamma \notin \overline{O_m \cap U_\delta}\}$, choose $x_\alpha \in C \cap (\overline{O_{n+m} \cap U_{\delta(\alpha)}}) - \overline{\{x_\gamma : \gamma<\alpha\}}$. Note $\delta(\alpha)$ can be defined because $C_m = \emptyset$;

the set $C\cap(\overline{O_{n+m}\cap U_{\delta(\alpha)}})-\overline{\{x_\alpha:\gamma<\alpha\}}$ is nonempty since otherwise the closure of the countable set $\{x_\gamma:\gamma<\alpha\}$ would not be separated from p.

$Y\cup\{p\}$ is discrete because $Y-\overline{\{x_\gamma:\gamma<\alpha\}}-\overline{O_m\cap U_{\delta(\alpha+1)}} = \{x_\alpha\}$. Since $x_{\beta+n}\in\overline{O_{n+m}\cap U_{\delta(\beta+n)}} \subset \overline{O_n\cap U_\beta}$, Y cannot be separated from p. So X is not $HCWT_2$.

DEFINITION. For a regular ordinal τ let $D(\tau)$ be the statement: every $HCWT_2$, regular, right separated space of type τ contains a discrete subspace of cardinality τ.

Note for τ a strong limit, i.e. $\alpha<\tau$ implies $2^\alpha<\tau$, that $D(\tau)$ is true even without the $HCWT_2$ hypothesis.

EXAMPLE 5. ([T]) There is a regular first countable space $-\omega^\omega$ with underlying set ${}^\omega\omega$ which contains a right separated subspace A of type b and no discrete subspace of cardinality b, where b is an uncountable regular ordinal $\leq 2^{\aleph_0}$. However, $-\omega^\omega$ is ccc. So if A is not an S space, $-\omega^\omega$ is not $HCWT_2$.

QUESTION. In example 5 can A always be chosen in ZFC to be $HCWT_2$? This seems to be quite a difficult problem. If it can be done, then in ZFC using the technique in example 4 there would be a nonregular, $HCWT_2$, lob space of character b.

For a globular space X and nonisolated $p\in X$, p has a neighborhood subbase $\bigcup_{i<n} \mathcal{O}_{\tau_i}$ for some nonzero $n<\omega$ where for each $i<n$ τ_i is a regular ordinal, $\tau_i<\tau_{i+1}$ if $i+1<n$, and $\mathcal{O}_{\tau_i} = \{O_{i,\alpha}\}_{\alpha<\tau_i}$ is a decreasing sequence of open sets.

Generalizing the proof of theorem 3 we have the following.

THEOREM 4. If for all regular ordinals τ $D(\tau)$ is consistent, then it is consistent that every globular $HCWT_2$ space is regular.

NOTE. Actually we only need $D(\tau)$ to be consistent for globular spaces.

REFERENCES

[D] E. K. van Douwen, Remote points, Dissertationes Math., vol. 188, 1981.

[N] P. J. Nyikos, Order-theoretic base axioms, in: G. M. Reed, ed., Surveys in General Topology, Academic Press, New York, 1980, 367-398.

[P_1] S. Purisch, The orderability and closed images of scattered spaces, to appear.

[P_2] S. Purisch, Monotonically normal scattered spaces, in preparation.

[R] J. Roitman, Basic S and L, in: K. Kunen and J. E. Vaughn, eds., Handbook of Set-Theoretic Topology, North Holland, Amsterdam, 1984, 295-326.

[T] S. Todorcevic, Remarks on cellularity in products, Compositio Math. 57 (1986), 357-372.

[W] M. L. Wage, A collectionwise Hausdorff, non-normal Moore space, Can. J. Math. 28(1976), 632-634.

CLASSES OF COMPACT SEQUENTIAL SPACES

Peter J. Nyikos
Department of Mathematics, University of South Carolina
Columbia, South Carolina 29208

Eric van Douwen was a regular contributor to the Problems Section of <u>Topology Proceedings</u>, of which I have been the Problems Editor since its inception. One of the problems he contributed to Volume 8 was the following:

<u>Problem</u>. In the class of compact Hausdorff spaces, are there any other implications besides the ones shown?

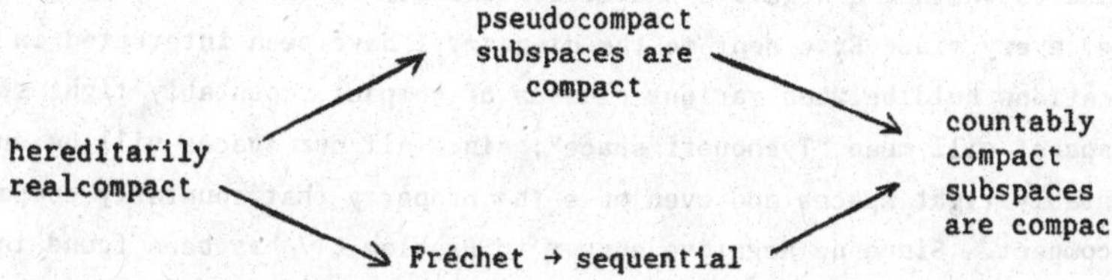

[For definitions, see Section 1.]

I happened to know that Zhou Hao-Xuan had shown [Zh] that if every subspace of a compact Hausdorff space is pseudocompact, the space is Fréchet. This makes it possible to straighten out the diagram to a linear sequence of implications. I also knew (as did Eric) that the one-point compactification of Ψ (Example 3.3) was sequential without being Fréchet. The problem of whether "countably compact subspaces are compact" implies "sequential" had already been posed by me in Volume 3 (C17) in an equivalent form (see below). So I just made two separate problems out of Eric's, whether the first two implications in the straightened-out diagram reversed, together with consistency results (discussed below) on each.

I found it very interesting that "Fréchet" and "sequential", which in Tychonoff spaces have little to do with the other three concepts, should come to be sandwiched in between two such similar-sounding properties. "Pseudocompact" and "countably compact" are often treated together, as in the Gillman and Jerison text [GJ], and so are "Fréchet" and "sequential", as in [F] and [AF]. What made it even more interesting is that, as I also knew at the time, "countably compact subspaces are compact" is in turn sandwiched in between "sequential" and "countably tight" (another concept treated in [AF]) in the class of compact Hausdorff spaces, and at the time it was still a famous unsolved problem whether there is a compact space of countable tightness that is <u>not</u> sequential. (A space is <u>countably tight</u> if whenever $x \in A$, there is a countable $B \subset A$ such that $x \in \bar{B}$.) This problem is now solved, thanks to the combined efforts of D.H. Fremlin, Z. Balogh, A. Dow, and myself (and the older consistency result of Ostaszewski [OS_1] whereby axiom $\diamondsuit$ implies an affirmative answer): under the Proper Forcing Axiom, and also in some models not requiring the consistency of anything more than ZF, every compact Hausdorff space of countable tightness is sequential [BDFN].

This did not dispose of "Problem C17," however. In [IN] we had shown that a compact Hausdorff space is sequential if, and only if, it is sequentially compact and countably compact subspaces are compact, so that the question of whether the last arrow in Eric's straightened-out diagram reverses becomes equivalent to:

Problem C17. If a compact Hausdorff space has the property that all countably compact subsets are compact, is the space sequentially compact?

We also showed [ibid] that the answer is yes if either MA or $2^{\omega} < 2^{\omega_1}$. These seemingly disparate axioms were combined by Eric into one [vD, 6.4]: $2^{t} > c$. (For the definition of t and many other cardinals important to set-theoretic topology, see [vD].) The issue is whether a negative answer is consistent, and that is still unsolved.

At any rate, every since Eric sent me the diagram, I have been interested in what sorts of implications hold between various classes of compact, countably tight spaces. From now on, "space" will mean "Tychonoff space", since all our spaces will be subspaces of compact countably tight spaces and even have the property that countably compact subspaces are compact. Since no negative answer to Problem C17 has been found in any model, I have taken the liberty of simplifying the title of this paper.

The first two sections are devoted to greatly expanding van Douwen's diagram, to the point where it seemed best to split it into three. Section 3 gives examples (and some theory as well) to justify the absence of arrows between the various classes, although some questions still remain in this regard. Section 4, devoted to some topological games of Gruenhage, will also justify some arrows that do appear. In Section 5, it is shown what happens when one is restricted to the compact scattered spaces, associated with superatomic Boolean algebras via Stone duality. In anticipation of this, as many examples in Section 3 as practical are scattered. Section 6 points the way to further expansions of van Douwen's diagram.

1. Additional classes and implications.

Most of the classes of compact spaces dealt with here fall into five informal classifications:

A. Classes defined by convergent sequences: Sequential, Fréchet, bisequential, $\aleph_0$-bisequential, (weakly) first countable, α_1.

B. Banach space classes: Eberlein, Gul'ko, and Corson compact spaces.

C. "Rings of continuous functions" classes: hereditarily realcompact, pseudocompact subspaces are compact, countably compact subspaces are compact.

D. Measure classes: Radon, hereditarily α-realcompact.

E. Hereditary covering and separation properties: (weakly) σ-metacompact, weakly θ-refinable, metalindelöf, Property wD.

Let us look more closely at each in turn.

Classification A. A subset A of a space X is sequentially closed in X if no sequence in A converges to a point outside A. A space is sequential if every sequentially closed subset is closed. A space is Fréchet-Urysohn (or simply Fréchet) if whenever $x \in \overline{A}$, there is a sequence from A converging to x. A space is bisequential if for every point x and every ultrafilter μ converging to x, there is a countable subset $\{A_n: n \in \omega\}$ of μ such that every neighborhood containing x contains all but finitely many A_n. A space is weakly first countable if to each point x one can associate a countable weak base, i.e. a countable collection $\{B_n(x): n \in \omega\}$ of subsets of X containing x such that a set S is open iff for each $x \in S$ there exists n such that $B_n(x) \subset S$. A compact space is $\aleph_0$-bisequential if it is countably tight and every countable subspace is bisequential.

A sheaf at x is a countable collection of sequences converging to x. A point of a space X is an α_i-point ($i=1,2,3,4$) iff for each sheaf $\{\sigma_n: n \in \omega\}$ at x, there is a sequence σ converging to x such that:

α_1: ran $\sigma_n \subset^*$ ran σ for all n; [As usual, we write $A \subset^* B$ if $A \setminus B$ is finite.]

α_2: ran $\sigma_n \cap$ ran σ is infinite (equivalently, nonempty) for all n.

α_3: ran $\sigma_n \cap$ ran σ is infinite for infinitely many n.

α_4: ran $\sigma_n \cap$ ran σ is nonempty for infinitely many n.

A space is an α_i-space if every point is an α_i-point.

The α_i properties are primarily of interest when the spaces in question are Fréchet, and that is the context in which we will consider them. A space is countably bisequential or strongly Fréchet if it is Fréchet and α_4, a w-space if it is Fréchet and α_2, and a v-space if it is Fréchet and α_1. Actually, except for "v-space", these were originally given completely different-looking definitions. For the original definition and equivalences, see [Mi_1], [A_1, 5.23] and [Sh], [No_1] respectively. An important theorem is that every compact Fréchet space is countably bisequential [Mi_1].

The α_i-properties are especially important in product theorems, as are ($\aleph_0$)-bisequential. For instance, the countable product of countably compact Fréchet α_i-spaces is again such a space if $i=1,2,3$ [No_1, 3.12], and any countable product of ($\aleph_0$)-bisequential spaces is again one [A_4]. (Caution: The first definition of "bisequential" in [A_4] is incorrect.)

The equivalence in the definition of α_2 above is an easy exercise [No_2]. Once this is granted, the implications $\alpha_i \Rightarrow \alpha_{i+1}$ are obvious; so are Fréchet => sequential, and first countable => bisequential => $\aleph_0$-bisenquential. Less easy to see is the fact [A_4, 6.23] that $\aleph_0$-bisequential implies both Fréchet and α_3. Despite some resemblance in the definitions, neither of "bisequential" or "weakly first countable" implies the other. In fact, a space is first countable iff it is weakly first countable and Fréchet [Si_1]. However, the following is still unsolved:

Problem 1. Is there a weakly first countable compact space that is not first countable?

An affirmative answer is consistent (Example 3.11).

Every weakly first countable space is sequential [Si_1], [Ny_2]. Thus for compact spaces we have the implications => in:

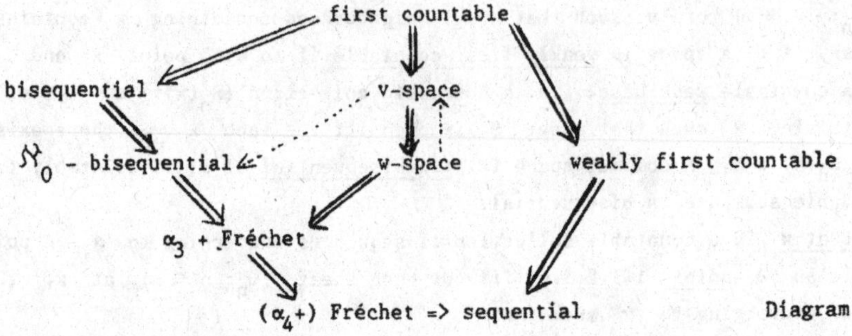

Diagram 1

A remarkable recent result of Alan Dow is:

1.1. Theorem. [Do] In the Laver forcing model [L] every w-space is a v-space.

For more on this and the following result, see Section 4.

1.2. Theorem. [Dow and Steprans] It is consistent that every countable v-space is first countable.

A corollary is that it is consistent that every compact v-space is $\aleph_0$-bisequential. I have indicated consistent implications in the above diagram by dotted lines. No other implications, consistent or otherwise, are possible, except those embodied in:

Problem 2. Is it consistent that every compact, α_3, Fréchet space is $\aleph_0$-bisequential? or that every compact w-space is $\aleph_0$-bisequential? (See Example 3.8.)

Classification B. A compact space is Corson compact if it embeds in a Σ-product of real lines. It is Eberlein compact if it embeds in a Banach space with the weak topology (not the one making it Banach). A compact space X is Gul'ko compact if the space $C_p(X)$ of continuous real-valued functions with the product topology is a Lindelöf Σ-space. (For the concept of a Σ-space see [Bu] or [Gr_2].) Related classes of compact spaces and many equivalent conditions may be found in [Ne], as well as the implications

Eberlein => Gul'ko => Corson

and examples showing the arrows do not reverse. See also the discussion of Classification E.

It is easy to see that every separable subset of a Corson compact space is metrizable, and not too difficult to show that a Corson compact space (in fact a Σ-product of first countable spaces) is Fréchet [Gr_1, 4.6]. Hence we also have:

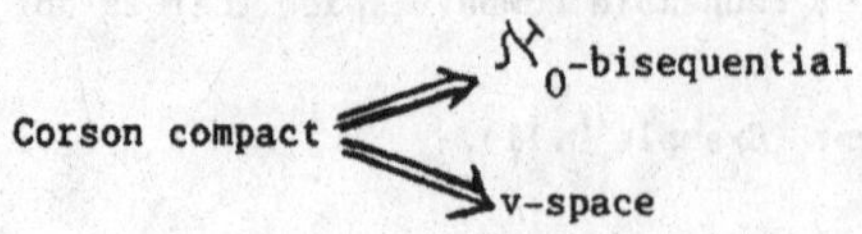

Classification C. I assume readers are familiar with the concept of a countably compact space, but here is a characterization which helps relate it to some of the other classes: a space is countably compact iff every filterbase of closed sets has the countable intersection property. A space is pseudocompact if every real-valued continuous function is bounded, and realcompact if it can be embedded as a closed subspace in a product of real lines.

Although it may grate on some set theorists' nerves, I will follow here the custom of calling a maximal centered subcollection F of a family A of sets an "A-ultrafilter", and calling F "fixed" if it has nonempty intersection, and "free" otherwise. [I never could understand why logicians persist in using the cumbersome expression "non-principal ultrafilter" when "free ultrafilter" is available.] A subset of a space X is a zero-set if it is of the form $f^{-1}\{0\}$ for some continuous real-valued function f. Letting Z stand for the collection of all zero-sets in X, we recall:

1.3. Theorem [GJ,]. A space is (A) compact iff every Z-ultrafilter is fixed; (B) pseudocompact iff every Z-ultrafilter has the c.i.p. (C) realcompact iff every Z-ultrafilter with the c.i.p. is fixed.

From this it is obvious that a space is compact iff it is pseudocompact and realcompact, accounting for one arrow in van Douwen's diagram. Another is accounted for by the fact (also clear from the above) that every countably compact space is pseudocompact. We have already provided references for the other implications in his diagram, as straightened out by Zhou's theorem.

Classification D. A Borel measure μ is inner-regular with respect to a certain class K if $\mu(B) = \sup\{\mu(K): K \subset B, K \in K\}$ for each Borel set B. A space X is Radon if every finite (i.e. $\mu(X) < +\infty$) measure on the Borel sets is Radon, i.e. is inner-regular with respect to the compact subsets. A space is α-realcompact (also known as closed-complete) if every "closed ultrafilter" (= C-ultrafilter where C is the collection of closed sets) with the countable intersection property is fixed. Every Radon space is hereditarily α-realcompact; this follows from 6.8, 7.4, and 8.12 of [GP] and the diagram on p. 992 of [GP], which gives a number of concepts intermediate between these (but some are equivalent to Radon for compact spaces, cf. [GP, 7.9]).

Classification E. A space is metalindelöf if every open cover has a point-countable open refinement. A collection A is σ-point-finite if it is the countable union of point-finite collections, and weakly σ-point-finite if $A = \cup\{A_n : n \in \omega\}$ where, for each point x, A is the union of all the A_n such that x is in at most finitely many members of A_n. A space is [weakly] σ-metacompact if every open cover has a [weakly] σ-point finite open refinement. A weakly θ-refinable space is one where every open cover has an open refinement $U = \cup\{U_n : n \in \omega\}$ where for each point x there is an n such that x is in at least one, but no more than finitely many, members of U_n. [If in addition each U_n can be a cover, then we have the definition of a θ-refinable space.]

It is easy to see that the following implications hold:

$$\sigma\text{-metacompact} \Rightarrow \text{weakly } \sigma\text{-metacompact} \begin{array}{l} \nearrow \text{metalindelöf} \\ \searrow \text{weakly } \theta\text{-refinable} \end{array}$$

Of course, every compact space is σ-metacompact, so that these properties only are interesting in our context if they are satisfied hereditarily. Now, as is so often the case with covering properties, it is enough to check that every open subspace has the respective property. The idea is that, if $\mathbf{U}$ is a cover of a subspace Y by sets open in Y, then each $V \in \mathbf{U}$ is of the form $W \cap Y$ for some W open in the whole space, and if we refine the collection $\mathbf{W}$ of all these expansions on the open subspace $\cup\mathbf{W}$ in the appropriate way, the traces of the refinement on Y will also behave as desired. This is especially worth noting in the context of compact spaces since every open subspace of a (locally) compact space is locally compact, and these hereditary covering properties are neither created nor destroyed in passing to one-point compactifications. It also leads to such simplifications as the following theorem [Ny_5], which uses alternate characterizations of weak θ-refinability in [BL] and [Bu].

1.4. Theorem. A locally compact space is weakly θ-refinable iff every open cover has a σ-relatively-discrete refinement by compact subsets. [A collection of subsets of a space X is σ-relatively-discrete if it is of the form $\cup\{\mathbf{A}_n : n \in \omega\}$ where each point of $\cup \mathbf{A}_n$ has a neighborhood meeting at most one member of $\mathbf{A}_n$.]

It also helps in the analysis of the hereditary metalindelöf property. In Section 4 we will see that "hereditarily metalindelöf" implies both "v-space" and "$\aleph_0$-bisequential" for compact spaces. See also [PY] for a quick proof that if x is a point of a compact space X and $X-\{\infty\}$ is metalindelöf and $x \in \overline{A}$, then there is a sequence from A converging to x.

Finally, a space X satisfies Property wD if for every countably infinite closed discrete subspace S there is an infinite discrete (in X) collection $\mathbf{U}$ of open sets, each of which meets S in exactly one point. (Recall that a collection $\mathbf{A}$ is discrete if each point has a neighborhood meeting at most one member of $\mathbf{A}$.

Property wD is the weakest in a hierarchy of properties extending to normal (and beyond to collectionwise normal, etc.) [vD, Section 12], [V_1]. It is included here because compact sequential spaces which satisfy wD hereditarily happen to fit nicely into Eric's diagram and illuminate some of the relationships. On the one hand, every realcompact space satisfies wD [V_1]; on the other, every pseudocompact space satisfying wD is countably compact. This is easy to see if one considers another characterization of pseudocompact spaces [E, 3.10.23]: every discrete family of open sets is finite. It might be said that normal $\Rightarrow$ wD is the "real" reason for the fact, initially surprising to many students, that every normal pseudocompact space is countably compact. At any rate, we can squeeze "sequential and satisfying wD hereditarily" in between the first two classes of Eric's straightened-out diagram, because of what happens on the other end:

sequential implies countably compact subsets are closed in any space, hence compact in a compact space.

2. More implications

The covering properties in Classification E all imply compactness in a countably compact space. This is part of a theme carried considerably further in [V_2, Section 6], [A_3] and [T]: "countably compact + ________ => compact." The simplest thing to put in the blank is "Lindelöf," but the weaker the property, the better.

So, if one of the Classification E covering properties is satisfied hereditarily in a compact space, it implies all countably compact subspaces are compact. But, except for σ-metacompactness, I do not know how much further we can go (see Problem 6, and Example 3.6 below). We do have the following 1984 result of Uspensky:

2.1. Theorem. [U] Every σ-metacompact, pseudocompact space is compact.

Gardner [GP, 10.2-3] showed a hereditarily weakly θ-refinable, locally compact space is Radon iff it has no discrete subspace of a real-valued measurable cardinal, and that a weakly θ-refinable space is α-realcompact iff it has no closed discrete subspace of a measurable cardinal.

Gruenhage [Gr_3] gave the following unexpected characterizations:

2.2. Theorem. The following are equivalent for a compact space X:

(i) X is Corson [resp. Eberlein] compact.

(ii) X^2 is hereditarily metalindelöf [resp. σ-metacompact]

(iii) $X^2 - \Delta$ is metalindelöf [resp. σ-metacompact]

He also showed [Gr_6] that X^2 is hereditarily weakly σ-metacompact if X is Gul'ko compact and conjectured that the converse is true, and also asked:

Problem 3. If $X^2 - \Delta$ is weakly σ-metacompact, is X^2 hereditarily weakly σ-metacompact? is X Gul'ko compact?

So far, we have the following implications, for compact spaces where H. stands for "hereditarily":

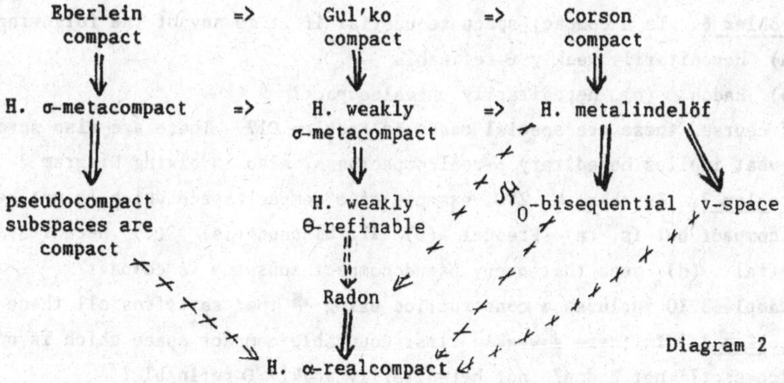

Diagram 2

Here ====> means the implication holds if there are no real-valued measurable cardinals, and +++> means it is consistent that the implication does not hold, but that it is not known whether it fails in ZFC. See Problem 7 below, and:

Problem 4. If a compact space X is hereditarily metalindelöf, and no discrete subspace is of measurable cardinality, is X hereditarily α-realcompact? Is it consistent that X is always Radon?

In [GP, Example 11.20] there is a compact, hereditarily metalindelöf non-Radon space constructed using CH, in which no discrete subspace is of a real-valued measurable cardinal. Gardner [Gd_2] has asked whether MA + ¬CH implies no such space exists.

I do not even know the answer to:

Problem 5. If X is metalindelöf, and no closed discrete subspace is of measurable cardinality, is X α-realcompact?

This is part of a theme related to the one mentioned at the beginning of this section: "No discrete subspace of measurable cardinality + ______________ => α-realcompact." If one looks at the characterizations of α-realcompact and countably compact in Section 2, it is clear that anything one can truthfully put in this blank will serve for the other one. But this theme has not progressed nearly so far: As I said, weak θ-refinability works, but it is the weakest covering property I have seen so far that does. But "metalindelöf" is a reasonable candidate to try since metalindelöf spaces that are not weakly θ-refinable are still in short supply. The only "real" ones were described by Gruenhage (Example 3.5) and they are α-realcompact; in fact, they are subspaces of compact Radon spaces if one assumes c is not real-valued measurable.

It might be said that this second theme is the "real" reason why normal θ-refinable spaces [Z] and normal, countably paracompact, weakly θ-refinable spaces [Ga_1] with no closed discrete subspaces of measurable cardinality are realcompact: a normal, countably paracompact space is realcompact iff it is α-realcompact [Dy]; also every θ-refinable space is countably metacompact [Gi] and every normal, countably metacompact space is countably paracompact [E,] and [Ru_1, 1.1, (i) => (ii)].

A big unknown as far as Diagram 2 is concerned is:

Problem 6. Is a compact space sequential if it is any of the following:

(a) hereditarily weakly θ-refinable

(b) Radon (c) hereditarily α-realcompact?

Of course, these are special cases of Problem C17. There are also several problems about what implies hereditary α-realcompactness, also involving Diagram 2.

Problem 7. Is there a ZFC example of a compact space which is not hereditarily α-realcompact but is (a) Fréchet (b) $\aleph_0$-bisequential (c) hereditarily wD and sequential (d) such that every pseudocompact subspace is compact?

Example 3.10 includes a construction using ♣ that satisfies all these properties.

Problem 8. Is there a weakly first countable compact space which is not hereditarily α-realcompact? not Radon? not hereditarily weakly θ-refinable?

Here I do not know of any consistency results, not even under large cardinal hypotheses, nor am I aware of bounds on the cardinality of weakly first countable compact spaces.

It is quite easy to show that every realcompact space is α-realcompact [Dy]. [Caution: the collection of all zero-sets in a closed ultrafilter with the c.i.p. is not always a Z-ultrafilter, but that is all right: compare the proof of 2.3 below.] For compact spaces satisfying these properties hereditarily, we have an interesting interpolant, independently noticed by Reznichenko and myself.

2.3. Theorem. Every compact, hereditarily realcompact space is bisequential.

Proof. Let Y be hereditarily realcompact and compact. Let $\mathbf{U}$ be a free ultrafilter on Y converging to a point y. Let $X = Y - \{y\}$ and let

$$\mathbf{F} = \{F \in \mathbf{U}: F \text{ is a zero-set of } X\}.$$

Then $\mathbf{F}$ extends to a unique Z-ultrafilter $\mathbf{H}$ in X. Indeed, if Z_i, $i=0,1$ are disjoint zero-sets such that Z_1 meets every member of $\mathbf{F}$, let $f: X \to [0,1]$ be a continuous function such that $f^{\rightarrow}Z_i = \{i\}$ for $i=0,1$ [GJ, 1.15]. The image of $\mathbf{U}|X$ is a base for an ultrafilter that can only converge to 1, and so $f^{\leftarrow}[1/2,1]$ is in $\mathbf{F}$ and is disjoint from Z_0. It is now easy to see that $\mathbf{H} = \{Z \in \mathbf{Z}: Z \cap F \neq \phi$ for all $F \in \mathbf{F}\}$.

$\mathbf{H}$ is free because y has a base of zero-set neighborhoods in Y and their traces are all in $\mathbf{H}$. So there is a countable descending family $\{H_n\}_{n=1}^{\infty}$ in $\mathbf{H}$ with empty intersection. For each n, let $H_n = f_n^{\leftarrow}\{0\}$ for a continuous $f_n: X \to [0,1]$. Let $F_n = \mathrm{cl}_Y \bigcap_{i=1}^{n} f_i^{\leftarrow}[0,\frac{1}{n}]$. Then $\bigcap_{n=1}^{\infty} F_n = \{y\}$ and each F_n is in $\mathbf{U}$. Using compactness of Y as in [E, proof of 3.3.4], we see that the filterbase $\{F_n\}_{n=1}^{\infty}$ converges to y.

Hereditary realcompactness is not affected by adding one point, so Theorem 2.3 extends to locally compact spaces.

2.4. Theorem. Every compact bisequential space is hereditarily α-realcompact.

Proof. Let Y be a compact space. Suppose there is a subspace X of Y on which there is a free closed ultrafilter $\mathbf{F}$ with the c.i.p. Let $y \in Y - X$ be its unique (by regularity) adherent point. Then $\mathbf{F}$ is a filterbase on Y, and any ultrafilter $\mathbf{U}$ extending it converges to y. Let $U_n \in \mathbf{U}$ for each $n \in \omega$. The sets $(\mathrm{cl}_Y U_n) \cap X$ are in $\mathbf{F}$, so we can pick $x \in X$ in their intersection.

Let N be a closed neighborhood of y that misses x. None of the U_n can be a subset of N. So Y is not bisequential.

Theorem 2.4 does not extend to locally compact spaces. The ordinal space ω_1 is bisequential, as is any first countable space, but not α-realcompact.

Bisequentiality also interposes between hereditarily realcompactness and another Classification C property:

2.5. Theorem [A_3, in effect]. In a compact bisequential space, every pseudocompact subspace is compact.

Proof. Theorem 6' of $[A_3]$ states: A (regular) space is bisequential if and only if for any $x \in X$ and any filterbase ξ with x as an adherent point, there is a sequence $\{P_n\}_{n=1}^{\infty}$ of regular closed subsets of X which converges to x and such that every P_n meets every member of ξ. Now if Y is a pseudocompact subspace of X with x in its closure, the set of all interiors of the P_n trace a sequence of relatively open sets on Y which we may take to be $\subset$-decreasing. Then [GJ,9.13] there is a point of Y in the intersection of the closures of the P_n, and this must be x. Thus Y is closed.

Also hereditary realcompactness interposes between bisequentiality and another classification A property:

2.6. Theorem [GJ, 8.15] Every first countable realcompact space is hereditarily realcompact.

And so we have the implications => in the following diagram, again for compact spaces. The other arrows have the same meaning as in Diagram 2.

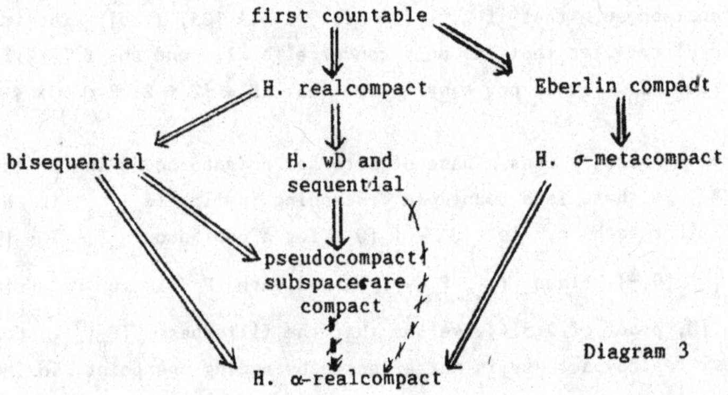

Diagram 3

Problem 9. If a compact space has no discrete subspace of measurable cardinality, is it bisequential if it is (a) Eberlein compact? (b) hereditarily σ-metacompact?

Note that the one-point compactification of a discrete space of measurable cardinality is not even hereditarily α-realcompact, but it is Eberlein compact. That is clear from Rosenthal's characterization of Eberlein compacta [Ro]: they are precisely the compact spaces K which admit a σ-point-finite, T_0-separating cover by open F_σ-sets. If one inserts "weakly" in front of "σ-point finite" here, one has Sokoloff's characterization of Gul'ko compact spaces [So]. And if one has "point-countable" instead, one has the Corson compact spaces [MR].

3. Counterexamples.

We begin this section with a space van Douwen felt could supplant the more complicated Tychonoff plank in most elementary texts.

3.1 <u>Thomas's plank</u>. Let W be an uncountable discrete space and let W + ∞ denote its one-point compactification. Let X = (W + ∞) × (ω + 1). Thomas's plank is the case where the underlying set of W is the set of real numbers.

X is not hereditarily normal. In fact, the subspace Y obtained by removing the corner point does not satisfy wD. Indeed, no infinite set of points <∞,n> can be expanded to a discrete collection of open sets. <u>A fortiori</u>, X is not hereditarily realcompact.

On the other hand, X is both bisequential and Eberlein compact. Indeed, part of the significance of the classes of ($\aleph_0$-)bisequential spaces is that they are countably productive and hereditary [Mi_1], [A_4]. The "Banach" classes of compacts are also countably productive: see [AL] for Eberlein compacta and [ ] for Gul'ko compacta. For Corson compacta it is trivial.

Since every separable subset of X is metrizable, it is a v-space. Since X is Fréchet but not first countable, it is not weakly first countable. If we make $|X| = \omega_1$, then X is Radon.

3.2. <u>Alexandroff's "two arrows"</u>. Let X = [0,1] × {0,1} with the lexicographical order topology. As is well known [E,] [SS], X is hereditarity separable and hereditarily Lindelöf, so that it satisfies all the Classification C and E properties. Since every discrete subspace is countable, X is Radon. Since X is first countable, it also satisfies the Classification A properties.

On the other hand, X satisfies none of the "Banach" properties since it is separable and nonmetrizable.

3.3. <u>The Mrówka-Isbell Ψ(+∞)</u>. Let A be an infinite, maximal almost disjoint family (MADF) of infinite subsets of ω. The underlying set of Ψ is ω ∪ A, topologized by letting the points of ω be isolated and letting {{A} ∪ (A-n): n ∈ ω} be a base for the neighborhoods of A for each A ∈ A.

As is well known [GJ, Exercise 5I] and easy to prove, Ψ is locally compact, and pseudocompact. Being a countable union of closed discrete subspaces, it is hereditarily (weakly) θ-refinable, and so is Ψ + ∞, the one-point compactification. On the other hand, no separable non-Lindelöf space can be metalindelöf, and it is even more obvious that {ω} ∪ {A ∪ {A}: A ∈ A} has no point-finite open refinement. Ψ + ∞ satisfies none of the "Banach" properties for the same reason as 3.2.

Ψ + ∞ is not Fréchet since no sequence from ω converges to ∞. But it is sequential: unless a subset of ω has compact closure, it has infinitely many points of A in its closure, and these have a sequence converging to ∞ since A ∪ {∞} is the one-point compactification of a discrete space.

From the discussion surrounding Yakovlev's space (3.11) it will be evident that Ψ + ∞ is not weakly first countable. It is Radon whenever c is not real-valued measurable. Special versions of it are Radon even in models where c is real-valued measurable. In fact, as far as I know, no negative answer is known to:

Problem 10. Is $\mathfrak{a}$, the least cardinality of an infinite MADF of subsets of ω, always less than the least real-valued measurable cardinal?

Eric van Douwen [vD] has called a space Ψ-like if it is locally compact, has a countably infinite dense set of isolated points, and the nonisolated points are a closed discrete subspace. If x is a nonisolated point and N_x is a compact neighborhood of x in which all other points are isolated, then the Hausdorff condition tells us each N_x is clopen and the sets $N_x \cap W$, where W is the set of isolated points, form an almost disjoint family. It is maximal iff the space is pseudocompact [vD, 11.6]. Conversely, any almost disjoint family of infinite subsets of a countable set gives rise to a Ψ-like space as a MADF gives rise to Ψ.

Every one-point compactification of a Ψ-like space is hereditarity weakly θ-refinable, being a countable union of (closed) discrete subspaces, but not hereditarily metalindelöf if the space is uncountable.

3.4. The Cantor tree + ∞. Let T denote the Cantor tree, i.e. the full binary tree of height $\omega + 1$ with the interval ("tree") topology. It is a Ψ-like space and is nonmetrizable, hence not metalindelöf. Unlike Ψ, it is hereditarily realcompact. In fact, it has a coarser compact metric topology, for which a base is the set of wedges $V_t = \{t' : t' \geq t\}$ where t is on a finite level of T, and their complements. And any space with a finer topology than a first countable realcompact space is hereditarily realcompact [GJ, 8.17].

Todorcevic' has pointed out that T with this latter topology, which I call "the coarse wedge topology", is "really" the tree of all initial segments (which he calls "paths") in the full binary tree of height ω, with the product topology. For a fuller discussion of this theme, see [Gr_5]. In [Ny_7] I show a very natural embedding of T with this topology in the plane, with the points of the top level topologically identified with the Cantor set.

Thus $T + \infty$, with T having the interval topology, is hereditarily realcompact, bisequential, etc. In [Ny_7] I show, using the Baire category theorem on the coarser topology, that it is not a w-space. Also [ibid.], if one removes all points from the top level of T except those corresponding to a λ'-set [Mi_2] and then takes the one-point compactification of what remains, one has a w-space.

3.5. The Todorcevic-Gruenhage space. Let S be a stationary, co-stationary subset of ω_1 and let T be the tree of all compact subsets of S, ordered by end extension. Let $\tilde{T}$ be obtained by adding a point at the end of each branch (maximal chain) of T. Give $\tilde{T}$ the coarse wedge topology (see 3.4). In [Gr_5] it is shown that $\tilde{T}$ is Corson compact, hence so is $\tilde{T}^2$, but $\tilde{T}^2 - \Delta$ is not weakly θ-refinable; also that $\tilde{T}^2$ is Radon if $\mathfrak{c}$ is not real-valued measurable. It is also shown that $\tilde{T}$ is hereditarily paracompact, hence it is hereditarily realcompact, etc. However, $\tilde{T}^2$ contains a copy of the Thomas plank (just take the subspace of points of level ≤ 1 and an appropriate-sized subset in each copy). On the other hand, Gruenhage mentions a first countable variant, and that has all countable powers first countable and so hereditarily realcompact, etc.

3.6. Reznichenko's space. This is a space with some amazingly strong properties, discovered in 1987. Uspensky privately communicated a description in English to Gruenhage. Since a full treatment is not likely to appear in print in English anytime soon, I thought it worthwhile to at least give the definition here.

Define sets A_α and finite sets F_α for $\alpha < \omega_1$ as follows. $A_0 = \omega$, $F_0 = [\omega]^1$. With A_β and F_β defined for all $\beta < \alpha$, let $F^*_\alpha = \cup\{F_\beta\colon \beta < \alpha\}$, $A^*_\alpha = \cup\{A_\beta\colon \beta < \alpha\}$. For each ω-sequence $s = \langle k^s_n\colon n \in \omega\rangle$ of disjoint members of F^*_α, define new elements $a_s \in A^*_\alpha$ with $a_s \neq a_{s'}$ if $s \neq s'$. Let

$$A_\alpha = \{a_s\colon S \text{ is à disjoint sequence in } F^*_\alpha\}$$

$$F_\alpha = \{k^s_n \cup \{a_s\}\colon n \in \omega,\ a_s \in A_\alpha\}.$$

Let $A = \cup\{A_\alpha\colon \alpha < \omega_1\}$, $F = \cup\{F_\alpha\colon \alpha < \omega_1\}$, with F regarded as a subset of 2^A in the natural way, with the product topology. Or rather, let us regard 2^A, topology and all, as the power set of A.

Note that the sets $T_n = \{k \in F\colon n \in k\}$ form a partition of F and each is a tree by inclusion. Each branch B of T_n converges to the point $\cup B$, and if these points are added, the resulting subspace C_n of 2^A is closed. Its topology is the coarse wedge ("path") topology. By a theorem of Gruenhage [Gr_5], C_n is Eberlein compact.

Let $C = \cup^\infty_{n=0} C_n$. It is not hard to see that $\overline{C}\setminus C = C \cup [A]^1 \cup \emptyset$, and of course $[A]^1 \cup \emptyset$ is the one-point compactification of the discrete subspace $[A]^1$. Reznichenko showed that $\bar{C}$ is Gul'ko compact. He also showed that $\beta(C \cup [A]^1) = \bar{C}$, i.e. the Stone-Cech compactification of $C \cup [A]^1$ is the one-point compactification! Thus $C \cup [A]^1$ is pseudocompact (and noncompact).

Of all the results in this survey, this was for me the most unexpected. I was familiar with the constructions of A. Berner [Be] and H.-X. Zhou [Zh], which gave spaces with similar properties under the axioms a=c and b=c respectively, and there seemed little hope of carrying out either construction or anything resembling it "in ZFC". The role of C_n was taken over in Zhou's construction by the Cantor set, in Berner's construction by a first countable compact space in which every set of cardinality $< c$ was nowhere dense, and in both cases the rest of the space was the one-point compactification of a discrete space, with the whole space minus the extra point being pseudocompact. Their spaces, however, were built by transfinite induction, like the well-known examples of Ostaszewski [Os_1], Juhász, Kunen, Rudin [JKR], and many others, including 3.8 and later examples below. These constructions leave room for a lot of optional details, whereas Reznichenko's space, as can be seen, is really just one space. (Of course, it is easy to construct variations on it.)

Another difference is that Zhou's space is separable, and Berner's seems inevitably to involve separable pseudocompact noncompact subspaces, while in Reznichenko's space, every countable subset has metrizable closure. Nevertheless, until I saw it, I would have guessed that the classes "Fréchet" and "every pseudocompact subspace is compact" were destined for another independence result.

3.7. Peter Simon's "barely Fréchet" compacta. This is a pair of one-point compactifications of Ψ-like spaces whose properties are intermediate between those of Ψ and the Cantor tree. On the one hand, their one-point compactifications fail to be α_3, but on the other hand they are Fréchet, and compactness gives α_4.

The product of the two spaces fails to be Fréchet; on the other hand, the product of an α_3-Fréchet and a countably compact Fréchet space is Fréchet [A_4, 5.16], thus both compactifications fail to be α_3. These properties of Simon's spaces answer quite a few questions in [A_4]: 5.21, 5.22.1, 5.22.4, 6.12, 6.13, 6.14.

There is a close interplay between the construction and $\beta\omega$. Let Y be a noncompact Ψ-like space, with ω its set of isolated points. The identity on ω induces a continuous $f\colon \beta\omega \to Y + \infty$, with ω^* taken onto $(Y-\omega) \cup \{\infty\}$. Malyhin [Ma] showed that $Y + \infty$ is Fréchet iff $f^{\leftarrow}(Y-\omega)$ is regular open in ω^*. Simon was able to find a MAD family P of subsets of ω which was the disjoint union of two subfamilies P_0 and P_1, such that $\cup\{A^*: A \in P_i\}$ was a regular open subset of ω^* for $i=0,1$. [Here A^* denotes the remainder of A, i.e. $(cl_{\beta\omega}A) - \omega$.] In other words, here we have two families of disjoint clopen sets, with the union of each family regular open, and the union of these two regular open sets dense in ω^*. It is the fact that the union of these two regular open sets is not regular open that is behind the product of the associated spaces not being Fréchet.

I have called this setup "Petr Simon's checkerboard", with the remainder of each $A \in P_0$ a "red square" and that of each $A \in P_1$ a "black square". One might paraphrase Simon's closing note in [Si_1] by saying that every infinite MAD family traces a "checkerboard" on the remainder of some infinite subset of ω.

3.8. A consistent ($\mathbf{b}=\mathbf{c}$) example of a compact w-space which is not $\aleph_0$-bisequential. We begin with a ZFC construction which, like the λ'-modification of the Cantor tree, is an uncountable Ψ-like space whose one-point compactification is a w-space. It is the example "For later use" in [vD, 12.2], but used for a different purpose there.

The set of isolated points is $\omega\times\omega$, and the ADF is the set of all columns $\{n\}\times\omega$, together with the set of the graphs of a $<^*$-unbounded, $<^*$-well-ordered family $\{f_\alpha : \alpha < \mathbf{b}\}$ of increasing functions $f_\alpha\colon \omega \to \omega$, where $<^*$ is the eventual domination order, i.e. $f <^* g$ means there exists n such that $f(i) < g(i)$ for all $i \geq n$. Let X be the resulting space and Y its one-point compactification. In [Ny_8] I show Y is a w-space (and a v-space if the f_α form a scale).

Problem 11. Can Y ever be $\aleph_0$-bisequential?

I do not know whether Y can even be rigged "in ZFC" to fail to be $\aleph_0$-bisequential, but it can be done if $\mathbf{b} = \mathbf{c}$. Specifically, we make $(\omega\times\omega) \cup \{\infty\}$ fail to be bisequential. Let

$F = \{A \subset \omega\times\omega \mid A$ meets each column in a cofinite set$\}$ and let

$F' = \{A-(n\times\omega) | A \in F\}$. Then F' is a filter on $\omega\times\omega$ which converges to ∞ no matter how $\langle f_\alpha: \alpha < b\rangle$ is chosen. We will choose it so that no ultrafilter extending F' has a countable subfamily converging to ∞. This is equivalent to saying that, for each descending countable family $B = \{B_n: n \in \omega\}$ of subsets of $\omega\times\omega$ such that $B_n \cap F$ is infinite for each $F \in F'$, there is an f_α whose graph meets each B_n.

Well, $B_n \cup F$ is infinite for each $F \in F'$ if, and only if, each B_n meets infinitely many columns in an infinite set. And if $b=c$ we can arrange this simply by listing all such families B by $\{B_\alpha: \alpha < c\}$ and making sure f_α hits every member of B_α.

By omitting the columns from the ADF, one obtains a Ψ-like space Z that satisfies wD hereditarily, and its one-point compactification does also, but if $b=c$ the same filter F shows it is not $\aleph_0$-bisequential.

It can be shown that Z is hereditarily pseudonormal "in ZFC" [vD, 11.4e].

Definition. A space is pseudonormal if, given any two disjoint closed sets C and D, one of which is countable, there are disjoint open sets U and V such that $C \subset U$ and $D \subset V$.

Now every pseudnormal space satisfies wD [vD, 12.1]. So Z satisfies wD hereditarily, and it is easy to see that $Z + \infty$ does also.

3.9. A "barely pseudonormal" Ψ-like space. I have an unpublished ZFC construction of a Hausdorff gap $\langle A,B\rangle$ where, as usual, $A = \{A_\alpha: \alpha < \omega_1\}$ is a $\subset^*$-ascending sequence and $B = \{B_\alpha: \alpha < \omega_1\}$ is a $\subset^*$-descending sequence of subsets of ω such that $A_\alpha \subset^* B_\beta$ for all α, β but there is no $A \subset \omega$ such that $A_\alpha \subset^* A \subset^* B_\beta$ for all α, β. The added twist in the construction is that if one lets $C_\alpha = A_{\alpha+1} \setminus A_\alpha$ for all α, then for each pair of uncountable subsets Γ, Δ of ω_1 and any $G, D \subset \omega$ such that $C_\gamma \subset^* G$ for all $\gamma \in \Gamma$, $C_\delta \subset^* D$ for all $\delta \in \Delta$, it must be that $G \cap D$ is infinite.

In the Ψ-like space X which uses the ADF $\{C_\alpha: \alpha < \omega_1\}$ as in 3.3, this translates to the fact that no two uncountable subsets have disjoint neighborhoods. On the other hand, if Γ is countable, we can let $\lambda = \sup \Gamma$, and then $C_\gamma \subset^* A_{\lambda+1}$ for all $\lambda \in \Gamma$, $C_\alpha \subset^* \omega - A_{\lambda+1}$ for all $\alpha > \lambda$, so that $A_{\lambda+1}$ has countable clopen closure containing all the points associated with Γ, and now it is elementary to get disjoint open sets around the points indexed by Γ and those indexed by $\omega_1 - \Gamma$, and so these sets will be clopen. Thus X is pseudonormal, but just barely!

It is easy to see that a pseudonormal space in which the nonisolated points form a (closed) discrete subspace is hereditarily pseudonormal. So X satisfies wD hereditarily and $X + \infty$ is easily seen to satisfy it also.

But X is not realcompact. Disjoint zero-sets can be put into disjoint open sets [GJ,] and so the co-countable zero sets form a Z-ultrafilter with the countable intersection property which is not fixed.

I have been unable to ascertain whether $X + \infty$ is bisequential. The answer could well vary from one model to the next, but the question easily reduces to that of whether

$\omega \cup \{\infty\}$ is bisequential, so that if it is not, then $X + \infty$ is not $\aleph_0$-bisequential either.

Problem 12. Is it consistent that a compact space satisfying wD hereditarily is $\aleph_0$-bisequential?

The following example shows why we cannot have a ZFC theorem here.

4. Some topological games of Gruenhage.

Gary Gruenhage has invented a number of related topological games which are good for analyzing the sequential structure of spaces. His original one is the one that gave w-spaces their name.

As usual, the game lasts for ω plays and involves two players. I will follow a terminology I picked up from a recent paper co-authored by Shelah and call one player "the hero" and the other "the villain". The idea is that the spaces in which we are really interested, at least in the context of the game, are where the hero has a winning strategy, or at least the villain does not have one. Or, to paraphrase Gruenhage [Gr_1], for every strategy of the villain, the hero has a counter-strategy that will defeat it. It may not be immediately obvious that this is equivalent to the lack of a winning strategy for the villain, but it becomes clear from the usual definition of a strategy [ibid.]. It is a "decision tree" in which the nodes are the various possibilities for each player, but the player, call him/her X, whose strategy it represents has only one choice at each node which represents a turn by X. Now almost every topological game, including the ones we consider here, has no draws: any game played out in full is a win for either one player or another, and the only issue is whether one or the other has a winning strategy of a certain sort (see Definition 4.8). Thus "all" the opponent has to do, on being informed of the strategy [a very valuable piece of information, which alters the nature of the game!] is to check whether any branch leads to a loss by X, and then make the choices at each turn represented by such a branch. If there is no such branch, why then that is exactly what is meant by the tree being a winning strategy!

Game 1. (The point-open game.) Given a point j (the "jail") of a space X, the hero attempts to fence the villain down into the jail in ω moves. On each move he picks an open set U_n containing the jail, and the villain must always pick a point p_n inside U_n. The hero wins if the sequence $\langle p_n \rangle$ converges to j, otherwise he loses. A slight modification is:

Game 1'. Here the villain gets the first move and with it he picks a point j; that is where the hero must situate the jail, and then we proceed just as in Game 1. Of course, the villain will try to put the jail in the remotest possible point. It would clearly be suicidal to put it on an isolated point, and also to put it on a point of first countability. On the other hand, if there is a nonisolated point to which no nontrivial sequence converges, all the villain has to do is put the jail there and then he does not need to worry about any strategies because the hero cannot possibly win. Let's look at some other cases.

4.1 Lemma. If X is not Fréchet, the villain has a winning strategy in Game 1′.

Proof. Let the villain pick a point j for which there is a set A such that $j \in \bar{A}$, but no sequence from A converges to x. Ever after, the villain need only pick points of A, and this he can always do.

It is also easy to see that the villain has a winning strategy if X is not α_2: the villain picks a non-α_2-point and lists infinitely many sequences converging to it that witness this, then makes sure that on successive moves he hits every one of those sequences at least once. A nontrivial result is that these are the "only" spaces on which the villain has a winning strategy:

4.2. Theorem [Sh, in effect]: The villain has a winning strategy in Game 1′ iff X is not a w-space.

This statement of the theorem uses the definition I gave of "w-space". Originally, 4.2 was the definition of "w-space" rather then a theorem. Also:

4.3. Definition. A W-space is a space in which the hero has a winning strategy in Game 1′. A W-point is a point which, when chosen for j, gives the hero a winning strategy (for either game).

Of course, a space X is a W-space iff each point is a W-point. One can also define "w-points" in the same spirit.

Of course, every first countable space is a W-space; so is the one-point compactification of a discrete space: all the hero has to do is pick an open set on the nth move that excludes all the points p_k the villain has picked up to that point. An interesting fact [Gr_1, 3.5, and 3.6] is that every W-space has the property that separable subspaces are first countable and hence countable subspaces are metrizable. This extends to arbitrary cardinalities: $w(X) \leq |X|$, $\chi(X) \leq d(X)$ [ibid.]. Of course, by 4.1 W-spaces are Fréchet, so that they are v-spaces, and also $\aleph_0$-bisequential. Like $\aleph_0$-bisequential spaces, they have the nice property of being preserved by countable products (also Σ-products, see [A_4] and [Gr_1] respectively). Gruenbage commented in [Gr_1] that he did not know of an example of a w-space that is not a W-space and was surprised to see such a well-behaved (Fréchet and countable subspaces 1st countable) class sandwiched in between them. Galvin subsequently found a space in this class that was not a W-space. The first ZFC example of a countable w-space that is not first countable was given by Isbell and published by Olson [Ol], both of whom thought $2^{\aleph_0} < 2^{\aleph_1}$ was required for it. I observed in 1983 that it could be done "in ZFC" [No_2]. The ZFC portion of Example 3.8 was the first compact w-space with a countable subspace, $(\omega\times\omega) \cup \{\infty\}$, which is not first countable. Related results will appear in [Ny_8]. The λ′ variant of Example 3.4 is the first which is also bisequential.

A strange feature of all these examples and any other that might appear is:

"The villain's revenge". If a countable space is a w-space, but is not first countable, it is impossible to tell in ZFC whether it is a v-space.

This is a corollary of Theorems 1.1. and 1.2. I have not been through either proof yet but Dow has made me see, using some of my own results [Ny_8] how the following gives Theorem 1.1.

4.4. Theorem [Do] Laver forcing preserves ω-splitting families of subsets of ω.

A family S of subsets of ω is called ω-splitting if, given any countable family $\{A_n : n \in \omega\}$ of infinite subsets of ω, there exists $S \in S$ such that $A_n \cap S$ and $A_n \setminus S$ are both infinite for all n.

As for Theorem 1.2, it is established using a variant of Miller's rational forcing.

Game 2. This is like Game 1 except that a subset of X is used as the jail instead of a point, and for the hero to win, the sequence $\langle p_n : n \in \omega \rangle$ must "converge to the jail" in the sense that all but finitely many terms must lie in any given open set containing the jail. (It is sufficient for it to converge to some point in the jail for the hero to win, but not necessary: consider the case where the jail is the interval $[-1,1]$ on the y-axis and the space is the union of the jail with the $\sin \frac{1}{x}$ curve.)

The concept of a W-set is now defined in the obvious way to generalize that of a W-point. Here are some strong results from [Gr_3]:

4.5. Theorem. A compact space X is Corson compact $\Leftrightarrow$ the diagonal is a W-set in X^2 $\Leftrightarrow$ every closed subset of X^2 is a W-set.

4.6. Theorem. A closed subset in a compact space of countable tightness is a W-set if, and only if, its complement is metalindelöf.

4.7. Corollary. Every compact hereditarily metalindelöf space is a W-space.

Hence every countable subset in a compact hereditarily metalindelöf space is metrizable, and we get two of the implications in Diagram 2. Related results are given in Section 5.

In [Gr_4] some fine-tuning of strategies and more games are discussed. Recall:

4.8. Definition. A strategy for a player is stationary if a play depends only on the opponent's preceding move. A Markov strategy is one which depends only on the opponent's last move and the number of the move.

Unless the jail is an isolated point in Game 1 or an open set in Game 2, the villain can foil any stationary strategy by the hero by the simple expedient of landing in jail in odd-numbered moves! For then (s)he can pick any point p in the hero's corresponding open set that is not in the jail and keep playing p in even-numbered moves. But if we alter the game slightly by forbidding the villain to use the jail, then the hero has a stationary winning strategy if the jail is a point of first countability. All (s)he has to do is pick a descending local base and then always pick an open set that excludes the last point played by the villain.

The hero also has a Markov winning strategy in Game 1 if j is a point of first countability. In fact there is what might be called a winning "chronological" strategy, in which the hero only needs to keep track of what the number of the move is.

The situation is different on the one-point compactification of an uncountable discrete space. Even if the villain is barred from the jail, the hero has no stationary winning strategy. For each point p that the villain plays, the hero has to stick to the complement of a finite set $P(p)$ all through the game. Now there is a point q which is excluded from infinitely many $P(p)$, and among these p's there is a p_0 which is outside $P(q)$, so that the villain need only alternate between p_0 and q. The hero does not even have a Markov winning strategy: for $P(p)$ substitute $\bigcup_{n=1}^{\infty} P(p,n)$ where $P(p,n)$ is the finite set whose complement the hero picks if the villain picks p on the nth move. The villain may need to pick different p_{2n}'s (always outside $P(q,2n-1)$) but the hero's strategy is foiled as before.

But the situation is different if we let $\{j\} = J$ in:

The game $G(J,X)$. This is Game 2 with the rules altered so that the villain must play p_n in the intersection of the first n open sets played by the hero. This has the effect of relieving the hero of some bookkeeping. In the one-point compactification, the hero now need only follow up a play of p_n with $X - \{p_n\}$ and thus have a stationary strategy.

It is only with respect to these special strategies that $G(J,X)$ differs from Game 2; the question of whether the hero or villain has a winning strategy, with "perfect information" (of the past, not the future!) is easily seen to be the same in both games. Thus Theorem 4.5 tells us that a compact space X is Corson compact iff the hero has a winning strategy in $G(\Delta,X^2)$. The new wrinkle is:

4.9. Theorem. [Gr_4] A compact space X is Eberlein compact if, and only if, the hero has a Markov winning strategy in $G(\Delta,X^2)$.

The $P(p)$'s and $P(p,n)$'s mentioned earlier suggest turning things inside out, giving us:

The game $G(Y)$. On the nth turn, the hero picks a compact set K_n, and then the villain picks $p_n \notin \bigcup\{K_i : i \leq n\}$. The hero wins if the p_n's have no cluster point in the whole space Y. Here we might say the hero is trying to drive the villain into exile. For instance, if Y is locally compact and noncompact, then a win for the hero is equivalent to a win in the game $G(\{\infty\}, Y + \infty)$: the hero drives the villain out to the point at infinity. The only difference is that in $G(Y)$ the villain cannot land on ∞ during the game. Had we only required $p_n \notin K_n$, this difference could really make itself felt, as we already saw.

But the real point of bringing in $G(Y)$ here is to mention:

4.10. Theorem [Gr_4] If Y is locally compact and noncompact, then:

(a) Y is metacompact if, and only if, the hero has a stationary winning strategy in $G(Y)$.

(b) Y is σ-metacompact if, and only if, the hero has a Markov winning strategy in $G(Y)$.

By the comments in Section 2 on Classification E, it is now easy to formulate necessary and sufficient conditions for a compact space to be hereditarily (σ-)metacompact.

Further uses of $G(Y)$ and a simple variation $G^*(Y)$ are given in [Gr$_4$].

5. Relationships among scattered compact spaces

One characterization of scattered spaces is that in every subspace, the set of relatively isolated points is dense. A majority of the examples in Section 3 were scattered, many of them even Ψ-like. Nevertheless, there is quite a lot of simplification in restricting ourselves to compact scattered (also called dispersed) spaces. They are sequentially compact [Ba$_1$], so from Theorem 1.1 there follows:

5.1. Corollary. A compact scattered space is sequential if, and only if, countably compact subsets are compact.

A pleasantly parallel result is:

5.2. Theorem. A compact scattered space is Fréchet if, and only if, pseudocompact subsets are compact.

Proof. We need only show necessity. Suppose A is a pseudocompact subset of a compact scattered Fréchet space X. If $x \in \overline{A}$, then by the above characterization, there is a sequence of points of A isolated in A and converging to x. These give a discrete sequence of (relatively) open singetons of $A\setminus\{x\}$, which is therefore not pseudocompact. So $x \in A$, which shows A is closed.

Of course, $\Psi + \infty$ is a compact scattered sequential space which is not Fréchet.

First countability is a very strong property in compact scattered spaces: it is equivalent to countability, hence to metrizability. This is easily shown by induction on the scattered height of X, i.e. the least α such that the αth Cantor-Bendixson derivative $X^{(\alpha)}$ is empty. This derivative is defined by induction, with $X^{(0)} = X$ and $X^{(\alpha+1)}$ = the derived set of $X^{(\alpha)}$; if α is a limit ordinal, $X^{(\alpha)} = \cap \{X^{(\beta)}: \beta < \alpha\}$. This can be done for any space, and a standard exercise is that a space is scattered iff $X^{(\alpha)} = \emptyset$ for some α, and that a compact scattered space has a last nonempty derivative $X^{(\gamma)}$ and that $X^{(\gamma)}$ is finite.

The greatest coalescence is in Classifications B and E. Recall:

5.3. Definition. A compact space is strong Eberlein compact if it is embeddable in a σ-product of two-point discrete spaces. A σ-product is a subset of a product consisting of all points which differ in only finitely many coordinates from a fixed point x_0.

An internal characterization is the existence of a point-finite T_0-separating open cover by clopen sets.

5.4. Theorem. The following are equivalent for a space X.

(i) X is strong Eberlein compact.

(ii) X is a compact scattered W-space.

(iii) X is a compact scattered hereditarily metalindelöf space.

(iv) X is a compact scattered hereditarily metacompact space.

(v) X is a scattered Eberlein compact space.

Proof. The equivalence of (i) and (ii) is shown in [Gr_3]. The rest follow quickly from other results in the same paper. (iii) implies (ii) by Corollary 4.7, which is taken from [Gr_3], and (i) and (ii) together give (v), and every Eberlein compact space is hereditarily metalindelöf as we have seen. Finally, (iv) obviously implies (iii), while the converse follows from Theorem 4.6 and:

5.5. Theorem. [Gr_3] In a compact scattered space X, a closed subset H is a W-set if, and only if, $X - H$ is metacompact.

Of course, we are also using the fact (Section 1) that these hereditary covering properties need only be checked for open subspaces. In connection with scattered spaces, there is additional information on these in [Ny_1].

The following problem is reminiscent of Problem 12.

Problem 13. Is it consistent that every compact, scattered space satisfying Property wD hereditarily is bisequential?

I have not had the time to investigate the use of locally countable prototypes of the tangent bundle here. In [Ny_4], connectedness is used to establish "in ZFC" that $T^+ + \infty$ is never bisequential.

If we leave wD out of the picture, we can condense the known relationships between compact scattered sequential spaces into one diagram:

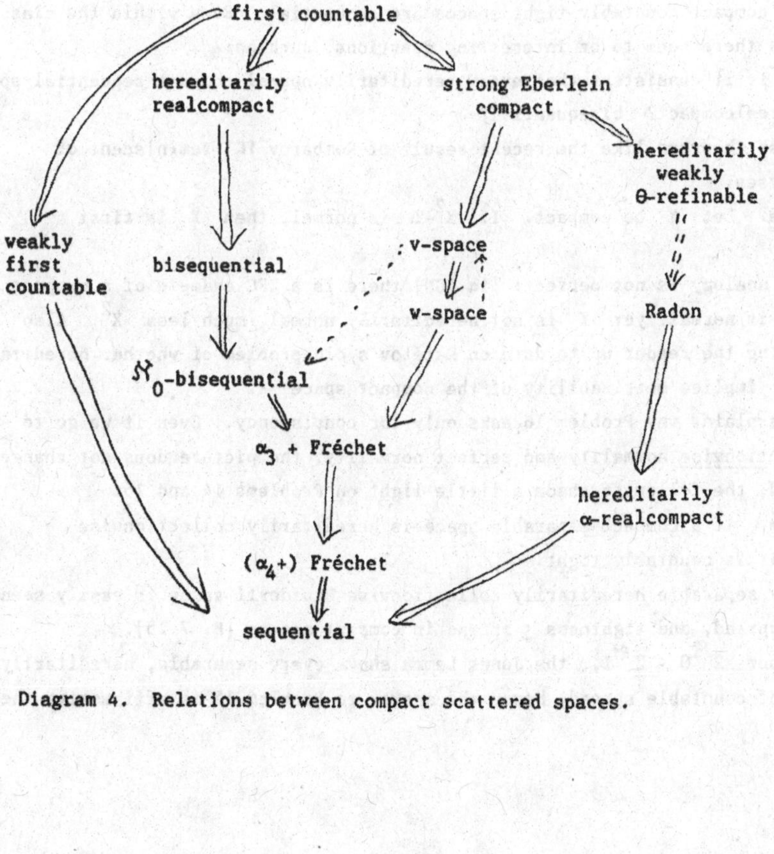

Diagram 4. Relations between compact scattered spaces.

6. A few additional classes

Three unpublished examples bring us up to date on Diagrams 1 through 4. There is a pair of 1981 constructions by D. H. Fremlin of compact spaces that are not Radon, one of which is first countable and the other scattered and hereditarily realcompact. The third space is a very recent (1988) construction by Alan Dow of a compact scattered sequential space which is not hereditarily α-realcompact, nor Fréchet. All are ZFC examples and so, except for the unsolved problems already listed, complete the story of the relationships between the classes.

Of course, we could easily expand the classifications. For instance, the class of Talagrand-compact spaces is known to properly contain the Eberlein compacta, and be properly contained in the Gul'ko compacta. It can be inserted in Diagram 2 without much fuss, because Reznichenko's space (3.6) is Talagrand-compact.

A much more radical expansion was hinted at in defining "hereditarily wD". Hereditary normality is a natural strengthening which has attracted much attention over the years. The following old questions may be ripe for attack:

Problem 14. Is it consistent that every compact hereditarily normal space is sequentially compact?

Problem 15. Is every separable, hereditarily normal, compact space countably tight?

An affirmative answer to Problem 15 would imply one to Problem 14 under PFA, and in all models where compact countably tight spaces are sequential. Even within the class of sequential spaces there seem to be interesting questions, such as:

Problem 16. Is it consistent that every hereditarily normal compact sequential space is hereditarily realcompact? bisequential?

There are also theorems like the recent result of Kombarov [K] reminiscent of Gruenhage's analyses:

6.1. Theorem. Let X be compact. If $X^2-\Delta$ is normal, then X is first countable.

However, the analogy is not perfect: in [GN] there is a ZFC example of a compact X such that $X^2-\Delta$ is normal, yet X is not hereditarily normal, much less X^2. Also there we will bring the reader up to date on Katětov's old problem of whether hereditary normality of X^2 implies metrizability of the compact space X.

Example 3.1 explains why Problem 16 asks only for consistency. Even if we go to hereditary collectionwise normality and perfect normality, the picture does not change. On the other hand, the following sheds a little light on Problems 14 and 15.

6.2. Theorem. If a compact separable space is hereditarily collectionwise Hausdorff, then it is countably tight.

Proof. Every separable hereditarily collectionwise Hausdorff space is easily seen to be of countable spread, and tightness $\leq$ spread in compact spaces [H, 7.15].

Note that under $2^{\aleph_0} < 2^{\aleph_1}$, the Jones Lemma shows every separable, hereditarily normal space is of countable spread, hence the answer to Problem 15 is affirmative there.

And Problem 14 then entangles us in S and L spaces as shown in the Problem Section of Topology Proceedings, v. 2 and also in [Ny_3].

A theme related to the one at the beginning of Section 2 is:

collectionwise normal + ________ => paracompact.

As shown in [Ba_2], "locally compact + metalindelöf" works here, giving the corollary that every compact, hereditarily collectionwise normal, hereditarily metalindelöf space is hereditarily paracompact and, if every discrete subspace is of nonmeasurable cardinal, hereditarily realcompact. This meshes nicely with another, recent result of Balogh: in a model obtained by adding "supercompactly many" Cohen or random reals, every normal, locally compact space is collectionwise normal.

If we go all the way to monotonically normal spaces, then "separable" implies both "hereditarily separable" and "hereditarily Lindelöf" [Os_2]. So Problems 14 and 15 have affirmative answers for them, but Problem 16 still appears to be open for them.

There are other classes which similarly mesh well with compact sequential spaces, but the reader will probably profit more from reading the many fine papers of the Russian school of general topology, than anything I could add at present.

REFERENCES

[A_1] A. V. Arkhangel'skii, The frequency spectrum of a topological space and the classification of spaces, Doklady Akad. Nauk SSSR 206 (1972) 265-268 = Soviet Math. Dokl. 13 (1972).

[A_2] ________, On invariants of character and weight type, Trudy Moskov. Mat. Obs. 38 (1979) = Trans. Moscow Math. Soc. 1980, Issue 2, 1-23.

[A_3] ________, The star method, new classes of spaces, and countable compactness, Soviet Math. Dokl. 21 (1980) 550-554.

[A_4] ________, The frequency spectrum of a topological space and the product operation, Trudy Moskov. Mat. Obs. 40 (1979) = Trans. Moscow Math. Soc. 1981, Issue 2, 163-200.

[AF] ________ and S. P. Franklin, Ordinal Invariants for Topological Spaces Michigan Math. J. 15 (1968) 313-323.

[AL] D. Amir and J. Lindenstrauss, The structure of weakly compact subsets in Banach spaces, Ann. Math. 88 (1968) 35-46.

[Ba_1] J. W. Baker, Ordinal subspaces of topological spaces, Gen. Top. Appl. 3 (1973) 85-91.

[Ba_2] Z. Balogh, Paracompactness in locally nice spaces, preprint.

[BDFN] Z. Balogh, A. Dow, D. H. Fremlin, and P. J. Nyikos, Countable tightness and proper forcing, AMS Bulletin, to appear.

[Be] A. Berner, βX can be Fréchet, AMS Proceedings. 80 (1980) 367-373.

[BL] H. R. Bennett and D. J. Lutzer, A note on weak θ-refinability Gen. Top. Appl. 2 (1972) 49-54.

[Bu] D. Burke, Covering properties, in: Handbook of Set-Theoretic Topology, K. Kunen and J. Vaughan ed., North-Holland, 1984, pp. 347-422.

[vD] E. K. van Douwen, The integers and topology, in: Handbook of Set-Theoretic Topology, K. Kunen and J. Vaughan ed., North-Holland, 1984, pp. 111-167.

[Do] A. Dow, Two classes of Fréchet-Urysohn spaces, preprint.

[Dy] N. Dykes, Generalizations of realcompact spaces, Pac. J. Math. 33 (1970) 571-581.

[E] R. Engelking, General Topology, Polish Scientific Publishers, Warsaw, 1977.

[F] S. P. Franklin, Spaces in which sequences suffice I and II, Fund. Math. 57 (1965) 107-115 and 61 (1967) 51-56.

[Ga_1] R. J. Gardner, The regularity of Borel measures and Borel measure-compactness, Proc. London Math. Soc. 30 (1975) 95-113.

[Ga_2] ________, Corson compact and Radon spaces, Houston J. Math 13 (1987) 37-46.
[Gi] R. F. Gittings, Some results on weak covering conditions, Canad. J. Math. 26 (1974) 1152-1156.
[GJ] L. Gillman and M. Jerison, Rings of Continuous Functions, D. van Nostrand Co., 1960.
[GN] G. Gruenhage and P. Nyikos, Hereditary normality in squares of compact spaces, preprint.
[GP] R. J. Gardner and W. Pfeffer, Borel measures, in: Handbook of Set-Theoretic Topology, K. Kunen and J. Vaughan ed., North-Holland, 1984, pp. 961-1043.
[Gr_1] G. Gruenhage, Infinite games and generalizations of first-countable spaces, Gen. Top. Appl. 6 (1976) 339-352.
[Gr_2] ________, Generalized metric spaces, in: Handbook of Set-Theoretic Topology, K. Kunen and J. Vaughan ed., North-Holland, 1984, pp. 423-501.
[Gr_3] ________, Covering properties in $X^2-\Delta$, W-sets, and compact subsets of Σ-products, Top. Appl. 17 (1984) 287-304.
[Gr_4] ________, Games, covering properties, and Eberlein compacts, Top. Appl. 23 (1986) 207-316.
[Gr_5] ________, A Corson compact space of Todorcevic, Fund. Math.
[Gr_6] ________, A note on Gul'ko compact spaces, AMS Proceedings.
[H] R. Hodel, Cardinal functions I, in: Handbook of Set-Theoretic Topology, K. Kunen and J. Vaughan ed., North-Holland, 1984, pp. 1-61.
[IN] M. Ismail and P. Nyikos, On spaces in which countably compact subsets are closed, and hereditary properties, Top. Appl. 11 (1980) 281-292.
[JKR] I. Juhász, K. Kunen, and M. E. Rudin, Two more hereditarily separable non-Lindelöf spaces, Canad. J. Math. 28 (1976) 998-1005.
[K] Kombarov,
[L] R. Laver, On the consistency of Borel's conjecture, Acta. Math. 137 (1976) 151-169.
[Ma] V. I. Malyhin, On countable spaces having no bicompactification of countable tightness, Dokl. Akad. Nauk. SSSR 203 (1972) 1001-1003 = Soviet Math. Sokl. 13 (1972) 1407-1411.
[Mi_1] E. Michael, A quintuple quotient quest, Gen. Top. Appl. 2 (1972) 91-138.
[Mi_2] A. Miller, Special subsets of the real line, in: Handbook of Set-Theoretic Topology, K. Kunen and J. Vaughan ed., North-Holland, 1984, 201-233.
[No_1] T. Nogura, Fréchetness of inverse limits and products, Top. Appl. 20 (1985) 59-66.
[No_2] ________, The product of $\langle\alpha_i\rangle$-spaces, Top. Appl. 21 (1985) 251-259.
[Ny_1] P. Nyikos, Covering properties on σ-scattered spaces, Top. Proceedings 2 (1977) 509-542.
[Ny_2] ________, Metrizability and the Fréchet-Urysohn property in topological groups, AMS Proceedings 83 (1981) 793-801.
[Ny_3] ________, Axioms, theorems, and problems related to the Jones Lemma, in: General Topology and Modern Analysis, L. F. McAuley and M. M. Rao ed., 441-449.
[Ny_4] ________, Various smoothings of the long line and their tangent bundles I: Basic ZFC results, Advances in Math., to appear.
[Ny_5] ________, Various smoothings of the long line and their tangent bundles II: Countable metacompactness and sections, Advances in Math., to appear.
[Ny_6] ________, Various smoothings of the long line and their tangent bundles III: Consistency and independence results, Advances in Math., to appear.
[Ny_7] ________, The Cantor tree and the Fréchet-Urysohn property, to appear in the proceedings of the Third New York Conference on Limits.
[Ny_8] ________, ${}^{\omega}\omega$ and the Fréchet-Urysohn and α_i-properties, preprint.
[Ol] R. C. Olson, Bi-quotient maps and countably bisequential spaces, Ge. Top. Appl. 4 (1974) 1-28.
[Os_1] A. Ostaszewski, On countably compact, perfectly normal spaces, J. London Math. Soc. (2) 14 (1976) 505-516.
[Os_2] ________, On spaces having a G_δ-diagonal, Coll. Math. Soc. János Bolyai (1978).
[PY] E. G. Pytkeev and N. N. Yakovlev, On bicompacta which are unions of spaces defined by means of coverings, Comment. Math. Univ. Carolinae 21 (1980) 247-261.
[Ro] H. P. Rosenthal, The heredity problem for weakly compactly generated Banach spaces Compositio Math. 28, Fasc. 1 (1974) 83-111.

[Ru_1] M. E. Rudin, Dowker spaces, in: Handbook of Set-Theoretic Topology, K. Kunen and J. Vaughan ed., North-Holland, 1984, pp. 761-780.

[Ru_2] __________, A perfectly normal manifold from *, Top. Appl., to appear.

[Sh] P. L. Sharma, Some characterizations of W-spaces and w-spaces, Gen. Top. Appl. 9 (1978) 289-293.

[Si_1] F. Siwiec, On defining a space by a weak base, Pac. J. Math. 52 (1974) 233-245.

[Si_2] P. Simon, A compact Fréchet space whose square is not Fréchet, Comment. Math. Univ. Carolinae 21 (1980) 749-753.

[So] G. A. Sokoloff, On some classes of compact spaces lying in Σ-products, Comment. Math. Univ. Carolinae 25 (1984).

[SS] L. A. Steen and J. A. Seebach, Counterexamples in Topology, Springer-Verlag, 1978.

[T] M. G. Tkachenko, On compactness of countably compact spaces having additional structure, Trudy Moskov. Matem. Obs. 46 (1983) = Trans. Moscow Math. Soc. 1984, Issue 2, 149-167.

[U] V. V. Uspensky, Pseudocompact spaces with a σ-point-finite base are metrizable, Comment. Math. Univ. Carolinae 25, 2 (1984) 261-264.

[V_1] J. E. Vaughan, Discrete sequences of points, Top. Proceedings 3 (1978) 237-265.

[V_2] __________, Countably compact and sequentially compact spaces, in: Handbook of Set-Theoretic Topology, K. Kunen and J. Vaughan ed., North-Holland, 1984, pp. 569-602.

[Y] N. N. Yakovlev, On the theory of σ-metrizable spaces, Dokl, Akad. Nauk SSSR 229 (1976) 1330-1331 = Soviet Math. Dokl. 17 (1976) 1217-1219.

[Z] P. L. Zenor, P. L. Zenor, Certain subsets of products of Θ-refinable spaces are realcompact, AMS Proceedings 40 (1973) 612-614.

[Zh] H.-X. Zhou, A conjecture on compact Fréchet space, AMS Proceedings 89 (1983) 326-328.

Special Non-Special $\aleph_1$-trees.

Chaz Schlindwein
University of Kansas
Lawrence, KS 66045 USA

In [Sch2], the consistency of "ZFC + continuum hypothesis + Suslin's hypothesis + not Every Aronszajn Tree is Special (¬EATS)" is demonstrated using proper forcing. In fact, the following stronger statement is shown consistent:

(*) ZFC + CH + there is an Aronszajn tree T^* and a stationary co-stationary set S^* such that T^* is S-*st*-special iff $S - S^*$ is non-stationary, and every Aronszajn tree is S^*-*st*-special.

See [Sh, p. 286] for the definition of S-*st*-special. The proof relies heavily on [Sh, chap. V]. Of course the consistency of CH with SH is due to Jensen [DJ]. In this paper, we build a model of (*) + every ω_1-tree is S^*-$*$-special (and therefore there are no Suslin or Kurepa trees) starting from an inaccessible cardinal. Consequently, every combination of ZFC±CH±SH±KH±EATS not ruled out by EATS$\Longrightarrow$SH, is attained. The main new ingredient of the proof, compared to [Sch2], is the use of a method of Todorcevic [To]. Familiarity with [Sch2] is not assumed, but familiarity with [Sh, chaps. III & V] is. We confuse sequences with tuples and structures with their universes. By "tree," we mean "normal tree."

Definition 1. *Suppose T is an ω_1-tree (i.e., height ω_1 and every level countable) and $S \subsetneq \mathrm{Lim} \cap \omega_1$. Then f S-$*$-specializes T iff $\mathrm{dom}(f) = T{\upharpoonright}S = \{x \in T \colon \mathrm{ht}(x) \in S\}$ and $f(x) < \mathrm{ht}(x)$ and whenever $x <_T u$ and $x <_T v$ and $u \not\leq_T v$ and $v \not\leq_T u$ then we do not have $f(x) = f(u) = f(v)$. If such an f exists, T is said to be S-$*$-special.*

It is easy to show that if T is Aronszajn, then T is S-$*$-special iff T is S-*st*-special. Consequently, if T is an ω_1-tree and S is stationary and T is S-$*$-special, then T is not Suslin. Also, under the same hypothesis, T is not Kurepa, for if f S-$*$-specializes T and $G(x)$ is the unique branch of T such that $S_x = \{y \in T \colon y > x \text{ and } f(x) = f(y)\} \subseteq G(x)$ when S_x is uncountable, then by Fodor's theorem each uncountable branch of T contains some uncountable S_x, so G maps onto the set of uncountable branches.

By [Sch2] (essentially proved by [Sh, chap. V & IX]) we have the following result; see [Sh, pp. 154, 164, 178–180, 293] for the relevant definitions:

Theorem 1. *Suppose* ZFC *proves that for every Aronszajn tree T^* and ω_1-tree T and stationary co-stationary $S^* \subseteq \mathrm{Lim} \cap \omega_1$ there is a poset $P = P(T, S^*)$ such that P is $< \omega_1$-proper, is $(\{A \subseteq [\aleph_1]^{\aleph_0}: \omega_1 - S^* \subseteq A\}, \mathcal{D})$-complete for some simple $\aleph_1$-complete completeness system $\mathcal{D}$, is $(\dot{T}^*, S^*)$-preserving, and $1 \Vdash_P$ "$\check{T}$ is $\check{S}^*$-$*$-special." Then* Con(ZFC + κ is inaccessible) *implies* Con((*) + every ω_1-tree is S^*-$*$-special).

We turn our attention to building the poset $P = P(T, S^*)$ as in the hypothesis of theorem 1. We may assume $T, T^* \in H_{\omega_2}$. Let $B = \{b: b \text{ is an uncountable branch of } T\}$; fix $\kappa > \mathrm{card}(B) + \aleph_1$, κ regular. $\mathcal{N}$ is a κ-chain iff $\mathcal{N}$ is an $\in$-chain of countable elementary substructures of H_κ (not necessarily continuous); $L_{\mathcal{N}} = \bigcup\{\mathcal{N}(\gamma): \gamma \in \mathrm{dom}(\mathcal{N})\}$.

Preliminary definition. $Q = \{\langle f, S, \mathcal{N}\rangle: S \subseteq \mathrm{Lim} \cap \omega_1,\ |S| \leq \aleph_0,\ f\ S\text{-}*\text{-specializes } T,\ \mathcal{N} \text{ is a } \kappa\text{-chain, and } \omega_1 \cap \mathcal{N}(\alpha) \in S \text{ whenever } \alpha \in \mathrm{dom}(\mathcal{N})\}$, ordered by $\langle f, S, \mathcal{N}\rangle \leq \langle f', S', \mathcal{N}'\rangle$ iff S end-extends S', $f \supseteq f'$, and $\mathcal{N}$ end-extends $\mathcal{N}'$. $T^n = \{\langle x_0, \ldots, x_{n-1}\rangle: (\exists \alpha)(\forall i)(x_i \in T_\alpha)\}$, ordered by $\bar{x} \leq \bar{y}$ iff $(\forall i)(x_i \leq y_i)$. Γ is a *promise* iff there is $n = n(\Gamma) \in \omega$ and $C = C(\Gamma) \subseteq \omega_1$ closed unbounded and $\bar{x} = \min(\Gamma) \in T^n_{o(\Gamma)}$ (where $o(\Gamma) = \bigcap C$) and $G = G(\Gamma) \subseteq n$ and $b_t = b_t(\Gamma) \in B$ $(t \in G)$, such that $\bar{x} \in \Gamma \subseteq T^n \restriction C$ and $\bar{y} \in \Gamma \implies \bar{y} \geq \bar{x}$ and if $\alpha < \beta$ are in C and $\bar{y} \in \Gamma \cap T^n_\alpha$ then there is $W \subseteq \Gamma \cap T^n_\beta$ such that $\bar{w} \in W \implies \bar{w} > \bar{y}$, and if $\bar{w}$ and $\bar{w}'$ are distinct elements of W then $\{\bar{w}(t): t \in n - G\}$ is disjoint from $\{\bar{w}'(t): t \in n - G\}$, and for all $\bar{y} \in \Gamma$ we have $\bar{y}(t) \in b_t$ for $t \in G$, and W is infinite unless $G = n$. $\bar{A}$ is a *finite rectangle* if $\bar{A}$ is a finite sequence with $\bar{A}(i) \subseteq \omega_1$ finite for $i \in \mathrm{dom}(\bar{A})$. $\heartsuit(\bar{w}, f, \bar{A})$ iff $(\forall i < \mathrm{lh}(\bar{w}) = \mathrm{lh}(\bar{A}))(\forall y \leq \bar{w}(i))(f(y) \notin \bar{A}(i))$. $\langle f, S\rangle$ *fulfills* Γ iff whenever $\bar{y} \in \Gamma \cap T^{<\omega}_\alpha$ and $\beta > \alpha$ and $\beta \in C(\Gamma)$ and $\bar{A}$ a finite rectangle, then there is $W \subseteq \Gamma \cap T^{<\omega}_\beta$ such that W is infinite unless $G = n$ and if $\bar{w}$ and $\bar{w}'$ are distinct elements of W then $\{\bar{w}(t): t \in n(\Gamma) - G(\Gamma)\} \cap \{\bar{w}'(t): t \in n(\Gamma) - G(\Gamma)\} = \emptyset$ and, for all $\bar{w} \in W$, $\bar{w} \geq \bar{y}$ and $(\heartsuit(\bar{y}, f, \bar{A}) \implies \heartsuit(\bar{w}, f, \bar{A}))$. For b an uncountable branch of T, we let $\heartsuit(b, f, F)$ iff $(\forall x \in b)\heartsuit(x, f, F)$.

Main Definition. $\langle f, S, \mathcal{N}, \Psi\rangle \in P$ *iff:* $\langle f, S, \mathcal{N}\rangle \in Q$ and Ψ is a countable set of promises that $\langle f, S\rangle$ fulfills, and if $\{\omega_1 \cap \mathcal{N}(\beta): \beta < \alpha\}$ is unbounded in $\gamma \in S^*$ then $\alpha \in \mathrm{dom}(\mathcal{N})$ and $\mathcal{N}(\alpha) = \bigcup_{\beta<\alpha} \mathcal{N}(\beta)$, and $T \in \mathcal{N}(0)$ (unless $\mathcal{N} = \emptyset$), and, for all $\alpha \in \mathrm{dom}(\mathcal{N})$, if $x \in \mathrm{dom}(f) - \mathcal{N}(\alpha)$, then the following are equivalent:

i) $(\exists y < x)(y \in \mathrm{dom}(f) \cap \mathcal{N}(\alpha) \text{ and } f(x) = f(y))$

ii) $(\exists b \in B \cap \mathcal{N}(\alpha))(x \in b)$.

For $p \in P$, we give f_p, S_p, $\mathcal{N}_p$, Ψ_p, and $\mathrm{ht}(p)$ their obvious meanings; we let $L_p = L_{\mathcal{N}_p}$ and

$\rho_p(x) =$ (least γ such that $x \in \mathcal{N}(\gamma)$) for $x \in L_p$, and for $b \in B \cap L_p$ we set $\delta_p(b) = \omega_1 \cap \mathcal{N}_p(\rho_p(b))$ and $\mu_p(b) =$ (unique $x \in b$ such that $\mathrm{ht}(x) = \delta_p(b)$) and $\sigma_p(b) = f_p(\mu_p(b))$ and $U_N = \bigcup(B \cap N)$ and $U_p = U_{L_p}$. We order P by $p \le q$ iff $\langle f_p, S_p, \mathcal{N}_p\rangle \le \langle f_q, S_q, \mathcal{N}_q\rangle$ and $\Psi_p \supseteq \Psi_q$ and $S_p - S_q \subseteq C(\Gamma)$ for all $\Gamma \in \Psi_q$.

This definition should be compared with the Main Definition of [Sh, p. 184] and with the poset of Todorcevic [To] for specializing ω_1-trees with finite conditions.

Notice that if $b \in B \cap L_p$ and $\heartsuit(b, f_p, F)$ and $q \le p$, then $\heartsuit(b, f_q, F)$.

Definition 4. *$(M, p, \delta, n, \bar{x}, \bar{A})$ is as usual iff $M \prec H_\lambda$ (λ large enough, regular); $|M| = \aleph_0$; P, $T, \ldots \in M$, $\delta = \omega_1 \cap M$, $\bar{x} \in T^n_\delta$; $p \in P \cap M$; $\bar{A}(i) \subseteq \omega_1$ is finite for $i < n$; and $\heartsuit(\bar{x}, f_p, \bar{A})$.*

Lemma 1. *Suppose $(M, p, \delta, n, \bar{x}, \bar{A})$ is as usual and $\theta \in H_\kappa \cap M$. Then there is $q \le p$ with $\theta \in L_q$ and $(M, q, \delta, n, \bar{x}, \bar{A})$ is as usual.*

Proof: First take $N \prec H_\kappa$ countable with $N \in M$ and p, θ, T, $\bar{A} \cap \delta^n$ in N. Build $q \le p$ with $\Psi_q = \Psi_p$ and $\mathcal{N}_q = \mathcal{N}_p{}^\frown\langle N\rangle$ as follows. First set $\delta_\omega = \omega_1 \cap N$ and then choose $\delta_m \nearrow \delta_\omega$ with sufficient care: namely, for every $\Gamma \in \Psi_p$ and every $\bar{y} \in \Gamma \cap T^{<\omega}_{\delta_m}$, there is an infinite $W \subseteq \Gamma \cap T^{<\omega}_{\delta_{m+1}}$ (or $W = \Gamma \cap T^{<\omega}_{\delta_{m+1}}$ if this is a singleton) such that whenever $\bar{w}$ and $\bar{w}'$ are distinct elements of W then $\{\bar{w}(t) : t \in n(\Gamma) - G(\Gamma)\}$ is disjoint from $\{\bar{w}'(t) : t \in n(\Gamma) - G(\Gamma)\}$, and also disjoint from U_p. Then list all $\langle \bar{y}^k, \Gamma_k, \bar{A}^*_k, m_k\rangle$ with $\Gamma_k \in \Psi_p$ and $\bar{y}^k \in \Gamma_k \cap T^{<\omega} \cap N$ and $\bar{A}^*_k \subseteq (\delta_\omega)^{<\omega}$ a finite rectangle and $m_k \le \omega$, with infinitely many repetitions, and so in order type ω one builds the required f_q (with $S_q = S_p \cup \{\delta_m : m \le \omega\}$) by defining, at stage k, f_q appropriately on some $\bar{z}^k > \bar{y}^k$ with $\mathrm{ht}(\bar{z}^k) = \delta_{m_k}$ with $\bar{z}^n(t) \notin U_p$ for $t \in n(\Gamma_k) - G(\Gamma_k)$ and each such $\bar{z}^k(t)$ incomparable with all previously considered notes from T (doing nothing if $\mathrm{ht}(\bar{y}^k) \ge \delta_{m_k}$). Also we have a listing of $T \restriction S_q$ in order-type ω which we use to ensure that $\mathrm{dom}(f_q) = T \restriction S_q$. The care in choosing δ_m for $m < \omega$ is needed because we are constrained to take $f_q(x) = \sigma_p(b)$ for $x \in b \in B \cap L_p$ (assuming $x \notin L_p$). Aside from that quibble, the construction is simply an elaborated version of Fact 6.6 [Sh, p. 184], and there is no problem (although we must distinguish between $x \in U_N$ and $x \notin U_N$ when $x \in T_{\delta_\omega}$; in the former case we must take $y < x$ not yet considered, with $y \in T \restriction \{\delta_m : m \in \omega\} - U_p$, and set $f_q(y) = f_q(x)$, at the time we consider x).

Definition 5. *We say that H is M-good iff $H \subseteq P \cap M$ and whenever $(M, p, \delta, n, \bar{x}, \bar{A})$ is as usual and $D \subseteq P$ is open dense and $D \in M$, then there is $q \le p$ such that $q \in H \cap D$ and $\heartsuit(\bar{x}, f_q, \bar{A})$.*

Lemma 2. *Suppose $M \prec H_\lambda$ (large enough λ, regular), $|M| = \aleph_0$, $P \in M$. Then $P \cap M$ is M-good.*

Proof: Let $(M, p, \delta, n, \bar{x}, \bar{A})$ and D be a counterexample. We may assume $\bar{A} \subseteq \delta^n$ hence $\bar{A} \in M$. Let $\alpha = \mathrm{ht}(p)$. Let $\bar{z} < \bar{x}$ have height α and let $\Delta = \{\bar{y} \in T^n : \bar{y}$ is comparable to $\bar{z}$ and there is no $q \leq p$ such that $q \in D$ and $\mathrm{ht}(q) \leq \mathrm{ht}(\bar{y})$ and $\heartsuit(\bar{y}, f_q, \bar{A})\}$. Notice that Δ is downward closed (i.e., $\bar{y} \leq \bar{y}' \in \Delta \Longrightarrow \bar{y} \in \Delta$). Every $\bar{y} < \bar{x}$ is in Δ, for otherwise the q witnessing $\bar{y} \notin \Delta$ could be taken to be in M, which would contradict the assumption that we are dealing with a counterexample to the lemma. Therefore $M \models$ "Δ is uncountable."

Claim. Suppose $\Delta' \subseteq T^k$ (some finite k) is uncountable and downward closed and every element of Δ' is comparable to $\bar{z}'$. Then there is a promise $\Gamma' \subseteq \Delta'$ with $\min(\Gamma') = \bar{z}'$.

Proof of claim: We build $G^* \subseteq k$ and uncountable branches $\langle b_t : t \in G^* \rangle$, and an uncountable downward closed Δ^* consisting of sequences $\langle \bar{w}^*(i) : i \in k - G^* \rangle$ such that for each $\bar{w}^* \in \Delta^*$, the unique $\bar{w} \in T^k_{\mathrm{ht}(\bar{w}^*)}$ extending $\bar{w}^*$ with $\bar{w}(t) \in b_t$ for $t \in G^*$ is a member of Δ', and $\{\bar{w}^*(i) : \bar{w}^* \in \Delta^*$ and $i \in k - G^*\}$ is Aronszajn or empty. The procedure is as follows: start with $G^* = \emptyset$ and $\Delta^* = \Delta'$. If $\{\bar{y}(t) : \bar{y} \in \Delta^*$ and $t \in \mathrm{dom}(\bar{y})\}$ is Aronszajn, we are done; otherwise, we take b an uncountable branch. Fix t such that $b \cap \{\bar{y}(t) : \bar{y} \in \Delta^*\}$ is uncountable. Since Δ^* is downward closed, $b \subseteq \{\bar{y}(t) : \bar{y} \in \Delta^*\}$. Let $b_t = b$ and put t into G^* and take the new Δ^* to be $\{\langle \bar{y}(t') : t' \in k - G^* \rangle$: the unique $\bar{y}'$ extending $\langle \bar{y}(t') : t' \in k - G^* \rangle$ with $\mathrm{dom}(\bar{y}') = k - G^* \cup \{t\}$ is in the old $\Delta^*\}$. Continue this procedure until we finish (i.e., we reach an Aronszajn tree or $G^* = k$). Let $\bar{z}^*$ be the restriction of $\bar{z}$ to $k - G^*$. Now, in the case $G^* \neq k$, the fact that Δ^* is a downward closed, uncountable subset of some Aronszajn $(T^*)^{k'}$ (some k'), and every element of Δ^* is comparable with $\bar{z}^*$, allows us to assert the existence of a promise $\Gamma^* \subseteq \Delta^*$ with $\min(\Gamma^*) = \bar{z}^*$ and $G(\Gamma^*) = \emptyset$ (see [Sh, pp. 188–189]). Let Γ' consist of all $\bar{w} \in T^k$ such that $\bar{w}$ extends some $\bar{w}' \in \Gamma^*$ and $\bar{w}(t) \in b_t$ for $t \in G^*$. Then Γ' is the required promise. QED claim.

Take $\Gamma \subseteq \Delta$ a promise with $\min(\Gamma) = \bar{z}$. Let $p' = \langle f_p, S_p, \mathcal{N}_p, \Psi_p \cup \{\Gamma\} \rangle$. Now suppose $r \leq p'$. We may take $\bar{w} \in \Gamma$ with $\mathrm{ht}(\bar{w}) = \mathrm{ht}(r)$ and $\heartsuit(\bar{w}, f_r, \bar{A})$. Now, since $\bar{w} \in \Delta$, there is no $q \leq p$ such that $q \in D$ and $\mathrm{ht}(q) \leq \mathrm{ht}(\bar{w})$ and $\heartsuit(\bar{w}, f_q, \bar{A})$. In particular, $r \notin D$. But this contradicts the fact that D is dense. QED lemma.

Lemma 3. *Suppose $(M, p, \delta, n, \bar{x}, \bar{A})$ is as usual. Then there is $q \leq p$ such that q is M-generic and $\mathrm{ht}(q) = \delta \in S_q$ and $\heartsuit(\bar{x}, f_q, \bar{A})$.*

Proof: Let $\langle D_m : m \in \omega \rangle$ enumerate the dense open sets of P in M, $\langle \theta_m : m \in \omega \rangle$ enumerates $N = H_\kappa \cap M$, and $\langle \bar{y}^m, \Gamma_m, \bar{A}^*_m \rangle$ enumerates all triples in M with Γ_m a promise and $\bar{y}^m \in \Gamma_m \cap T^{<\omega} \cap N$ and $\bar{A}^*_m \subseteq \delta^{<\omega}$ a finite rectangle with $\mathrm{lh}(\bar{A}^*_m) = n(\Gamma_m)$, with each triple listed infinitely often. We build $p \geq p_0 \geq p_1 \geq \ldots$ with $p_{2n} \in D_n$ and $\theta_n \in L_{p_{2n+1}}$ and $p_n \in M$;

simultaneously we construct $g: T_\delta \mapsto \delta$ such that in the end $q = \langle \bigcup f_{p_n} \cup g, \bigcup S_{p_n} \cup \{\delta\}, \bigcup \mathcal{N}_{p_n}{}^\frown \langle N \rangle, \bigcup \Psi_{p_n} \rangle \in P$ and $\heartsuit(\bar{x}, f_q, \bar{A})$. Because of lemmas 1 and 2, there is essentially no problem in doing this. Notice that at stage n we may assign finite sets F_i to finitely many nodes x_i in T_δ and demand that $g(x_i) \notin F_i$ and $f_{p_m}(y) \notin F_i$ for all $y < x_i$ and all $m \in \omega$ (if $x_i \in U_{p_n}$ then we can only do this if $\heartsuit(x_i, f_{p_n}, F_i)$); this is how we take steps to ensure that each Γ which appears in some Ψ_{p_k} will be fulfilled by $\langle f_q, S_q \rangle$ at level δ. Also, when we put $x \in T_\delta$ into dom(g), we check whether $x \in U_N$; if so, we immediately strengthen our current p_n to $p'_n \in M$ such that the unique $b \in B \cap N$ with $x \in b$ will be in $L_{p'_n}$ and we set $g(x) = \sigma_{p'_n}(b)$; if not, i.e., $x \notin U_N$, take $g(x) < \delta$ unequal to $f_{p_n}(y)$ for all $y < x$ and demand that $f_{p_m}(y) \neq g(x)$ for all $y < x$ and all $m \in \omega$. This completes the construction.

Lemma 4. $\mathbf{1} \Vdash$ "$\check{T}$ is $\check{S}^*$-$*$-special."

Proof: Let $\dot{S}$ be the name for $\bigcup\{\omega_1 \cap \mathcal{N}_p(\alpha): p \in \dot{G}$ and $\alpha \in \text{dom}(\mathcal{N}_p)\}$. Notice $\mathbf{1} \Vdash$ "$\check{T}$ is $\dot{S}$-$*$-special." Let $\dot{C}$ be such that $\mathbf{1} \Vdash$ "$\dot{C} = \{\alpha < \check{\omega}_1: \sup(\dot{S} \cap \alpha) = \alpha\}$." Thus $\mathbf{1} \Vdash$ "$\dot{C}$ is closed unbounded." Suppose $p \Vdash$ "$\dot{\alpha} \in \dot{C} \cap \check{S}^*$." Take $q \leq p$, $\beta < \omega_1$ with $q \Vdash$ "$\dot{\alpha} = \check{\beta}$." Since $q \Vdash$ "$\check{\beta} \in \dot{C}$," we have $\sup(\{\omega_1 \cap \mathcal{N}_q(\alpha): \alpha \in \text{dom}(\mathcal{N}_q)\} \cap \beta) = \beta \in S^*$. By the Main Definition, $\omega_1 \cap \mathcal{N}_q(\gamma) = \beta$ for some γ. Thus $q \Vdash$ "$\check{\beta} \in \dot{S}$." We conclude $\mathbf{1} \Vdash$ "$\dot{C} \cap \check{S}^* \subseteq \dot{S}$." This establishes the lemma.

Lemma 5. *P is $(E, \mathcal{D})$-complete for some simple $\aleph_1$-complete completeness system $\mathcal{D}$, where $E = \{A \subseteq [\aleph_1]^{\aleph_0}: \omega_1 - S^* \subseteq A\}$.*

Proof: First note that lemma 3 holds with T_δ replaced by any countable collection of branches of $T_{<\delta}$ by the same argument. Then an especially tedious coding argument gives the completeness system (and the first-order formula defining it); see [Sh, p. 250, lines 8–13] for a thumbnail sketch.

Lemma 6. *P is $<\omega_1$-proper.*

The proof of this lemma is identical to the proof of lemma 13 in [Sch2], so we give only an outline. Take $\langle N_\alpha: \alpha \leq \rho \rangle$ a countable tower of countable elementary substructures of H_λ; we prove by induction that $H \cap N_{\beta+1}$ is $N_{\beta+1}$-good, where $H = \{p \in P: (\forall \xi)(p \notin N_\xi \implies p \text{ is } N_\xi\text{-generic})\}$. For successor β, for $p \in H \cap N_{\beta+1}$, by induction we may assume $p \notin N_\beta$ and by lemma 2 we are done. For limit β, repeat the proof of lemma 3 using the induction hypothesis rather than the M-goodness of $M = N_\beta$; we thereby see that again we may assume $p \notin N_\beta$, and again we finish by lemma 2.

The following lemma corresponds to lemma 12 of [Sch2].

Lemma 7. *Suppose $M \prec H_\lambda$ (large regular λ), $|M| = \aleph_0$, $P, T \in M$, $p \in P \cap M$, and $\delta = \omega_1 \cap M \notin S^*$. Then there is q which is preserving for (p, P, T^*, S^*) (see [Sh, p. 293] for the definition), and hence P is (T^*, S^*)-preserving.*

Proof: Let D_m, $\bar{y}^m$, Γ_m, $\bar{A}^*_m$ $(m \in \omega)$ be as in the proof of lemma 3, and let $\langle x_m, \dot{A}_m\rangle$ enumerate all pairs $\langle x, \dot{A}\rangle$ such that $\dot{A} \in M$ is a P-name and $x \in T_\delta$ and $(\forall X \in M)(x \in X \Longrightarrow (\exists y < x)(y \in X))$. We build $p \geq p_0 \geq p_1 \geq \ldots$ such that $p_i \in M$ and $p_{2n+1} \in D_n$ and $\{p_m : m \in \omega\}$ is bounded below by some $q \in P$, and in choosing p_{2n+2} we either have $p_{2n+2} \Vdash$ "$\check{y} \in \dot{A}_n$" for some $y < x_n$, or we have taken certain steps, described below, towards ensuring $q \Vdash$ "$\check{x}_n \notin \dot{A}_n$." Unlike the situation in lemma 3, we need not have f_q defined on T_δ; however, we still must take steps to ensure that each $\Gamma \in \bigcup_{m\in\omega} \Psi_{p_m}$ is fulfilled at level δ. Thus, at stage n, we assign finitely many finite sets $F_0, \ldots F_{m_n} \subseteq \delta$ to nodes $x_0, \ldots, x_{m_n}$ in T_δ and declare that $f_{p_k}(y) \notin F_i$ for all $y < x_i$ and all $k < \omega$. We do this according to our ordering of $\langle \bar{y}^m, \Gamma_m, \bar{A}^*_m\rangle$ (doing nothing, of course, if $\Gamma \notin \Psi_{p_n}$ or $\mathrm{ht}(\bar{y}^n) > \mathrm{ht}(p_n)$ or $\neg\heartsuit(\bar{y}^n, f_{p_n}, \bar{A}^*_n)$) thereby causing each promise in $\bigcup_{n\in\omega} \Psi_{p_n}$ to be fulfilled at level δ. The restrictions thus imposed do not hinder us from choosing $p_{2n+1} \in D_n$ by lemma 3; however, they introduce a further complication towards choosing p_{2n+2}. What we require is the following claim:

Claim: Suppose $\overline{w} \in T^m_\delta$ and $\bar{A} \subseteq \delta^m$ a finite rectangle and $\heartsuit(\overline{w}, f_{p_{2n+1}}, \bar{A})$, and $\overline{w}^\# < \overline{w}$ and $\mathrm{ht}(\overline{w}^\#) = \mathrm{ht}(p_{2n+1})$. Then either:

(*) $(\exists q \leq p_{2n+1})(\exists y < x_n)(\mathrm{ht}(q) < \delta$ and $\heartsuit(\overline{w}, f_q, \bar{A})$ and $q \Vdash$ "$\check{y} \in \dot{A}_n$"); or

(**) $x_n \in Y = \{y \in T^* : (\exists \overline{w}^* \in T^m_{\mathrm{ht}(y)})(\overline{w}^*$ is comparable to $\overline{w}^\#$ and $(\forall q \leq p_{2n+1})$(if $\mathrm{ht}(q) = \mathrm{ht}(y)$ and $\heartsuit(\overline{w}^*, f_q, \bar{A})$ then there is a promise Γ such that $o(\Gamma) = \mathrm{ht}(q)$ and $\langle f_q, S_q, \mathcal{N}_q, \Psi_q \cup \{\Gamma\}\rangle \Vdash$ "$\check{y} \notin \dot{A}_n$")$\}$.

We show that the claim implies the lemma. Suppose (*) holds and $q \leq p_{2n+1}$ and $y < x_n$ witness (*). Then certainly $(\exists q' \leq p_{2n+1})(\mathrm{ht}(q') = \mathrm{ht}(q)$ and $\heartsuit(\overline{w}', f_{q'}, \bar{A})$ where $\overline{w}' \leq \overline{w}$ and $\mathrm{ht}(\overline{w}') = \mathrm{ht}(q')$ and $q' \Vdash$ "$\check{y} \in \dot{A}_n$"); and since $\mathrm{ht}(q')$ and $\overline{w}'$ are in M, we may take p_{2n+2} to be such a q' in M. If instead (**) holds, we set $p_{2n+2} = p_{2n+1}$ and we take $\overline{w}^*$ to be a witness to (**) and demand that $f_{p_k}(y) \notin \bar{A}(i)$ whenever $y < \overline{w}^*(i)$ and $k \in \omega$. Having done so, we know by (**) that there is Γ_n such that $\langle \bigcup_{k\in\omega} f_{p_k}, \bigcup_{k\in\omega} S_{p_k}, \bigcup_{k\in\omega} \mathcal{N}_{p_k}, \bigcup_{k\in\omega} \Psi_{p_k} \cup \{\Gamma_n\}\rangle \Vdash$ "$\check{x}_n \notin \dot{A}_n$;" thus, taking $q = \langle \bigcup f_{p_k}, \bigcup S_{p_k}, \bigcup \mathcal{N}_{p_k}, \bigcup \Psi_{p_k} \cup \{\Gamma_n$: (**) holds at stage $2n+1\}\rangle$, we satisfy the demands of the lemma. Thus, it suffices to prove the claim.

Proof of claim: Suppose (*) fails and $y < x_n$. We show that $y \in Y$; by choice of x_n (i.e., $(\forall X \in N)(x_n \in X \Longrightarrow (\exists y < x_n) y \in X)$) this suffices. In fact, we show that the unique $\overline{w}^* \leq \overline{w}$

with $\text{ht}(\overline{w}^*) = \text{ht}(y)$ witnesses $y \in Y$. For suppose $q^* \leq p_{2n+1}$ and $\text{ht}(q^*) = \text{ht}(y)$ and $\heartsuit(\overline{w}^*, f_{q^*}, \bar{A})$ but there is no Γ with $o(\Gamma) = \text{ht}(q^*)$ and $\langle f_{q^*}, S_{q^*}, \mathcal{N}_{q^*}, \Psi_{q^*} \cup \{\Gamma\}\rangle \Vdash$ "$\check{y} \notin \dot{A}_n$"}. We may assume that $q^* \in M$. Let $\Delta = \{\overline{w}' \in T^m : \overline{w}'$ is comparable with $\overline{w}^*$ and $(\forall q^+ \leq q^*)$(if $\text{ht}(q^+) < \text{ht}(\overline{w}')$ and $\heartsuit(\overline{w}', f_{q^+}, \bar{A})$ then $q^+ \not\Vdash$ "$\check{y} \in \dot{A}_n$")}. Notice that every $\overline{w}' < \overline{w}$ is in Δ, hence $M \models$ "Δ is uncountable." Also, Δ is clearly downward closed. Let $\Gamma \subseteq \Delta$ be a promise with $\min(\Gamma) = \overline{w}^*$. I claim that $q^+ = \langle f_{q^*}, S_{q^*}, \mathcal{N}_{q^*}, \Psi_{q^*} \cup \{\Gamma^*\}\rangle \Vdash$ "$\check{y} \notin \dot{A}_n$," which is the desired contradiction. Suppose that $q^+ \not\Vdash$ "$\check{y} \notin \dot{A}_n$." Take $q' \leq q^+$ such that $q' \Vdash \check{y} \in \dot{A}_n$." Since $\langle f_{q'}, S_{q'}\rangle$ fulfills Γ, we may take $\overline{w}' \in \Gamma$ with $\text{ht}(\overline{w}') > \text{ht}(q')$ such that $\heartsuit(\overline{w}', f_{q'}, \bar{A})$. Since $\overline{w}' \in \Delta$, there is no $r \leq q^*$ with $\text{ht}(r) < \text{ht}(\overline{w}')$ and $\heartsuit(\overline{w}', f_r, \bar{A})$ and $r \Vdash$ "$\check{y} \in \dot{A}_n$;" but q' witnesses the opposite, a contradiction. Thus $q^+ \Vdash$ "$\check{y} \notin \dot{A}_n$" and we are done.

We have proved:

Theorem 2. Con(ZFC + κ is inaccessible) *iff* Con(ZFC + CH + there is an Aronszajn tree T^* and a stationary co-stationary S^* such that T^* is S-$*$-special iff $S - S^*$ is non-stationary, and every ω_1-tree is S^*-$*$-special).

References

[Ba] Baumgartner, James, "Iterated Forcing," in *Surveys in Set Theory*, A. R. D. Mathias (ed.), London Mathematical Society Lecture Note Series, vol. 87, Cambridge University Press, Cambridge, 1983.

[DJ] Devlin, K. J., and H. Johnsbråten, *The Souslin Problem*, Lecture Notes in Mathematics, vol. 405, Springer-Verlag, New York, 1974.

[Sch1] Schlindwein, Chaz, *Club Sandwich Forcing*, PhD. thesis, University of California, Berkeley, 19xx.

[Sch2] Schlindwein, C., "Proper Forcing, Aronszajn Trees and the Continuum," 19xx.

[Sh] Shelah, Saharon, *Proper Forcing*, Lecture Notes in Mathematics, vol. 940, Springer-Verlag, Berlin, 1982.

[To] Todorcevic, Stevo, "A Note on the Proper Forcing Axiom," in *Axiomatic Set Theory*, J. Baumgartner, D. A. Martin, S. Shelah (eds.), Contemporary Mathematics, vol. 31, Amer. Math. Soc., Providence, 1984.

Consistency of positive partition theorems for graphs and models

by

Saharon Shelah

Institute of Mathematics
The Hebrew University
Jerusalem, Israel

Department of Mathematics
Rutgers University
New Brunswick N.J. U.S.A.

Recently A. Hajnal, P. Komjath [1] have dealt with the partition relation $H \to (G)^2_\sigma$: if we colour the edges of a graph H by σ colours, there is an induced subgraph isomorphic to G which is monochromatic (i.e. all edges get the same colour). They prove (generalizing a proof from Shelah [2]) that it is consistent (with ZFC) that there is a graph G of cardinality $\aleph_1$ such that for no graph $H : H \to (G)^2_2$.

They ask whether the negation is consistent. We give here an affirmative answer (even for much stronger partition relations). We first prove it using a class of measurable cardinals (in §1, §2). In §3, §4 we eliminate this.

We can also generalize $M \to (N)^{\langle\beta(*)}_\theta$ to $M \to [N]^{\langle\beta(*)}_{\theta,\theta_1}$ (we get an isomorphism of N in which only $\leq \theta_1$ colours occurs). Our positive independence result ([3],[4]) like $2^{\aleph_0} \to [\aleph_1]^2_{n,2}$ are generalized naturally. This will be discussed elsewhere. Later are given generalizations with finite conclusion, but infinite number of colours ; and we improve the bounds for $\to_{wsp}$.

This research was partially supported by the United States Israel Binational Science Foundation (Number 289). We thank Juris Steprans for handling the paper.

§ 1. The consistency of the partition theorem from a measurable cardinal.

1.1 Notation: K^α_σ (for $\alpha \le \omega$, σ a cardinal) is the class of triples $M = (A,<,F)$ where $<$ is a well ordering of the nonempty set A and $F : [A]^{<\alpha} \to \sigma$ such that $F(\emptyset) = 0$. We let $[A]^n = \{u : u \subseteq A, |u| = n\}$, $[A]^{<\alpha} = \bigcup\{A^{[n]} : n < \alpha\}$, and do not distinguish strictly between F and $\langle F^{[n]} : n < \alpha\rangle$. If A is a set of ordinals, $<$ will be the usual order and we omit it.

We write $M = (|M|, <^M, F^M)$, and use M for $|M|$ sometimes. Now $f : M \to N$ is an (K^α_σ)-embedding if $(\forall x,y \in M)[x <^M y \Leftrightarrow f(x) <^N f(y)]$ and $(\forall u \in [|M|]^{<\alpha})[F^M(u) = F^N(f''(u))]$ and we write $M \subseteq N$ (M a submodel of N) if the identity is an embedding.

1.1A Explanation: We are thinking of M as a model, $F^M(u)$ as the quantifier free type of u, more formally, if $u = \{a_0,\dots,a_{n-1}\}$, $a_0<\dots<a_{n-1}$ we call

$$\bigwedge_{v \subseteq u} F(\dots,x_i,\dots)_{i\in v} = F(\dots,a_i,\dots)_{i\in v} \wedge x_0<\dots<x_{n-1}$$

the quantifier free type tp_{qf} of u (this notation happens when we do not have a fixed model).

Below in 1.2, d is thought of as a coloring of M.

1.2 Definition: For $M,N \in K^\alpha_\sigma$, $\beta \le \omega$ and cardinal θ, $M \to (N)^{<\beta}_\theta$ if: for every function $d : [M]^{<\beta} \to \theta$, there is an embedding f of N into M such that for every non empty $u \in [N]^{<\beta}$, $d(f''(u))$ depend only on the quantifier free type of u in N.

The arrow $\to_{sp}$ is defined in 2.1. First we define the weaker notion $\to^{eh}$ where eh stands for end homogeneity.

1.3 Definition: For $M,N \in K^\alpha_\sigma$, $\beta < \omega$ and cardinal θ (eh stands for end homogeneous) $M \to^{eh} (N)^{<\beta}_\theta$ if: for every function $d : [M]^{<\beta} \to \theta$, there is an embedding f of N into M such that: if $m < \beta$, $\langle a_0,\dots,a_{m-1},a_m\rangle$ a

$<^N$-increasing sequence of members of N, and

$$tp_{qf}(\langle a_0,\ldots,a_{m-2},a_{m-1}\rangle,\emptyset,N) = tp_{qf}(\langle a_0,\ldots,a_{m-2},a_m\rangle,\emptyset,N)$$

then

$$d(f(a_0),\ldots,f(a_{m-2}) , f(a_{m-1})) = d(f(a_0),\ldots,f(a_{m-2}) , f(a_m)).$$

1.3A Remark: There are obvious monotonicity properties. Here qf stands for quantifier free.

1.4 Fact: If $\lambda_0 \le \lambda_1 \le \ldots \le \lambda_{n-1}$ and L is a finite vocabulary with predicates only and if for $m < n$

$(*)_m$ for every model $N \in K^\alpha_\sigma$ of cardinality λ_m
for some model $M \in K^\alpha_\sigma$ of cardinality λ_{m+1}
$M \rightarrow^{eh} (N)^{<n}_M$ holds.

then for every model $N \in K^\alpha_\sigma$ of cardinality λ_0 there is an model $M \in K^\alpha_\sigma$ of cardinality λ_{n-1} such that $M \rightarrow (N)^{<n}_M$.

Proof: Trivial.

1.4A Remark: We can define canonization relations - say, in how many variables the coloring does not depend. See [EHMR]: [Sh95].

1.5 Lemma: Suppose $\mu = \mu^{<\mu}$, $\sigma < \mu \le \theta < \kappa \le \lambda$, $\alpha,\beta \le \omega$, and $\kappa \rightarrow_{sp} (\mu+1)^{<\mu,<\beta}_{\mu,\theta}$ (see Definition 2.1), let P be the forcing notion defined by: $p \in P$ iff p is a partial function from $[\lambda]^{<\alpha}$ to σ of cardinality $< \mu$ and $p(\emptyset) = 0$ and ordered by inclusion.

Then in V^P for some $M \in K^\alpha_\sigma$, $|M| = \kappa$, and for every $N \in K^\alpha_\sigma$ of order type $\le \mu+1$ we have: $M \rightarrow^{eh} (N)^{<\beta}_\theta$.

1.5A Remark: (1) P is really just adding λ Cohen subsets of μ.
(2) If we let $|N| < \mu$, the proof is somewhat simplified.

Proof of Lemma 1.5 : Let $\underset{\sim}{M} = (\kappa,<,F^{\underset{\sim}{M}})$, where $F^{\underset{\sim}{M}}(u) = j$ iff for some $f \in G_P$ $f(u) = j$. Let $\underset{\sim}{d} : [\kappa]^{<\beta} \to \theta$, and $\underset{\sim}{N} \in K^{\alpha}_{\sigma}$ be P-names. Without loss of generality $|\underset{\sim}{N}|$ is an ordinal $\epsilon+1 \le \mu + 1$, and (as $\underset{\sim}{N}$ depends on at most μ of the Cohen subsets and doing P in two stages) without loss of generality $\underset{\sim}{N} = N$. Let χ be a large enough regular cardinal and let $<^*_{\chi}$ a well-ordering of $H(\chi)$. Let $p^* \in P$. By the hypothesis (see Definition 2.1 and 2.3) (for second phrase in (I) use for $i < \mu$ the function

$$F(x_0,\ldots,x_j,\ldots)_{j<i} \overset{\text{def}}{=} \langle x_j : j < i \rangle) .$$

(*) There is $B \subseteq \kappa$, B of order type $\epsilon + 1$ and $\langle N_u : u \in [B]^{<\beta} \rangle$ and $H_{u,v}$, $(u,v \in [B]^{<\beta}$, $|u| = |v|)$ such that:

(I) $N_u \prec (H(\chi),\in,<^*_{\chi})$, $(\forall a \in [N_u]^{<\mu})[a \in N_u]$ (hence $\mu \subseteq N_u$)

and

$$\{\langle u,\sigma,\kappa,\lambda,P,\underset{\sim}{d},N\rangle\} \in N_u$$

(II) (b) - (h) of 2.1

(III) $p^* \in P \cap N_{\emptyset}$.

Let $B = \{\xi_i : i \le \epsilon\}$ and $[i < j \Rightarrow \xi_i < \xi_j]$. First assume $\epsilon < \mu$. Let g be the unique order preserving function from $\epsilon + 1$ onto B, i.e. $g(i) = \xi_i$. Let for $i \le \epsilon$

$$I_i = [\{\xi_j : j < i \text{ or } j = \epsilon\}]^{<\beta} .$$

Next let for $u \in I_i$, $K(u) = \{(u,h) : h$ is a function from $[u]^{<\alpha}$ to $\sigma, h(\emptyset) = 0\}$

$$J_i = \{(u,h) : u \in I_i \text{ and } (u,h) \in K(u)\} .$$

Note that $[i < j \Rightarrow I_i \subseteq I_j]$, and $[i < j \Rightarrow J_i \subseteq J_j]$. We say $(u_1,h_1) \le (u_2,h_2)$ if $u_1 \subseteq u_2, h_1 \subseteq h_2$.

[**Explanation:** Note that we have already decided that the desired embedding will take i to $g(i)$, so the universe of the image of N will be $I \overset{\text{def}}{=} \{\xi_i : i \le \epsilon\}$. What we have to do is to find a condition in P forcing that the embedding is as required. Now ξ_{ϵ} is simultaneously a good "approximation" to ξ_j over $\{\xi_i : i < j\}$ and we define a condition for $\{\xi_j : j < i$ or $g(j) = \epsilon\}$ by induction on i, though in N, $j_1 \neq j_2$ may realize different quantifier free type over $\{i : i < j_1 \cap j_2\}$. We are saved by dealing simultaneously with conditions $p_{(u,h)}$ for $(u,h) \in J_i$.]

We now define by induction on $i \leq \epsilon$, $\langle p^i_{(u,h)} : (u,h) \in J_i \rangle$ such that:

(a) $p^i_{(u,h)} \in P \cap N_u$, and $p^* \leq p^i_{(u,h)}$

(b) $h \subseteq p^i_{(u,h)}$

(c) if $j < i$, $(u,h) \in J_j$, then $p^j_{(u,h)} \leq p^i_{(u,h)}$.

(d) if $(u_1,h_1) \leq (u_2,h_2)$ then $p^i_{(u_1,h_1)} = p^i_{(u_2,h_2)} \restriction (\lambda \cup N_{u_1})$.

(e) $p^i_{(u,h)}$ forces a value to $\underset{\sim}{d}(v)$ for every $v \subseteq u$.

(f) $p^{i+1}_{u\cup\{\xi_i\},h_i} \geq H_{u\cup\{\epsilon\},u\cup\{\xi_i\}}(p^i_{(u\cup\{\epsilon\},h_\epsilon)})$ if $u \subseteq i$ and h_i, h_ϵ are functions such that $(u \cup \{\xi_i\},h_i) \in J_{i+1}$, $(u \cup \{\xi_i\},<,h_i) \cong (u \cup \{\epsilon\},<,h_\epsilon)$.

We shall carry out the definition in detail.

[Explanation: Condition (a) is in order to have control over the conditions and to utilize the indiscernibility.

Condition (b), note it is the role of $p^i_{(u,h)}$ to ensure our being able to deal with the case $h = (F^N \circ (g^{-1})) \restriction u$.

Condition (c) should be clear.

Condition (d) enables us to form the condition $\bigcup_{x\in X} p_{(u_x,h_x)}$ for suitable X.

Condition (e) as we want to form a condition forcing the "right" values of $\underset{\sim}{d}$, we certainly have to have approximations forcing some values for them.

Condition (f) This comes for the end-homogeneity, we want to say that ξ_i, ϵ are similar over $\{\xi_j : j < i\}$, of course the minute we want to force a value to $\underset{\sim}{d}(\{\xi_i,\xi_\epsilon\})$ this similarity cannot be maintained.]

The Induction.

Case A: $i = 0$.

So $I_i = \{\epsilon\}^{<\beta}$, so $I_i = \{\emptyset, \{\epsilon\}\}$ (except when $\beta \leq 1$ which is not so interesting).

For every $(u,h) \in J_i$ we have to define $p^i_{(u,h)}$. Let $u_0 = \emptyset, u_1 = \{\epsilon\}$.

Let us enumerate the functions $h : u_1 \to \sigma$, $h(\emptyset) = 0 : \{h_\gamma : \gamma < \gamma_0\}$, (so $\gamma_0 < \mu$). We define p_γ by induction on γ such that:

(i) $p_\gamma \in N_{\{\epsilon\}} \cap P$

(ii) $h_\gamma \subseteq p_\gamma$

(iii) for every $\beta < \gamma$, $p_\beta \restriction N_\emptyset \leq p_\gamma$

(iv) p_γ forces a value to $\underset{\sim}{d}(\{\epsilon\})$.

There is no problem in doing this by $\oplus_1, \oplus_2$ below. In the end let

$$p^i_{(u_0,\emptyset)} = \bigcup_{\gamma<\gamma_0} p_\gamma \restriction N_\emptyset$$

$$p^i_{(u_1,h_\gamma)} = p_\gamma \cup p^i_{(u_0,\emptyset)}$$

where

$\oplus_1$ if $q_\gamma \in P$ for $\gamma < \gamma(*) < \mu$, then $\bigcup_{\gamma<\gamma(*)} q_\gamma$ is in P and is the least upper bound of $\{q_\gamma : \gamma < \gamma(*)\}$ if and only if for any $\gamma_1, \gamma_2 < \gamma(*)$ the functions $q_{\gamma_1}, q_{\gamma_2}$ are compatible

$\oplus_2$ if $q_1 \in N_{u_1} \cap P$, $q_2 \in N_{u_2} \cap P$ then: q_1, q_2 are compatible if and only if $q_1 \restriction N_{u_1 \cap u_2}$, $q_2 \restriction N_{u_1 \cap u_2}$ are compatible [by 2.1 d].

Case B: i limit.

For any $(u,h) \in J_i$ there is $j_{(u,h)} < i$ such that $(u,h) \in J_{j_{(u,h)}}$

We let $p^i_{(u,h)} \overset{\text{def}}{=} \bigcup\{p^j_{(u,h)} : j < i \text{ and } (u,h) \in J_j\}$. There are no problems in checking the conditions (note: $p^i_{(u,h)} \in N_u$ because

$$[a \in N_u \wedge |a| < \mu \Longrightarrow a \in N_u] \text{ by I)}.$$

Case C: $i = j + 1$.

Let us enumerate $J_i = \{(u_\gamma, h_\gamma) : \gamma < \gamma(*)\}$.

Note that $\gamma(*) < \mu$. We define by induction on $\gamma \le \gamma(*)$ a sequence $\langle q^\gamma_{(u,h)} : (u,h) \in J_i \rangle$ such that (compare with (a) - (f) above):

(a)′ $q^\gamma_{(u,h)} \in P \cap N_u$

(b)′ $h \subsetneq q^\gamma_{(u,h)}$

(c)′ $\beta < \gamma$ implies $q^\beta_{(u,h)} \le q^\gamma_{(u,h)}$

(d)′ if $(u_1,h_1) \le (u_2,h_2)$ then

$$q^\gamma_{(u_1,h_1)} = q^\gamma_{(u_2,h_2)} \restriction N_{u_1}$$

(e)′ if $\beta < \gamma$ then $q^\gamma_{(u_\beta,h_\beta)}$ forces a value to $\underset{\sim}{d}(v)$ for every $v \subsetneq u$.

(f)′ the parallel of (f) .

Subcase C (a): $\gamma = 0$. Define $q^\gamma_{(u,h)}$ as follows:

(α) it is $p^j_{(u,h)}$ if $u \in I_j$

(β) it is $H_{u_1,u}(p^j_{(u,h)})$ if $u \in I_i$, $\xi_i \in u$, $\varepsilon \notin u$, $v \overset{\text{def}}{=} u \setminus \{\xi_i\} \in I_j$; and we let:

$$u_1 \overset{\text{def}}{=} v \cup \{\varepsilon\} ,$$

(γ) it is $p^j_{(u_0,h\restriction u_0)} \cup H_{u_0,u_1}(p^j_{(u_0,h_0)})$ if $u = v \cup \{\xi_j, \varepsilon\}$, $\xi_j \notin v$, $\varepsilon \notin v$, and we let $u_1 \overset{\text{def}}{=} v \cup \{\xi_j\}$, $u_0 = v \cup \{\varepsilon\}$.

Subcase C (b): γ is limit.

Use unions $q^\gamma_{(u,h)} = \bigcup_{\beta < \gamma} q^\beta_{(u,h)}$.

Subcase C (c): $\gamma = \beta + 1$.

Note that the only demand on γ in $\langle q^\gamma_{(u,h)} : (u,h) \in J_i \rangle$ which is not clearly satisfied by $\langle q^\beta_{(u,h)} : (u,h) \in J_i \rangle$ is (e)′ for β . We first choose $r = r^\gamma$, such that:

(i) $q^{\beta}_{(u_\beta,h_\beta)} \le r \in N_{u_\beta} \cap P$

(ii) r forces a value to $\underset{\sim}{d}(v)$ for every $v \subseteq u_\beta$.

Clearly such r exists. Now for every $(u,h) \in J_i$ let

$q^{\gamma}_{(u,h)} \overset{\text{def}}{=} q^{\beta}_{(u,h)} \cup\bigcup\{r \restriction N_v : v \subseteq u \cap u_\beta, h \restriction v = h_\beta \restriction v\}$ [note that it is possible that $v_1, v_2 \subseteq u \cap u_\beta$ and $h \restriction v_1 = h_\beta \restriction v_1$, $h \restriction v_2 = h_\beta \restriction v_2$ but $h \restriction (v_1 \cup v_2) \ne h_\beta \restriction (v_1 \cup v_2)$.]

Now $q^{\gamma}_{(u,h)} \in P$ by $\oplus_1, \oplus_2$ and as $q^{\beta}_{(u,h)} \restriction N_v = q^{\beta}_{(v,h\restriction v)} \le q^{\beta}_{(u_\beta,h_\beta)} \le r$ whenever $(v \subseteq u \cap u_\beta, h \restriction v = h_\beta \restriction v)$. It is easy to check that $\langle q^{\gamma}_{(u,h)} : (u,h) \in J_1\rangle$ is as required: i.e. conditions (a)′ - (f)′ are satisfied.

* * *

So we have defined $\langle q^{\gamma}_{(u,h)} : (u,h) \in J_\gamma\rangle$ for $\gamma \le \gamma(*)$ as required, and we can finish Case C: let $p^{i}_{(u,h)} \overset{\text{def}}{=} q^{\gamma(*)}_{(u,h)}$ for $(u,h) \in J_i$.

* * *

So we have finished the definition of $\langle p^{i}_{(u,h)} : (u,h) \in J_i\rangle$ for $i \le \varepsilon$.

Lastly let

$$p^{**} = \cup\{p^{\varepsilon}_{(u,h)} : (u,h) \in J_\varepsilon , u \subseteq \varepsilon , \text{ and } h(v) = F^N(g^{-1}(v)) \text{ for } v \in [u]^{<\beta}\} .$$

Clearly the union is well defined and forces what we need except when $\alpha > \beta$, then we have to add information to make g an embedding of N to M . So we have finished the case $\varepsilon < \mu$.

Secondly, we assume $\varepsilon = \mu$. We cannot use the definition above as the union will not be a condition (too large cardinality). But we can work in $V[G_P]$, and choose by induction on $i < \mu$, an ordinal α_i , such that $\sup_{j<i} \alpha_j < \alpha_i <$ Max B and $p^{i}_{(u,h)}$ such that $u \subseteq \{\alpha_j : j < i\} \cup \{\varepsilon\}$ and $(u,h) \in J_\varepsilon$ satisfying the parallel of (a) - (f) above such that: if $u \subseteq \{\alpha_j : j < i\} \cup \{\varepsilon\}$, and $(u,h) \in J_\varepsilon$ and $h(v) = F^N(g^{-1}(v)$ for $v \in [u]^{<\omega}$ then $p^{i}_{(u,h)} \in \underset{\sim}{G}_P$. End of proof of lemma 1.5.

1.6 Conclusion: Assume that there is a class of measurable cardinals. Then in some generic extension

$\forall m,n < \omega\ \forall \theta\ \forall N \in K_\sigma^{<n} (\exists M)[M \in K_\sigma^{<n} \wedge |M| \leq \beth_{m+1}((|N| + \sigma + \theta)^+) \wedge M \to (M)_\theta^m]$.

Proof: Iterate the forcing (with e.g. Easton support) $Q_{\delta+n}$ (δ limit or zero) is adding $\kappa_{\delta+n+1}$ Cohen subsets to $\kappa_{\delta+n}$, where $\kappa_0 = \aleph_0$, for limit ordinal δ, $\kappa_\delta = \bigcup_{\alpha<\delta} \kappa_\alpha$, if κ_δ is singular $\kappa_{\delta+1} = \kappa_\delta^+$. In all other cases $\kappa_{\alpha+1}$ is the first measurable $> \kappa_\alpha$. By 1.5 we get enough instances of $\to^{eh}$. Iterating their use by 1.4 we get the desired conclusion.

§ 2. On $\to_{sp}$.

2.1 Definition: We define $\lambda \to_{sp} (\kappa)_{\mu,\theta}^{<\sigma,<n}$ where $\lambda,\kappa,\sigma,\theta$ are cardinals and $n \leq \omega$. It says that if N is an algebra with universe λ and with $\leq \mu$ operations each with $< \sigma$ places, then there is $A \in [\lambda]^\kappa$ and N_u for $u \in [A]^{<n}$ such that:

(a) N_u is a subalgebra of N

(b) N_u has cardinality μ

(c) $N_u \cap A = u$

(d) $N_u \cap N_v = N_{u \cap v}$ (the main point!)

(e) for $u,v \in [A]^{<n}$ of the same cardinality, $N_u \cong N_v$ the unique isomorphism from N_u onto N_v, order preserving, exists, we call it $H_{u,v}$.

(f) $H_{u,v}$ maps u to v

(g) $H_{u,u}$ = the identity, $H_{u_2,u_3} \circ H_{u_1,u_2} = H_{u_1,u_3}$ and for $u_1 \subseteq u$, $H_{u,v} \restriction N_{u_1} \subseteq H_{v_1,u_1}$ where $v_1 \subseteq v$ is such that $H_{u,v}$ maps u_1 onto v_1 (so equality holds),

(h) for $u \in [A]^{<n}$, $N_u \cap \theta \subseteq N_\emptyset$.

2.2 Definition: (1) We define $\lambda \to_{spn} (\kappa)_{\mu,\theta}^{<\sigma,<n}$ similarly adding

(i) if $v \subseteq u \in [A]^{<n}$, $(\exists \xi) v = u \cap \xi$ then $|N_v|$ is an initial segment

of $|N_u|$;

(2) We omit θ when $\theta = \aleph_0$, i.e. omit (h) in 2.1 .

2.3 Observation: (1) If $\lambda \to_{spn}(\kappa)^{<\sigma,<n}_{\mu,\aleph_0}$ and $\theta < \lambda$ then $\lambda \to_{sp}(\kappa)^{<\sigma,<n}_{\mu,\theta}$.

(2) Those arrows have obvious monotonicity properties: we can decrease κ,σ,n,θ . For $\to_{sp}$ we can increase λ .

(3) In 2.1, 2.2 we can use as N any algebra such that $\lambda \subsetneq |N|$. (See [Sh3] and § 3).

2.4 Fact: (1) If λ is measurable, $\kappa \leq \lambda$, $\mu + \theta + \sigma < \lambda$, $n \leq \omega$ then $\lambda \to_{spn}(\kappa)^{<\sigma,<n}_{\mu,\theta}$.

(2) If λ is minimal such that $\lambda \to (\kappa)^{<\omega}_{\theta}$, $\theta \geq \mu$ then $\lambda \to_{spn}(\kappa)^{<\omega,<\omega}_{\mu,\theta}$.

2.5 Lemma: If $\zeta \geq 3$, $\lambda \to_{sp}(\zeta)^{<\omega,<3}_{\omega,\mu}$ then $\lambda \to (\zeta)^{<\omega}_{\mu}$.

Proof: Let $\chi > 2^\lambda$ be a regular cardinal and let $<^*_\chi$ be a well order of $H(\chi)$. So we have $\langle M_u : u \in [\zeta]^{\leq 2} \rangle$ such that:

(a) $M_u \prec (H(\chi),\in,<^*_\chi)$,

(b) $M_u \cap M_v = M_{u\cap v}$

(c) $\lambda \in M_u$

(d) if $|u| = |v|$, M_u,M_v are isomorphic and let $H_{u,v}$ denote the (unique) isomorphism

(e) if $u = \{i_1,i_2\}$, $v = \{j_1,j_2\}$, $i_1 < i_2$ and $j_1 < j_2$ then $H_{\{i_1\},\{j_1\}} \subseteq H_{u,v}$, $H_{\{i_2\},\{j_2\}} \subseteq H_{u,v}$, $id_{M_0} \subseteq H_{\{i_1\},\{j_1\}}$.

(f) $M_{\{i\}} \cap \lambda \neq M_\emptyset \cap \lambda$. (This follows from (c) of 2.1.)

Let $\alpha_i = \alpha(i) = \mathrm{Min}(M_{\{i\}} \cap \lambda - M_\emptyset)$. Clearly $H_{\{i\},\{j\}}(\alpha_i) = \alpha_j$ (use (d),(e)) and $\mu < \alpha_i$. Also $\alpha_i \neq \alpha_j$ (as $M_{\{i\}} \cap M_{\{j\}} = M_\emptyset$) for $i \neq j$.

So for all $i < j < \zeta$, "$\alpha_i < \alpha_j$" has the same truth value. Since

$\alpha_i \neq \alpha_j$ if $\zeta \geq \omega$, as the ordinals are well ordered:

(A) $\langle \alpha_i : i < \zeta \rangle$ is strictly increasing. If $\zeta < \omega$ we could inverse the indexing and also have (A).

Next we shall prove

(B) If $i < j$ and $\bar{c} \in M_{\{i\}}$, then $\bar{c}$ and $H_{\{i\},\{j\}}(\bar{c})$ realize the same type over $\{\gamma : \gamma < \alpha_i\}$ (in $(H(\chi), \in, <^*_\chi)$.)

[Proof: Let $\varphi(\bar{x},\bar{y})$ be a formula, $lg(\bar{x}) = lg(\bar{c}), lg(\bar{y}) = n$. Let $<_{gd}$ be the following order (of Godel) on n-tuples of ordinals: $\bar{\beta} <_{gd} \bar{\gamma}$ if and only if $\mathrm{Max}(\bar{\beta}) < \mathrm{Max}(\bar{\gamma})$ or $\mathrm{Max}(\bar{\beta}) = \mathrm{Max}(\bar{\gamma})$ and $\bar{\beta}$ is smaller than $\bar{\gamma}$ in the lexicographic order.

Let $F_\varphi(\bar{x}_1,\bar{x}_2)$ = the $<_{gd}$-first sequence $\bar{y}$ (n-tuple of ordinals) such that: $\varphi(\bar{x}_1,\bar{y}) \equiv \neg\varphi(\bar{x}_2,\bar{y})$.

Clearly F_φ is definable in $(H(\chi),\in,<^*_\chi)$ hence each M_u is closed under F_φ.

Let $\bar{c}_j = H_{\{i\},\{j\}}(\bar{c})$ for each $j < \zeta$ and assume that $F(\bar{c}_{j_1},\bar{c}_{j_2})$ is defined for some ($\equiv$ all) $j_1 < j_2 < 3$: otherwise (B) is immediate. So $F(\bar{c}_{j_1},\bar{c}_{j_2}) \in M_{\{j_1,j_2\}}$. However by a classical trick, if $j_1 < j_2 < j_3$ then the set $\{F_\varphi(\bar{c}_{j_1},\bar{c}_{j_2}), F_\varphi(\bar{c}_{j_1},\bar{c}_{j_3}), F_\varphi(\bar{c}_{j_2},\bar{c}_{j_3})\}$ has only two members. Assume e.g. that the first two are equal, so

$F_\varphi(\bar{c}_{j_1},\bar{c}_{j_2}) = F_\varphi(\bar{c}_{j_1},\bar{c}_{j_3}) \in M_{\{j_1,j_2\}} \cap M_{\{j_1,j_3\}} = M_{\{j_1\}}$. Generally (according to which of the three possible equalities holds) $F_\varphi(\bar{c}_{j_1},\bar{c}_{j_2})$ belongs to $M_{\{j_1\}}$ or to $M_{\{j_2\}}$. As clearly (for $\ell = 1,2$, as $\bar{c}_{j_m} = F_{\{i\},\{j_2\}}(\bar{c})$)

$F_\varphi(\bar{c}_{j_1},\bar{c}_{j_2}) \geq \sup(\alpha_{j_1} \cap \alpha_{j_2} \cap M_\emptyset) = \sup(\alpha_{j_\ell} \cap M_{\{j_\ell\}})$ we can deduce

$F_\varphi(\bar{c}_{j_1},\bar{c}_{j_2}) \geq \mathrm{Min}\{\alpha_{j_1},\alpha_{j_2}\}$. As φ was any formula, we have finished the proof of (B)].

(C) α_i is strongly inaccessible.

[Proof: Note that all α_i realize the same type in $(H(\chi),\in,<^*_\chi)$.

If each α_i is singular, there is $f_1 \in M_\emptyset$ such that for $\delta < \lambda$ singular, $f_1(\delta)$ is a club of δ of order type $cf(\delta)$. As $cf(\alpha_i) < \alpha_i$ and $cf(\alpha_i) \in M_{\{i\}}$ clearly $cf(\alpha_i) \in M_\emptyset$ hence for some $\theta \in M_\emptyset$, $(\forall i < \zeta)[cf(\alpha_i) = \theta]$. Let $f_2 \in M_\emptyset$ be such that for $\delta < \lambda$ of cofinality $\theta, f_2(\delta)$ is a one-to-one function from θ onto $f_1(\delta)$. So easily $f_1(\alpha_i) \cap M_{\{i\}} = f_1(\alpha_i) \cap M_\emptyset = \{f_2(\alpha_i)(\gamma) : \gamma \in \theta \cap M_\emptyset\}$; w.l.o.g. f_1, f_2 are definable over $\emptyset$ (in the model $(H(\chi),\in,<^*_\chi)$) . Now if $i_1 < i_2$, we get a contradiction to (B).

Next if α_i are not strong limit, then there is $\mu < \alpha_i$, $\mu \in M_{\{i\}}$, $2^\mu \geq \alpha_i$. So $\mu \in M_\emptyset$, and by the $H_{\{i\},\{j\}}$'s , $2^\mu \geq \alpha_j$ for each j , so in $M_\emptyset$ there is a (definable from $\emptyset$) one-to-one function from 2^μ to $\mathcal{P}(\mu)$ and we get contradiction to (B).]

(D) W.l.o.g. $M_{\{i,j\}}$ is the Skolem hull of $M_{\{i\}} \cup M_{\{j\}}$.

(E) For $i < j < \zeta$ the intersection of the Skolem hull of $M_{\{j\}} \cup (\alpha_j \cap M_{\{i\}})$ with α_j is included in $M_{\{i\}}$.

[Proof; If not, there are $\bar{c} \in \alpha_j \cap M_{\{i\}}$, $\bar{d} \in M_{\{j\}}$, $y = G(\bar{c},\bar{d})$, G definable, $y \in \alpha_j \setminus M_{\{i\}}$.

Let w.l.o.g. $j < j_1 < \zeta$ (we use that w.l.o.g. $i = 0$, $j = 1$ and remember $\zeta \geq 3$. As $\bar{d}$ and $\bar{d}' \overset{\text{def}}{=} H_{\{j\},\{j_1\}}(\bar{d})$ realize the same type over $\{\gamma : \gamma < \alpha_j\}$, clearly $y = G(\bar{c},\bar{d}')$ too, so

$$y \in M_{\{i,j\}} \cap M_{\{i,j_1\}} = M_{\{i\}} .]$$

We shall code, for each formula $\varphi(x_0,\ldots,x_{n-1},y)$, φ-types of n-tuples over $\{\gamma : \gamma < \alpha_i\}$ by an ordinal $< 2^{\alpha_i}$.

(F) For each φ , and $n \geq 1$, there is $\bar{c}_{\varphi,n,i} \in M_{\{i\}}$ such that:

(1) $\bar{c}_{\varphi,n,i}$ codes the φ-type of $\langle \alpha_i, \alpha_{i_1}, \ldots, \alpha_{i_{n-1}} \rangle$ over $\{\gamma : \gamma < \alpha_i\}$ whenever $i < i_1 < \ldots < i_{n-1} < \zeta$.

(2) $H_{\{i\},\{j\}}(\bar{c}_{\varphi,n,i}) = \bar{c}_{\varphi,n,j}$.

[**Proof**: For $n = 1$ this is easy.

For $n + 1$ if $i < i_1$, $\bar{c}_{\varphi,n,i}$ can be computed from α_i, $\bar{c}_{\varphi,n,i_1}$ (just think of the meaning) so $\bar{c}_{\varphi,n+1,i} = G(\alpha_i, \bar{c}_{\varphi,n,i_1})$ where G a definable function (over $\emptyset$).

However, by coding such things naturally $\bar{c}_{\varphi,n+1,i}$ is an ordinal $< 2^{\alpha_i}$, hence $< \alpha_1$ (by (C)). So it necessarily belongs to $M_{\{i\}}$ by (E), so (0),(1) holds.

By the way $\bar{c}_{\varphi,n+1,i}$ was defined, also (2) holds.]

If there is $F : [\lambda]^{<\omega} \to \mu$ which is a counterexample to the desired conclusion of 2.5, then such F belongs to $M_\emptyset$ and is definable over $\emptyset$ (in $(H(\chi), \in, <^*_\chi)$!), and $\langle \alpha_i : i < \zeta \rangle$ contradict its choice (by (F) above), so that the lemma 2.5 follows.

§ 3 Refining the combinatorics

3.1. Definition:

(1) For $x \in \{sp, spn\}$ we define

$$\lambda \longrightarrow_{ex(k)} (\kappa)^{<\sigma,n}_{\mu,\theta} \text{ like } \lambda \longrightarrow_{x} (\kappa)^{<\sigma,n}_{\mu,\theta}$$

(see definitions 2.1 and 2.2) except that we replace (e), (f), (g) by

$(e)^e$ if $u,v \in [A]^{<n}$ and $u \sim_k v$ (which means that for some w, w is an initial segment of u and v and $|u \backslash w| = |v \backslash w| \leq k$) then $N_u \cong N_v$ and let $H_{u,v}$ be the unique isomorphism

$(f)^e$ $H_{u,v}$ when defined maps u onto v; $H_{u,u} = id$

$(g)^e$ if $u_1 \sim_k u_2 \sim_k u_3$, $u_i \in [A]^{<n(*)}$

then $H_{u_2,u_3} \circ H_{u_1,u_2} = H_{u_1,h_3}$ and for any $u'_1 \subset u_1$ if $u'_2 = H_{u_1,u_2}(u'_1)$ then $H_{u_1,u_2} \restriction N_{u'_1} \subseteq H_{u'_1,u'_2}$ (so equality holds).

(2) If $k = 1$ we omit it.

(3) For $x \in \{sp, spn, esp, espn\}$ we define

$$\lambda \underset{wx}{\longrightarrow} (\kappa)^{<\sigma,<n}_{\mu,\theta} \quad \text{just like} \quad \lambda \underset{x}{\longrightarrow} (\kappa)^{<\sigma,<n}_{\mu,\theta}$$

replacing (d) by $(d)^w$ $N_u \cap N_v \subset N_{u \cap v}$ and if $\alpha < \beta$ are from β, $u \in [\beta]^{<n}$ and $(\alpha,\beta) \cap u = \emptyset$ then $(\alpha,\beta) \cap M_u = \emptyset$.

(Note that now in (g) equality does not follow.)

(4) For any of the x for which $\underset{x}{\longrightarrow}$ was defined $\lambda \underset{xv}{\longrightarrow} (\kappa)^{<\sigma,<n}_{\mu,\theta}$ is defined as above except that also $d : [\lambda]^{<n(*)} \to \theta$ is given and (h) is replaced by:

$(h)^v$ For each ℓ, $d \restriction [A]^{\ell}$ is constant when e does not appear in x; and $h(u)$ $(u \in [A]^{<n})$ does not depend on $\max(u)$ when it appears and, more generally, $exv(k)$ does not depend on the last k members of u (i.e. if $u_1,u_2 \in [A^{\ell}]$ and w is a common initial segment of u_1,u_2 $|u_{\ell} - w| \leq k$ then $h(u_1) = h(u_2)$).

3.2. Observation:

(1) We have $\lambda \underset{x}{\longrightarrow} (\kappa)^{<\sigma,<n}_{\mu,\theta} \Rightarrow \lambda \underset{y}{\longrightarrow} (\kappa)^{<\sigma,<n}_{\mu,\theta}$ where: y is x when we omit n or x is y when we omit e or w or v.

(2) If (e appears in x and) $\lambda_2 \underset{x(k)}{\longrightarrow} (\lambda_1)^{<\sigma,<n}_{\mu,\theta}$ and $\lambda_1 \underset{x(\ell)}{\longrightarrow} (\lambda_0)^{<\sigma,<n}_{\mu,\theta}$ then $\lambda_2 \underset{x(k+\ell)}{\longrightarrow} (\lambda_0)^{<\sigma,<n}_{\mu,\theta}$.

(3) If $\lambda \underset{x(k)}{\longrightarrow} (\kappa)^{<\sigma,<n}_{\mu,\theta}$ and $k \geq n-1$ then $\lambda \underset{y}{\longrightarrow} (\kappa)^{<\sigma,<n}_{\mu,\theta}$ where y is x with e omitted.

(4) If $\lambda_2 \underset{x(k)}{\longrightarrow} (\lambda_1)^{<\sigma,<n}_{\mu,\theta}$, and $\ell = n-1-k$, y is x with e omitted and $\lambda_1 \to (\lambda_0)^{\ell}_{2^M}$ then $\lambda_2 \underset{y}{\longrightarrow} (\lambda_0)^{<\sigma,<n}_{\mu,\theta}$.

3.2A Remark. By 3.2(4) even $\lambda \underset{x(0)}{\longrightarrow} (\kappa)^{<\sigma,<n}_{\mu,\theta}$, when $n \geq 3$, is quite strong (when w does not appear in x).

3.3. Definition. $\lambda \longrightarrow^{+} (\kappa)^{<\omega}_{\mu}$ means: for each club $C \subseteq \lambda$ and for $n < \omega$, $i < \mu$, $F_{n,i} : [\lambda]^n \to \lambda$ there is $A \in [C]^{\kappa}$ such that if $\alpha_0 < \ldots < \alpha_{n-1}$ belongs to A and $m < n$, $i < \mu$ and $F_{n,i}(\{\alpha_0, \ldots, \alpha_{n-1}\}) < \alpha_m$ then it does not depend on $\alpha_m, \ldots, \alpha_{n-1}$.

3.3.A. Remark: Replacing "a club $C \subseteq \lambda$" by "a final segment $C \subseteq \lambda$" does not change anything except that in the later version, if $\lambda = \bigcup_{i<1} \lambda_i$, each λ_i satisfying the first definition the λ satisfies the second definition.

3.4. Fact. If $\mu \leq \theta < \lambda$, $\lambda \overset{+}{\longrightarrow} (\kappa)^{<\omega}_{\mu}$ then $\lambda \underset{\text{spn}}{\longrightarrow} (\kappa)^{<\omega,<\omega}_{\mu,\theta}$ (κ can be, in fact, any limit ordinal, $\omega\kappa = \kappa$).

3.5. Lemma. If $\lambda \underset{\text{spn}}{\longrightarrow} (\xi)^{<\omega,<3}_{\omega,\mu}$, $\xi \geq 3$ then $\lambda \overset{+}{\longrightarrow} (\xi)^{<\omega}_{\mu}$.

Proof. Similar to the proof of 2.5 but by the definition of $\underset{\text{spn}}{\longrightarrow}$ we know $\sup(N_\emptyset \cap \lambda) < \alpha_0 = \lambda$. In the end, if there is a sequence $\langle C, \langle F_{n,i} : n < \omega, i < \mu \rangle \rangle$ contradicting the conclusion, wlog it is definable over $\emptyset$ as $C \in M_\emptyset$ and easily $\alpha_i \in C$ and continue as before.

3.6. Lemma. (1) For every $n < \omega$, there is $k = k^1_n < \omega$ (e.g. $k = (2n-1)^2$) such that : if $\kappa^{<\sigma} = \kappa$ then $_k(\kappa)^+ \underset{\text{wsp}}{\longrightarrow} (\kappa^+)^{<\sigma,<n}_{\kappa,\kappa}$.

(2) $\forall n < \omega \; \exists k = k^2_n < \omega$ such that : if $\sigma, \mu, \kappa < \lambda$ and λ is $(\alpha+k)$-Mahlo strongly inaccessible cardinal then $\lambda \underset{\text{wspn}}{\longrightarrow} (\kappa)^{<\sigma,<n}_{\mu,\mu}$.

Remark. 1. Using part 1 for 4.1 note that $\underset{\text{wsp}}{\longrightarrow}$ is stronger than $\underset{\text{wesp}}{\longrightarrow}$.

2. Part 2 is used for e.g. consistency of $2^{\mu} \to [\mu^+]^2_{0,3}$.

3. We do not try here to get the best bound (but see 3.8 and see [4]).

Proof. (1) Let $\lambda_0 = \kappa$, $\lambda_\ell + 1 = {}_{2n+1}(\lambda_\ell)$. Suppose N^* is an algebra with universe λ_n and at most κ functions each with $< \sigma$ places. We define, by induction on $m \leq n$, a set A_m and N^m_u $(u \in [A_m]^{<n})$:

I (i) $A_0 = \lambda_n$.

(ii) $A_{m+1} \subset A_m$.

(iii) $|A_m| = \lambda_{n-m}$.

II (i) N^m_u (for $u \in [A_m]^{<n}$) is a submodel of N^* of cardinality $\le \kappa$.

(ii) The answer to "is the γ_1-th element of N^m_v equal to the γ_2-th element of N^m_u?" where $u, v \in [A_{m+1}]^{<n}$ depends just on γ_1, γ_2 and the isomorphism type of $(u \cup v, u, v, < \restriction (u \cup v))$.

(iii) If $u, v \in [A_m]^{<n}$ then

$$|N^m_u| \cap |N^m_v| \subseteq N^{m+1}_{u \cap v}.$$

(iv) If $u \in [A_{m+1}]^{<n}$, $|u| \ge n - 1 - m$ then $N^m_u = N^{m+1}_u$.

For $m = 0$ let $A_0 = \lambda_n$, N^m_u is the subalgebra of N^* generated by the set $u \cup \kappa$. If $m < n - 1$, choose $A_{m+1} \subset A_m$ $|A_{m+1}| = \lambda_{n-m-1}$ such that II.(ii) holds (using $\lambda_{n-m} \to (\lambda_{n-m-1})^{2(n-1)}_{2^\kappa}$). Now for $u \in [A_{m+1}]^{<n}$, let N^{m+1}_u be: if $|u| < n - 1 - m$, the Skolem hull of $\bigcup\{N^m_w \cap N^m_v : w, v \in [A_{m+1}]^n$ and $u = w \cap v\}$ and if $|u| \ge n - 1 - m$, N^m_u. The cardinality of N^{m+1}_u is $\le \kappa$ by II (ii) i.e. if $X \in N^m_w \cap N^m_v$ and $w_1, v_1 \in [A_{m+1}]^{<n}$,

$$|w_1| = |w|, |v_1| = |v|, w_1 \cap v_1 = w \cap v$$

and $(\forall \alpha \in w \cup v)(\exists \beta \in w_1 \cup v_1)\ [|w \cap \alpha| = |w_1 \cap \beta| \wedge |v \cap \alpha| = |v_1 \cap \beta|]$ then $X \in N^m_{w_1} \cap N^m_{v_1}$.

(2) We need 3.7 below instead of Erdos-Rado and then the proof is similar to that of part 1.

3.7. Lemma: If λ is $(\alpha+n)$-Mahlo and strongly inaccessible and N^* is an algebra with universe λ and $< \lambda$ operations each with arity $< \lambda$ and A_0 is unbounded in λ <u>then</u> for every $\mu < \lambda$ there is $\kappa : \mu < \kappa < \lambda$, κ is α-Mahlo and strongly inaccessible and there is $A \subset A_0 \cap \kappa$ unbounded in κ:

(*) If for $\ell = 1,2$ $\beta < \alpha^\ell_0 < \ldots \alpha^\ell_{n-1}$ are from A then

$$\langle \alpha^1_0, \ldots \alpha^1_{n-1} \rangle, \langle \alpha^2_0, \ldots, \alpha^2_{n-1} \rangle$$

realize the same type over $\{\gamma : \gamma < \beta\}$ is N^*.

3.7A Historical Remark: We proved 3.7 in 1968 as part of some research on transfer theorems in model theory. As Schmerl was doing parallel research, it appeared in [ScSh20] but somehow this version does not appear - only the version with a finite conclusion. Subsequently Schmerl found a better lower bound for λ (how Mahlo it should be) and proved that it was exact. Hajnal independently proved 3.7 and the author wrongly told him it had appeared in [ScSh20].

Proof: We prove it by induction on n.
For $n = 0$ there is nothing to prove. For $n > 1$ use the induction hypothesis to find $\kappa < \lambda$ which is $(\alpha+1)$-Mahlo and $A_0 \subset \kappa$ as there for $n-1$. Expand $M \restriction \kappa$ by a predicate for A_0 and (as $n > 1$) apply the induction hypothesis for $n = 1$. For $n = 1$, let $C = \{\kappa < \lambda : \kappa$ is a strong limit and for each $\mu < \kappa$, there is $(N, A \restriction N) \prec_{L_{\mu,\mu}} (M, A)$ such that $\alpha \subsetneq |N| \subsetneq \kappa\}$. Clearly C is a club of λ, so there is $\kappa \in C$ which is α-Mahlo.

Choose $\gamma \in A - \kappa$, define a function $f : \kappa \to \kappa$ by $f(\alpha) = \min\{\gamma' \in A \cap \kappa : \gamma'$ realizes the type of γ over $\{i : i \leq \alpha\}\}$. Let $C' = \{\beta < \kappa : (\forall \alpha < \beta) f(\alpha) < \beta\}$. Clearly C' is a club of κ and $A_0 = \{f(\beta) : \beta \in C'\}$ is as required.

3.8. Lemma

Suppose $\lambda = \theta^+, \theta^\kappa = \theta$, $\kappa^\mu = \kappa$, $\mu^{<\sigma} = \mu$, τ is a vocabulary such that $|\tau| \leq \mu$ and each member of τ has arity $< \sigma$. *If* M is a τ-model with universe λ *then* we can find δ, α, B, $\langle M_s : s \in [B]^{\leq 2} \rangle$, $\langle M^-_{\{i\}} : i \in B \rangle$ $\langle H_{s,t} : |s| = |t| ; s,t \in [B]^{\leq 2} \rangle$ and W such that:

(a) $\delta < \lambda$, $\mathrm{cf}\delta = \mu^+$.

(b) B is a subset of δ of order type μ^+ (we could get $\mu^+ + 1$, actually but then $M_{\{\max B\}}$ is not defined).

(c) $M_s \prec_{L_{\mu,\sigma}} M$ for $s \in [B]^2$ and $M^-_{\{i\}} \prec_{L_{\mu,\sigma}} M_{\{i\}} \prec_{L_{\mu,\sigma}} M$ for $i \in B$ and

$M_\emptyset \prec_{L_{\mu,\sigma}} M$.

(d) $M_s \cap B = s$, $M_{\{i\}} \cap B = M^-_{\{i\}} \cap B = \{i\}$, $M_\emptyset \cap B = \emptyset$.

(e) For $s,t \in [B]^{\leq 2}$, $|s| = |t|$: $H_{s,t}$ is an isomorphism from M_s onto M_t (and $H_{s,s} = id_{M_s}$, $H_{s,t} = H^{-1}_{t,s}$ and $H_{s_0,s_2} = H_{s_1,s_2} \circ H_{s_0,s_1}$) and $H_{\{i\},\{j\}}$ maps $M^-_{\{i\}}$ onto $M^-_{\{j\}}$.

(f) All $H_{s,t}$ are compatible; $H_{s,t}$ maps s onto t .

(g) $M_s \cap M_t \subseteq M_{s\cap t}$.

(h) $\forall i < j < k$ from B

 (α) $M_{\{i,j\}} \cap M_{\{i,k\}} = M_{\{i\}}$,

 (β) $M_{\{i,j\}} \cap M_{\{j,k\}} = M^-_{\{j\}}$,

 (γ) $M_{\{i,k\}} \cap M_{\{j,k\}} = M^-_{\{k\}}$.

(i) For $i < j$ from B , j is the first element of $M_{\{i,j\}} \setminus M_{\{i\}}$.

(j) $M_\emptyset \subset M^-_{\{k\}}$, $M_\emptyset \subset M_s$ for $k \in B$, $s \in [B]^{\leq 2}$

(k) (α) $W \subset \lambda$, $\forall \xi \in W$ $cf\xi = \mu$

 (β) $\delta = \max W$

 (γ) If $H_{s,t}(\beta) = \gamma$ and $k \in W$ then $\beta < k \equiv \gamma < k$

 (δ) If $\beta \in M_{\{i\}} - M_\emptyset, i \in B$, $\kappa = \min\{\xi : \xi \in W , \beta < \xi\}$ then $\xi \neq \beta$ and $\langle H_{\{i\},\{j\}}(\beta) : j \in B\rangle$ is increasing converging to δ .

 (ϵ) If $\beta_1,\beta_2 \in M_{\{i\}} - M_\emptyset$, $\xi_e = \min\{\xi \in W : \beta_e \leq \xi\}$, $\xi_1 \neq \xi_2$, and $i < j \in B$ then $H_{\{i\},\{j\}}(\beta_1) > \beta_2$.

3.8A Remark: (1) We can instead " $\lambda = \theta^+$ " assume λ is inaccessible $\forall\alpha < \lambda\ [|\alpha|^\kappa < \lambda]$. Similarly for μ .

(2) For simplicity, θ,κ,μ,σ are regular and $<$ is a relation of M .

(3) We can replace $L_{\mu,\sigma}$ by any fragment of $L_{\mu^+,\sigma}$ of cardinality μ .

Proof: Let $M_0 \prec_{L_{\theta,\kappa^+}} M$ where $|M_0|$ is an ordinal $\delta_a < \lambda$ of cofinality θ , (or at least κ^+) so $\|M_0\| = \theta$. Let $N_a \prec_{L_{\kappa,\mu^+}} M$, $\delta_a \in N_a$, $|N_a| = \kappa$. By the choice of M_0 there is a model $N_b \prec_{L_{\kappa,\mu^+}} M_0$ and an isomorphism f from

N_a onto N_b over $N = M\restriction(|N_a| \cap |M_0|)$. Let $\delta_b = f(\delta_a)$. Let $N^* \prec_{L_{\mu,\sigma}} M$, $|N^*| = \mu$ be such that $\delta_a \in N^*$ and $\{N, N_a, N_b, f\} \in N^*$ in some coding. We let $M^-_{\{\delta_a\}} = N^* \restriction |N_a|$ and $M^-_{\{\delta_b\}} = N^* \restriction |N_b|$; $M_{\{\delta_b\}} = N^* \restriction |M_0|$, $M_{\{\delta_b,\delta_a\}}$ = the Skolem hull in M of $M^-_{\{\delta_a\}} \cup M_{\{\delta_b\}}$. Let

$h : |M_{\{\delta_b\}}| \to |N_a| : h(\beta) = \min\{\gamma : \gamma \in |N_a|, \beta < \gamma\}$. Let

$W = \text{range}(h) - \{\delta_a\}$. Let α be minimal element of N such that

$(\forall\beta)[\beta \in M_{\{\delta_b\}} \wedge (\exists\gamma \in N)\ \beta < \gamma \Rightarrow \beta < \alpha]$ i.e. $\alpha = \sup W$.

Now we define by induction on $\zeta < \mu^+$, δ_j, $M_{\{\delta_\zeta\}}$, $M^-_{\{\delta_\zeta\}}$, $M_{\{\delta_\xi,\delta_\zeta\}}$ for $(\xi < \zeta)$, $M_{\{\delta_\zeta,\delta_a\}}$ and $H_{\{\xi\},\{\zeta\}}$, $H_{\{\xi,\delta_a\},\{\zeta,\delta_a\}}$ for $\xi < \zeta$ (understand δ_ζ to be the ζ-th member of B) such that: the relevant cases of the desired conclusion holds, and $M_{\{\delta_\zeta\}} \subset N$, for $\xi < \zeta$, $M_{\{\delta_\xi,\delta_\zeta\}} \subset N$, $M_{\{\delta_\zeta,\delta_a\}} \subset N_a$, etc. and lemma 3.8 is proved.

3.9. Lemma: Suppose GCH for simplicity $\mu = \mu^{<\sigma}$, $\kappa = \kappa^\mu$, $\lambda \geq \kappa^{++}$, $\sigma < \mu < \kappa < \kappa^+ < \lambda$ are regular. There is a forcing notion P such that:

I A. P is strategically κ^+-complete.

B. P preserves cardinalities and cofinalities.

C. $|P| = \lambda$.

II (In V^P)

(*) There are $S^* \subset S \subset \lambda$, $\{C_\delta : \delta \in S\}$ and for $\delta \in S^*$, τ_δ, $\langle M^*_{\delta,s} : s \in [B^*_\delta]^{\leq 2}\rangle$, $\langle M^-_{\delta,s} : s \in [B_\delta]^1\rangle$, $H^\alpha = \langle H^\delta_{s,t} : s,t \in [B_\delta]^{\leq 2}$, $|s| = |t|\rangle$, W_δ, ξ_δ, ζ_δ, τ.

(A) The relevant conclusion of 1.1 holds for each $\delta \in S^*$ with B_δ an unbounded subset of $\xi_\delta < \kappa^{++}$, $\xi_0 = \min W < \kappa^+$.

(B) If $\xi_{\delta(1)} = \zeta_{\delta(2)}$ then there is a function $H_{\delta(n),\delta(2)}$ from $\bigcup_s M_{\delta(1),s} \cup W_{\delta(1)} \cup C_{\delta(1)}$ onto $\bigcup_s M_{\delta(2),s} \cup W_{\delta(2)} \cup C_{\delta(2)}$ which is order-preserving and preserves all relevant properties and the domain and range are disjoint.

(C) $\delta \in S^* \Rightarrow \mathrm{cf}\,\delta = \mu$ (follows from (A)) and for $\delta \in S$ C_δ is a club of δ of cardinality $\le \mu$ and

(D) if $\delta \in S^*$, α is an accumulation point of C_δ then $\alpha \in S \wedge C_\alpha = C_\delta \cap \alpha$ (follows from (B)).

(E) For $\delta \in S^*$, $W_\delta \subset C_\delta$.

Proof. If $\lambda = \kappa^{++}$ we shall force by approximations of cardinality κ. If we succeed to force for λ, we can force for λ^+ by approximations of cardinality κ^+. For $\lambda = \kappa^{+n}$, we iterate this, for $\lambda > \kappa^{+\omega}$ we have to take care of the singular case.

§ 4. Eliminating the Measurables

4.1. Lemma.

(1) Suppose $\mu = \mu^{<\mu}$, $\mu \le \theta < \kappa \le \lambda$, $\alpha \le \omega$, $\beta(*) < \omega$ and

$\kappa \underset{\text{wesp}}{\longrightarrow} (\mu+1)^{\mu,<\beta(*)}_{\mu,\theta}$ (see def. 3.1).

Let P be the forcing action as in 1.5. Then in V^P for some $M \in K^\alpha_\sigma$ of cardinality κ for every $N \in K^\alpha_V$ of power $\le \mu$, $M \xrightarrow{\text{eh}} (N)^{<\beta(*)}_\theta$.

(2) In applying $\underset{\text{wesp}}{\longrightarrow}$ we can weaken it replacing (d) in 3.1 (3) by

(d)$^-$: if $u \cup v \cup \{\alpha,\beta\} \subset A$, $(\forall i \in u \cup v)[i < \alpha \wedge i < \beta] \wedge [|u|,|v| < n-1]$ then $N_{u\cup\{\alpha\}} \cap N_{v\cup\{\beta\}} \cap \lambda \subset N_u \cap N_v$. However we still need $N_u \cap A = u$! but $\lambda = {}_{\beta(*)-2}(\kappa^{++})$ suffices for κ^+.

Proof: We indicate the changes in the proof of 1.5. Of course, we replace "(b) to (h) of 2.1" by the appropriate variants from definition 3.1 (3). Defining $p^i_{(u,h)}$ for $(u,h) \in J_i$ by induction on i we change (d) to:

(d) *If* (u_1,h_1), (u_2,h_2) (both from J), are compatible (i.e. $h_1 \restriction (u_1 \cap u_2) = h_2 \restriction (u_1 \cap u_2)$) *then*

$$p^i_{(u_1,h_1)} \restriction N_{u_1} \cap N_{u_2} \cap \lambda = p^i_{(u_2,h_2)} \restriction N_{u_1} \cap N_{u_2} \cap \lambda$$

and in case (C) we change (d)′ to

(d)′ If $(u_1,h_1),(u_2,h_2)$ are compatible

$$q^{\gamma}_{(u_1,h_1)} \restriction (N_{u_1} \cap N_{u_2} \cap \lambda) = q^{\gamma}_{(u_2,h_2)} \restriction (N_{u_1} \cap N_{u_2} \cap \lambda) .$$

In subcase (C)(a), we use (d) above (this influence (γ) there) and in the proof of subcase (C)(c), we let, for $(u,h) \in J_i$; $q^{\gamma}_{(u,h)}(w)$ is defined iff $w \in \mathrm{dom}\, q^{\beta}_{(u,h)}$ or $r(w)$ is defined and $w \in [N_u \cap \kappa]^{<\alpha(*)}$. The value of $q^{\gamma}_{(u,h)}(w)$ is $q^{\beta}_{(u,h)}(w)$ when defined and $r(w)$ otherwise. Let us check (a)′ - (f)′ .

(a)′ Trivially $q^{\gamma}_{(u,h)} \in P$ as $q^{\gamma}_{(u,h)} \subset N_u$ (as a set of pairs, by its definition) clearly $q^{\gamma}_{(u,h)} \in N_u$ by the demand $((\forall a \in [N_u]^{<\mu})[a \in N_u]$ from I in the beginning of the proof of 1.5).

(b)′ as $h \subset q^{\beta}_{(u,h)} \subset q^{\gamma}_{(u,h)}$

(c)′ $\beta' < \gamma$ implies $\beta' < \beta$ or $\beta' = \beta$ and check

(d)′ assuming $(u_1,h_1),(u_2,h_2)$ are compatible we have

$$q^{\beta}_{(u_1,h_1)} \restriction (N_{u_1} \cap N_{u_2} \cap \lambda) = q^{\beta}_{(u_2,h_2)} (N_{u_1} \cap N_{u_2} \cap \lambda) .$$

As clearly, $\mathrm{dom}\, q^{\gamma}_{(u_e,h_e)} = \mathrm{dom}\, q^{\beta}_{(u_e,h_e)} \cup (\mathrm{Dom}\,(r) \cap N_{u_e})$ the equality of the domains is easy, similarly check equalities of values.

(e)′ (f)′ immediate.

4.2. Conclusion: Assume, for simplicity only, that V satisfies GCH . Then in some generic extension, not collapsing cardinals nor changing cofinalities,

(a) $2^{\aleph_\alpha} < \aleph_{\alpha+\omega}$ for every α

(b) for every $n < \omega$ and model $N \in K^{<n}_{\sigma}$ and $m < \omega$ and θ for some $k < \omega$ and model M , $|M| < {}_k(\|N\| + \sigma + \theta)$ and $M \to (N)^m_\theta$. (By 3.6 (1) (see remark) and 4.1.)

Proof: Like 1.6 using 4.1 instead of 2.5.

§ 5 $K_4 \subseteq G \to (3)^2_{\aleph_0}$

The question we address is an old one of Erdos and Hajnal. K_n is the complete graph with n vertices.

<u>Question</u>: Is there a graph G which embeds no K_4 such that $G \to (3)^2_{\aleph_0}$?

We get here the consistency of a slightly stronger statement. We still deal with graphs although the proof says something more general. More on the case we are interested in (forbidden infinite subgraphs) will appear later.

5.1. Lemma: Suppose $\mu < \lambda < \kappa$, κ is measurable (or just $\kappa > l(\lambda)$ or $\lambda \to_{wsp} (2k(*))^{\omega,<3}_{\mu,\mu}$), $2 \le m < \omega_1$ and $\lambda = \lambda^{<\lambda}$. For some λ^+-c.c. λ-complete forcing notion P of power κ, $\Vdash_P$" $2^\lambda = \kappa$ and for some graph G of power κ,

(i) $G \to (K_{k(*)})^2_\mu$

(ii) G embeds no $K_{k(*)+1}$."

Proof: The forcing P introduces just the graph G. Let $|G|$, the set of vertices of G be

$$[\kappa]^m_{inc} = \{(\alpha_0, \ldots, \alpha_{m-1}) : \alpha_0 < \ldots < \alpha_{m-1} < \kappa\}.$$

We say $\eta = (\alpha_0, \ldots, \alpha_{m-1})$, $\nu = (\beta_0, \ldots, \beta_{m-1})$ from $[\kappa]^m_{inc}$ are potentially connected if $\alpha_0 < \beta_0 < \alpha_1 < \beta_1 < \ldots < \alpha_{m-1} < \beta_{m-1}$ (or interchange them).

Let $P = \{G : K_{k(*)+1}$ is not embeddable into G and G is a graph as above on $[\mathrm{dom}(G)]^m_{inc}$ where $\mathrm{dom}\, G$ is a subset of κ of power $< \lambda\}$. We say $G_1 < G_2$ if and only if $G_1 = G_2 \restriction [\mathrm{dom}\, G_1]^m_{inc}$. Clearly P is λ-complete, $P \vDash \lambda$-c.c., $\Vdash_P$ "$2^\lambda = \kappa$" and $|P| = \kappa$. Let $\underset{\sim}{Gr}$ be the P-name of $\bigcup\{L : L \in \underset{\sim}{G}_p\}$. It is a graph of the right form. Let $\underset{\sim}{d}$ be a P-name of a function from the set of edges of $\underset{\sim}{Gr}$ to μ and $p \in P$. Let χ be large enough. By the choice of κ and the partition theorem, we can find $U \subseteq \kappa$ such that $|U| = \lambda$ (U is really larger but this does not help). Let $I^\alpha = \{s \subseteq U : |s| \le 2^m\}$ and let $\{M_s : s \in I^\alpha\}$ be such that $U \cap M_s = s$,

$(\forall\alpha,\beta)[\alpha < \beta \wedge \alpha \in U \wedge \beta \in U \Rightarrow (\alpha,\beta) \cap M_\emptyset = \emptyset]$; $M_s \cap M_t = M_{s\cap t}$ (or just $M_s \cap M_t \subset M_{s\cap t}$) $(\forall a \subset M_s)(|a| < \lambda \Rightarrow a \in M_s)$ and $\|M_s\| = \lambda$ and for $s,t \in I^\alpha$, $|s| = |t|$ we have

$H_{s,t} : M_s \to M_t$, an isomorphism onto, so that $H_{s,t}(s) = t$ all the diagrams commute and $\Lambda_s\ [p,P,\lambda,\mu,\kappa,\underset{\sim}{G}_r,\underset{\sim}{d} \in M_s]$.

Now we want to find $p \le q \in P$ such that q forces a monochromatic $K_{k(*)}$. Let $\eta_e = (\alpha_0^e,\alpha_1^e,\ldots,\alpha_{m-1}^e) \in [U]^m$ for $e < k(*)$ such that

$$\alpha_0^0 < \alpha_0^1 < \alpha_0^2 < \ldots \alpha_0^{k(*)-1} < \alpha_1^0 < \alpha_1^1 < \ldots \alpha_1^{k(*)-1} < \ldots < \alpha_i^e \in U ,$$

$$t_e = \text{range } \eta_e .$$

We shall find a condition $q \ge p$. If $q \in P \cap M_{t_0}$, $p \le q$ then we can find $r \in M_{t_0 \cup t_1} \cap P$ and $\xi < \mu$ such that 1. – 6. below holds, where

1. $r \Vdash (\eta_0,\eta_1) \in$ edges of $\underset{\sim}{Gr}$
2. $r \restriction M_{t_0} \ge q$
3. $r \restriction M_{t_1} \ge h_{t_0,t_1}(q)$
4. $r \Vdash \underset{\sim}{d}\ (\eta_0,\eta_1) = \xi$
5. $r \Vdash \forall x,y \in$ vertices (q), if $x,y \notin M_\emptyset$

   $[\{x,y\} \ne \{\eta_0\} \Rightarrow \langle x,h_{t_0,t_1}(y)\rangle \notin$ edges of $\underset{\sim}{Gr}]$.

6. if $r \restriction M_{t_0} \le q' \in P \cap M_{t_0}$ and $q' \restriction M_\emptyset = q'' \restriction M_\emptyset$ and $r \restriction M_{t_1} \le q'' \in P \cap M_{t_1}$ then we can find r' such that $q',q'' \le r' \in M_{t_0 \cup t_1}$ and r' satisfies 1 – 4 and: [$x \in$ vertices $q' - M_\emptyset$ and $y \in$ vertices $q'' - M_\emptyset$ and $(x,y) \in$ edges $r' \Rightarrow (xy) = (\eta_0,\eta_1)$ and $r' \Vdash$ "$\underset{\sim}{d}(\eta_0,\eta_1) = \xi$" .

As P is λ-complete also $P \cap M_{t_0}$ is μ^+-complete so there are q_0,ξ_0 , such that $p \le q_0 \in P \cap M_{t_0}$ and: $\forall q : q_0 \le q \in P \cap M_{t_0}$ we can find r as above for $\xi = \xi_0$. Note $\Vdash$ "the distance in $\underset{\sim}{Gr}$ of η_0 from vertices in $\underset{\sim}{Gr} \cap M_\emptyset \cap \alpha_0$ is $\ge m$".

Now we can find $\xi_0 < \mu$ and $\langle q_\ell^0 : \ell < k(*)\rangle$ such that

(i) $q^0_\ell \in M_{t_\ell}$

(ii) $q^0_\ell \restriction M_\emptyset = q^0_0 \restriction M_\emptyset$

(iii) $h_{t_0,t_i}(q) \le q^0_\ell$

(iv) for $\ell_1 < \ell_2 < k(*)$ we have: if $q^0_{\ell_1} \le q' \in M_{t_{\ell_0}}$ and $q^0_{\ell_2} \le q'' \in M_{t_{\ell_1}}$ and $q' \restriction M_\emptyset = q'' \restriction M_\emptyset$ then we can find r as above.

[Why? We define, by induction on $i < k(*)$, $\langle q^{0,i}_\ell : \ell \le i\rangle$ such that $\langle q^{0,i}_\ell : \ell \le i\rangle$ satisfies (i),(ii),(iii),(iv) above with the natural restrictions. For $i = 0$, $q^{0,0}_0 = q_0$. For $i = j+1$ apply the assertion above (before 1. - 6.) so with $h_{t_\ell,t_0}(q^{o,j}_j)$ here standing for q there; get there r and let

$$q^{0,i}_i = h_{t_0 \cup t_1, t_j \cup t_i}(r \restriction M_{t_1})$$

$$q^{0,i}_j = h_{t_0 \cup t_1, t_j \cup t_i}(r \restriction M_{t_0}),$$

and for $\ell < j$, $q^{0,i}_\ell = q^{0,j}_\ell$.

In the end let $q^0_\ell = q^{o,k(*)-1}_\ell$.

Let $\{(\beta_\ell,\gamma_\ell) : \ell \le \binom{k(*)}{2} = m\}$ list the increasing pairs. Now we define by induction on $\ell \le \binom{k(*)}{2}$

$$\{q^\ell_\beta : \beta \le k(*)\}, \quad r_{\beta_\ell,\gamma_\ell} \text{ such that:}$$

1. $q^{\ell_1}_\beta \le q^{\ell_2}_\beta$ for $\ell_1 \le \ell_2$
2. $q^\ell_\beta \in M_{t_\beta}$
3. $q^\ell_{\beta_1} \restriction M_\emptyset = q^\ell_{\beta_2} \restriction M_\emptyset$
4. $r_{\beta_\ell,\gamma_\ell} \restriction M_{t_{\beta_\ell}} \le q^{\ell+1}_{\beta_\ell}$
5. $r_{\beta,\gamma} \restriction M_{t_{\gamma_\ell}} \le q^\ell_{\gamma_\ell}$
6. $r_{\beta,\gamma} \Vdash \underset{\sim}{d}(\eta_\beta,\eta_\gamma) = \xi_0$

7. If e_{ℓ_i} is an edge of $r_{\beta_{\ell_i},\gamma_{\ell_i}}$ not in

$(M_{t_{\beta_{\ell_i}}} \times M_{t_{\beta_{\ell_i}}}) \cup (M_{t_{\gamma_{\ell_i}}} \times M_{t_{\gamma_{\ell_i}}}) \cup \{(\eta_{\beta_{\ell_i}}, \eta_{\gamma_{\ell_i}})\}$ for $i \in 0,1$ and $\ell_0 \neq \ell_1$

then e_0, e_1 have no vertex in common.

8. If $\gamma \notin \{\alpha_\ell, \beta_\ell\}$ then $\mathrm{edges}(q_\gamma^{\ell+1}) = \mathrm{edges}(q_\gamma^\ell) \cup \mathrm{edges}(q_{\alpha_\ell}^{\ell+1} \restriction M_\emptyset)$.

There is no problem in this - q_0 is tailor-made for this.

Now we define q:

$$\mathrm{dom}\ q = \bigcup_{\beta,\ell} \mathrm{dom}\ q_\beta^\ell \cup \bigcup_{\ell \le \binom{k(*)}{2}} \mathrm{dom}\ r_{\alpha_\ell,\beta_\ell}$$

edges of q = union of the set of edges of q_β^ℓ, $r_{\alpha_\ell,\beta_\ell}$.

(Note that any node in $\mathrm{dom}\ r_{\beta_i,\gamma_i} \setminus (M_{t_{\beta_i}} \cup M_{y_{\gamma_i}})$ is connected.)

(Note that the q_β^ℓ, $r_{\alpha_\ell,\beta_\ell}$ are pairwise compatible.)

The least trivial is to show $K_{k(*)+1}$ is not embeddable into q. Let Ξ be a set of $k(*)+1$ vertices.

Assume that Ξ is a complete graph (in q) and we shall derive a contradiction.

If we omit the edges $\{(\eta_i,\eta_j)\colon\ i < j < k(*)\}$ from q, the resulting graph is obtained by successive edgeless amalgamation (look at the restriction to $\bigcup_{\ell \le i} \mathrm{Dom}\ q_\ell^i \cup_{j<i} \mathrm{Dom}\ r_{\beta_j,\gamma_j}$, for $i \le k(*)$). Hence it has no subgraph isomorphic to $K_{k(*)+1}$. So necessarily for some $i(1)$, $\eta_{i(1)} \in \Xi$. Now by the definition of "potential edge" and as $(m \ge 2$ and) the interval $(\eta_{i(1)}(0), \eta_{i(1)}(1))$ is disjoint to $M_\emptyset$, we have: $\eta_{i(1)}$ is not connected to any vertex from $M_\emptyset$. So $\Xi \cap M_\emptyset = \emptyset$. Now consider the sequence

$$\left\langle \bigcup_{\ell<k(*)} \mathrm{Dom}\ q_\ell^i \cup \bigcup_{j<i} \mathrm{Dom}\ r_{\beta_0,\gamma_j} \setminus M_\emptyset : i \le k(*) \right\rangle$$

and the restrictions of the graph q to them. Easily the first is in P, and in each step we use edgeless amalgamation (we could have started with this argument) so we finish.

Concluding Remarks:

5.2 Easy variants: We can have $G \rightarrow (H)^2_\mu$ such that the family of finite subgraphs of G is S (up to isomorphism) where for some n :

1. $S \neq \emptyset$

2. S closed under edgeless amalgamation

3. If $L_1, \ldots, L_{|H|} \in S$; $i \neq j \Rightarrow L_i \cap L_j = L$;

   $x_i \in L_i$ and the distance of x_i from L in L_i is $\geq n$ then

   $L^* \in S$ where: $\text{vertices}(L^*) = \bigcup_{i=1}^{n} \text{vertices}(L_i)$

   $\text{edges}(L^*) = \bigcup_{i=1}^{n} \text{edges}(L_i) \cup \text{edges}(L^* \restriction \{x_i : i = 1, \ldots, |H|\})$

   $L^* \restriction \{x_1, \ldots, x_{|H|}\} \cong H$.

5.3 Easy Remark: Instead of graphs we can have a model where relations are a partition of the singleton and of the pairs.

5.4 Note that the proof of 5.1 tells us that in 4.2 for $n = 3$ (i.e. coloring of singletons and pairs) we do not need 1.4 but can directly prove hence lowering the required cardinal.

5.5 On generalizing 5.2 to relation and colorings with more places see later works.

References

[1] A. Hajnal and P. Komjath, Embedding graphs and colored graphs.

[2] S. Shelah, Notes on combinatorial set theory. *Israel J. Math.* 14 (1973) 262-277.

[3] S. Shelah, Was Sierpinski right ? I. *Israel J. Math.* 62 (1988) 355-380.

[4] S. Shelah, Was Sierpinski right ? II. Preprint.

[5] J. Nesestril and V. Rodl, Partition (Ramsey) theory, a survey. Colloq. Math. Soc. Janos Bolyai, Vol. 18, North Holland, Amsterdam, (1978), 75-192.

[Sh95] S. Shelah, Canonization theorems and applications, J. of Symb. Logic 40 (1981) 345-353.

[EHMR] P. Erdos, A. Hajnal, A. Mate and R. Rado, Combinatorial Set Theory: Partition relations for cardinals. Disquisitiones Mathematicae Hungaricae 13, Akademiai Kiado Budapest, 1984, North Holland.

[Sc Sh20] J. Schmerl and S. Shelah, On power-like models of hyperinaccessible cardinals, J. of Symb. Logic 37 (1972) 531-537.

Topological Problems for Set-theorists

Franklin D. Tall[1]
University of Toronto

This short note is aimed at set-theorists who have heard there are interesting applications of set theory to topology but are perhaps deterred by an overabundance of terminology e.g., is every weakly $\delta\theta$-refinable space with a regular $\bar{G}_\delta$-diagonal submetrizable? We shall introduce some of the classic problems of set-theoretic topology, explaining the concepts involved, briefly outlining what is known, and referring the reader to relevant literature. We make no attempt to be comprehensive. We assume no more topology than the reader can be expected to recall from the one course taken in graduate school many years ago.

Our criteria for selection (with occasional exceptions) are that a problem should

a) have been around for at least a decade,
b) have been worked on by more than one strong researcher,
c) be well-known to set-theoretic topologists,
d) require a minimum of topological knowledge to state and work on,
e) be apparently set-theoretic in nature.

In addition, we have attempted to select problems from a variety of areas within set-theoretic topology. Those who wish to see large numbers of additional problems may consult the Problems Section in back issues of *Topology Proceedings*.

I should like to thank Jim Baumgartner for inviting me to speak in his seminar on this topic, which led me to compile this assortment. Input from the members of the Toronto Set-theoretic Topology Seminar has also been helpful. At the suggestion of the referee, I have omitted well-known problems which have been surveyed elsewhere: see [To$_2$] and [R], [Wi] and [vD], [Ru], [N$_3$] and [V] for L & S spaces, box products, Dowker spaces, countably compact spaces respectively.

1. The Cardinality of Lindelöf T_2 spaces with points G_δ, and a related problem.

Arhangel'skii proved that Lindelöf T_2 first countable spaces have cardinality $\leq 2^{\aleph_0}$. A natural question is whether first countability can be weakened to "points G_δ". (This condition is equivalent to first countability in compact T_2 spaces.) It is not difficult to show that even if only "T_1" is assumed, such spaces cannot have weakly compact cardinality, or cardinality $\geq$ the first measurable. Shelah showed it consistent with GCH that there exist an example of size $\aleph_2$, and, assuming the consistency of a weak compact, the consistency with CH that there exists no example of size $\aleph_2$. One certainly expects that a supercompact should suffice to obtain the consistency of there being no examples of size greater than $\aleph_1$, but as usual, the difficulty is in proving a suitable preservation lemma. All one needs to know about this problem appears in [J].

Shelah's example (reworked and improved in [HJ$_2$]) is a graph constructed by forcing or by a morass with built in $\diamond$.

<u>Definition</u>. $L(X) = \min\{\kappa$: every open cover of X has a subcover of cardinality $\leq\kappa\} + \aleph_0$.

This <u>cardinal function</u> (i.e. function from the class of topological spaces into the class of cardinals, invariant under homeomorphism) is known as the <u>Lindelöf degree</u>.

Shelah's example (or rather, two versions of it) partially solve another problem on Lindelöf spaces: how big is $L(X \times Y)$, where X and Y are Lindelöf? There is an easy example of a (hereditarily) Lindelöf space X such that $L(X^2) = 2^{\aleph_0}$: X is the <u>Sorgenfrey line</u>, i.e. the real line equipped with a basis of all $[a,b)$. The line $y = -x$ in the plane is closed discrete; its complement plus the open sets witnessing discreteness form an open cover with no subcover of size $< 2^{\aleph_0}$. One might conjecture that that is as bad as matters get, that if X and Y are Lindelöf, $L(X \times Y) \leq 2^{\aleph_0}$. This may well be consistently true, but it is consistently false. See [J_2] for the example where $L(X \times Y) = \aleph_2 > 2^{\aleph_0} = \aleph_1$. Indeed, the only upper bound known for $L(X \times Y)$ at present is the first strongly compact cardinal. (Proof: generalize Tychonoff's Theorem.) Surely this can be improved.

2. <u>Linearly Lindelöf Spaces</u>.

<u>Definition</u>. A point x is a <u>complete accumulation point</u> of a set Y if for each open U containing x, $|U \cap Y| = |Y|$.

It is not difficult to show that a space is compact if and only if each infinite set has a complete accumulation point.

<u>Definition</u>. A space is <u>finally compact in the sense of complete accumulation points</u> (FCCAP), if every uncountable set has a complete accumulation point.

It is not difficult to show that Lindelöf implies FCCAP, but the natural attempt to prove the converse encounters difficulty with covers of size $\aleph_\omega$. Indeed, there is a completely regular counterexample [M]. However the assumption that every sequence $\{F_n\}_{n<\omega}$ of closed sets such that $F_n \supseteq F_{n+1}$ and $\bigcap_{n<\omega} F_n = \emptyset$ can be fattened to a sequence $\{U_n\}_{n<\omega}$ of open sets such that $U_n \supseteq U_{n+1}$, $U_n \supseteq F_n$, and $\bigcap_{n<\omega} U_n = \emptyset$ is sufficient to reduce covers with size of countable cofinality to ones with regular size, and hence get that FCCAP implies Lindelöf. The open question is whether there is a <u>normal</u> FCCAP space that is not Lindelöf. For those who know the terminology, this is asking for a special kind of <u>Dowker space</u> [Ru]. It is interesting that FCCAP has an equivalent form that appears even closer to Lindelöf.

<u>Definition</u>. A space is <u>linearly Lindelöf</u> if every well-ordered by inclusion open cover has a countable subcover.

To work on this problem, the minimal topological background can be found in [M] and [H].

3. <u>Non-Archimedean spaces</u>.

<u>Definition</u>. A space is <u>non-Archimedean</u> if it has a basis $\mathcal{B}$ s.t. $B, B' \in \mathcal{B}$ implies $B \subsetneq B'$ or $B' \subsetneq B$ or $B \cap B' = \emptyset$. A collection $\mathcal{U}$ of subsets of a space is <u>point-countable</u> if each point is in at most countably many elements of $\mathcal{U}$. A space is <u>perfectly normal</u> if it is normal and closed sets are G_δ's.

The linear order obtained by squashing a Souslin tree is an example of a perfectly normal non-Archimedean non-metrizable space with a point-countable base of size $\aleph_1$.

Todorčević [To$_1$] proved that, assuming MA plus the non-existence of weak Kurepa [also known as Canadian] trees, every perfectly normal non-Archimedean space of weight $\leq\aleph_1$ is metrizable. The question is whether MM, for example, allows us to drop the weight restriction, even adding "point-countable base". For further information, also see [N$_1$] and [N$_2$].

4. Reflection problems.

In recent years there has been a growing interest in problems that ask whether, if all small subspaces of X have a certain property, then X does. Here is a typical one:

Problem. If X is first countable and every subspace of size $\aleph_1$ is metrizable, is X metrizable?

A non-reflecting stationary set of ω-cofinal ordinals is a counterexample [HJ$_1$]. The non-existence of such sets (which holds in the model obtained by Lévy-collapsing a supercompact to ω_2 [B] implies there is no counterexample in which each point has a neighbourhood of size $\leq\aleph_1$ [Do]. Replacing "$\aleph_1$" by "$<2^{\aleph_0}$" yields a positive answer if supercompact many Cohen reals are adjoined [DTW]; in the given problem we are missing a preservation lemma.

Another problem of the same sort presumably requires less topological knowledge.

Definition. A space is κ-collectionwise Hausdorff if for each closed discrete subspace $\{x_\alpha\}_{\alpha<\lambda}$, $\lambda\leq\kappa$, there exist disjoint open sets U_α, with $x_\alpha\in U_\alpha$.

Problem. Is it consistent that every first countable $\aleph_1$-collectionwise Hausdorff space is κ-collectionwise Hausdorff for all κ?

The answer is again "no" if there is a non-reflecting stationary set of ω-cofinal ordinals, but "yes" if the space is locally separable. See [F$_1$], [F$_3$], [S], [T$_2$], [T$_3$].

5. Omitting Cardinals.

There is a problem-generating machine manufactured in Hungary, into which one inputs a cardinal function and two cardinals and outputs a problem. We mention only one such: does every Lindelöf space of cardinality $2^{2^{\aleph_0}}$ have a Lindelöf subspace of cardinality $2^{\aleph_0}$? See [JW].

6. Partition problems.

W. Weiss and others have extended the partition calculus to topological spaces, thereby creating a host of new theorems and problems. For a survey, see [We]. One of the more interesting questions is the following:

<u>Problem</u>. Can every Hausdorff space be partitioned into two pieces, neither of which includes a Cantor set? V = L implies "yes" [HJW], but perhaps it's just true in ZFC.

7. <u>Can ω^* be homeomorphic to ω_1^*?</u>

This problem and the next one are purely set-theoretic but arise frequently in topological contexts. ω^* is the Stone-Cech compactification of ω, with the natural copy of ω that lies within removed. Similarly for ω_1^*. By Stone duality, the question of whether the two spaces can be homeomorphic is equivalent to asking whether the Boolean algebras $\mathcal{P}(\omega)/\mathrm{Fin}$, $\mathcal{P}(\omega_1)/\mathrm{Fin}$ can be isomorphic. In all familiar models the answer is no; in particular MA and $2^{\aleph_0} < 2^{\aleph_1}$ provide negative answers. Although many people have worked on this, little is known. The completions of the two algebras are isomorphic [BW], but that doesn't seem to help. For pairs of distinct infinite cardinals κ,λ, it is known that ω^* and ω_1^* are the only infinite κ^* and λ^* that could conceivably be homeomorphic [BF].

8. <u>Can there be a small dominating family in ${}^{\omega_1}\omega$?</u>

The structure of ${}^{\omega}\omega$ is by now reasonably well understood, but the same cannot be said for ${}^{\kappa}\omega$, κ an uncountable cardinal. In particular, it would be interesting to known if it's consistent that there be a family $\mathcal{F}$ of fewer than $2^{\aleph_1}$ functions from ω_1 to ω, such that any function from ω_1 to ω is exceeded everywhere by some member of $\mathcal{F}$. An example of the topological import of this question can be found in [W_2]. Steprāns [S] and Jech-Prikry [JP] showed that there is no such small dominating family if $\mathrm{cf}(2^{\aleph_0}) < \min(2^{\aleph_1}, \aleph_{\omega_1})$, while the existence of such a family implies the existence of a measurable cardinal in an inner model.

9. <u>Are locally compact normal metacompact spaces paracompact?</u>

There used to be many problems in general topology which asked for the implications among rather basic properties. Virtually all have been settled now, often being shown to be undecidable. One of the few that remains is the one mentioned above. Recall

<u>Definition</u>. A space is <u>metacompact</u> (<u>paracompact</u>) if every open cover has a point-finite (locally finite) open refinement.

Several things are known. To obtain a positive answer, it suffices to prove collectionwise Hausdorffness. If "metacompact" is strengthened so that for each open cover there is an $n \in \omega$ such that it has a point-n refinement, again the answer is positive [D]. Under V = L, the answer is also positive [W_1]. If there is a counterexample, there is one which is a subspace of $PR(C_\kappa)$ for some κ [D]. C_κ is κ with the cofinite topology. $PR(C_\kappa)$ is the topology on the collection of non-empty finite subsets of κ with basis of sets $[A,U] = \{B \in [\kappa]^{<\omega}: A \subseteq B \subseteq U\}$, where $A \in [\kappa]^{<\omega}$ and U is open in C_κ.

10. Compact spaces with hereditarily normal powers.

A curious theorem of Katevov [K] states that compact spaces with hereditarily normal cubes are metrizable. One naturally wonders whether "squares" suffices. Under MA (with or without ~CH), the answer is "no", but whether the question is undecidable is unknown. [GN] surveys the situation.

In conclusion, I hope that this list of easily stated problems will stimulate even more set theorists to lend their talents to the problem-strewn field of set-theoretic topology.

Footnote

1. The author acknowledges support from Grant A-7354 of the Natural Sciences and Engineering Research Council of Canada.

References

[A] C.E. Aull, Some base axioms for topology involving enumerability, 54-61 in Gen. Top. and its relations to Modern Anal. and Alg. (Proc. Kanpur Top. Conf., 1968), Academia, Prague, 1971.

[AT] U. Abraham and S. Todorčević, Martin's Axiom and first countable S- and L-spaces, 327-346 in Handbook of Set-theoretic Topology, ed. K. Kunen and J.E. Vaughan, North-Holland, Amsterdam, 1984.

[B] J.E. Baumgartner, A new class of order types, Ann. Math. Logic 9(3) (1976) 187-222.

[BF] B. Balcar and R. Frankiewicz, To distinguish topologically the space m^*, II, Bull. Acad. Polon. Sci. Sér. Mat. Astronom. Phys. 26 (1978) 521-523.

[BW] S. Broverman and W. Weiss, Spaces co-absolute with $\beta N-N$, Top. Appl. 12 (1981) 127-133.

[D] M. Daniels, Normal locally compact boundedly metacompact spaces are paracompact - an application of Pixley-Roy spaces, Canad. J. Math. 35(1983) 827-833.

[Do] A. Dow, An introduction to applications of elementary submodels to topology, Top. Proc., to appear.

[vD] E.K. van Douwen, Covering and separation properties of box products, 55-130 in Surveys in General Topology, ed. G.M. Reed, Academic Press, New York, 1980.

[vDTW] E.K. van Douwen, F.D. Tall, W.A.R. Weiss, Nonmetrizable hereditarily Lindelöf spaces with point-countable bases from CH, Proc. Amer. Math. Soc. 64 (1977) 139–145.

[DTW] A. Dow, F.D. Tall, W. Weiss, New proofs of the consistency of the normal Moore space conjecture, Top. Appl., to appear.

[F_1] W.G. Fleissner, Separation properties in Moore spaces. Fund. Math. 98 (1978) 279–286.

[F_2] W.G. Fleissner, The normal Moore space conjecture and large cardinals, 733–760 in Handbook of Set-theoretic Topology, ed. K. Kunen and J.E. Vaughan, North-Holland, Amsterdam, 1984.

[F_3] W.G. Fleissner, Left-separated spaces with point-countable bases, Trans. Amer. Math. Soc. 294 (1986) 665–678.

[GN] G. Gruenhage and P.J. Nyikos, Normality in X^2 for compact X, preprint.

[H] R.W. Heath, Screenability, pointwise paracompactness and metrization of Moore spaces, Canad. J. Math. 16 (1964) 763–770.

[HJ_1] A. Hajnal and I. Juhász, On spaces in which every small subspace is metrizable, Bull. Acad. Polon. Sci. Sér. Math. Astronom. Phys. 24 (1976) 727–731.

[HJ_2] A. Hajnal and I. Juhász, Lindelöf spaces á la Shelah, Coll. Math. Soc. J. Bolyai 23 (1978) 555–567.

[HJW] A. Hajnal, I. Juhász, W. Weiss, Ramsey type theorems for topological spaces, in preparation.

[Ho] N.R. Howes, Ordered coverings and their relationship to some unsolved problems in topology, 60–68 in Proc. Washington State U. Conf. on Gen. Top., March 1970, Pullman, Washington, 1970.

[J] I. Juhász, Cardinal Functions II, 63–110 in Handbook of Set-theoretic Topology, ed. K. Kunen and J.E. Vaughan, North-Holland, Amsterdam, 1984.

[JP] T. Jech and K. Prikry, Cofinality of the partial ordering of functions from ω_1 into ω under eventual domination, Math. Proc. Camb. Phil. Soc. 95 (1984) 25–32.

[JW] I. Juhász and W. Weiss, A Lindelöf scattered space that omits $\underline{c}$, Top. Appl. (to appear).

[K] M. Katetov, Complete normality of Cartesian products, Fund. Math. 36 (1948) 271–274.

[L] L.B. Lawrence, The box product of countably many copies of the rationals is consistently paracompact, preprint.

[M] A. Miščenko, Finally compact spaces, Sov. Math. Dokl. 145 (1962) 1199–1202.

[N_1] P.J. Nyikos, Some surprising base properties in topology, 427–450 in Studies in Topology, ed. N. Stavrakis, Academic Press (New York), 1975.

[N_2] P.J. Nyikos, Order-theoretic basis axioms, 367–397 in Surveys in General Topology, ed. G.M. Reed, Academic Press, New York, 1980.

[N_3] P. Nyikos, Progress on countably compact spaces, 379–410 in General Topology and its Relations to Modern Analysis and Algebra VI, Proc. 6th Prague Top. Symp. 1986, Heldermann Verlag, Berlin, 1988.

[R] J. Roitman, Basic S and L, 295–326 in Handbook of Set-theoretic Topology, ed. K. Kunen and J.E. Vaughan, North-Holland, Amsterdam, 1984.

[Ru] M.E. Rudin, Dowker Spaces, 761–780 in Handbook of Set-theoretic Topology, ed. K. Kunen and J.E. Vaughan, North-Holland, Amsterdam, 1984.

[S] S. Shelah, Remarks on λ-collectionwise Hausdorff spaces, Top. Proc. 2 (1977) 583–592.

[St] J. Steprāns, Some results in set theory, Thesis, University of Toronto, 1982.

[T_1] F.D. Tall, Normality versus collectionwise normality, 685–732 in Handbook of Set-theoretic Topology, ed. K. Kunen and J.E. Vaughan, North-Holland, Amsterdam, 1984.

[T_2] F.D. Tall, Topological Applications of Supercompact and Huge Cardinals, 545–558 in General Topology and its Relations to Modern Analysis and Algebra VI, Proc. 6th Prague Top. Symp. 1986, Heldermann Verlag, Berlin, 1988.

[T_3] F.D. Tall, Topological applications of generic huge embeddings, Trans. Amer. Math. Soc., to appear

[To_1] S. Todorčević, Some consequences of MA + ~wKH, Top. Appl. 12 (1981) 187–282.

[To_2] S. Todorčević, Partition problems in general topology, American Mathematical Society, Providence, 1988.

[V] J.E. Vaughan, Countably compact and sequentially compact spaces, 569–602 in Handbook of Set-theoretic Topology, ed. K. Kunen and J.E. Vaughan, North-Holland Amsterdam, 1984.

[W_1] S. Watson, Locally compact normal spaces in the constructible universe, Canad. J. Math. 34 (1982) 1091–1096.

[W_2] S. Watson, Separation in countably paracompact spaces, Trans. Amer. Math. Soc. 290 (1985) 831–842.

[We] W. Weiss, Partitioning topological spaces, in Mathematics of Ramsey Theory, ed. J. Nesetril and V. Rödl, to appear.

[Wi] S.W. Williams, Box products, 169–200 in Handbook of Set-theoretic Topology, ed. K. Kunen and J.E. Vaughan, North-Holland, Amsterdam, 1984.

A Beginning For Structural Properties of Ideals on $P_\kappa\lambda$

William S. Zwicker
Department of Mathematics
Schenectady, New York 12308
U.S.A.

§0. Introduction

This paper flows from the observation that coding sets, introduced in [Z1], provide a means by which one can begin to carry out for $P_\kappa\lambda$ a program analogous to that which Baumgartner, Taylor, and Wagon developed for an uncountable cardinal κ in their paper, "Structural Properties of Ideals" [BTW]. Essential background is provided by D. Carr's papers ([C1],[C2],[C3]), which also develop the basic notation for ideals on $P_\kappa\lambda$, and of course by Jech [J1].

Our approach will be to view $P_\kappa\lambda$ as a set partially ordered by $\subsetneq$; κ and λ will always be cardinals with κ regular and $\kappa \leq \lambda$. All ideals on $P_\kappa\lambda$ will be assumed to be κ-additive extensions of $I_{\kappa\lambda}$, the ideal of "not unbounded" subsets of $P_\kappa\lambda$, where "unbounded" means cofinal. Thus for any $X \in I_{\kappa\lambda}$, X has a "nubound" - an element of $P_\kappa\lambda$ that no element of X properly contains. Ideals on κ will likewise be assumed to be κ-additive extensions of I_κ, the ideal of bounded subsets of κ.

The first conundrum facing a theory of ideals on $P_\kappa\lambda$ is that any reasonable definition of "selective" (which has occurred to any one!) fails to be satisfied by the ideal $NS_{\kappa\lambda}$ of non-stationary subsets of $P_\kappa\lambda$. Since $NS_{\kappa\lambda}$ is the prototypical normal ideal on $P_\kappa\lambda$, and one certainly want normal ideals to be selective, this *is* a problem. One solution is to strengthen the definition of "normal" on $P_\kappa\lambda$ so that it does imply selective; the cost is that $NS_{\kappa\lambda}$ cannot have the stronger property. This approach begins to seem more natural when one recalls that while every normal measure on κ has the partition property, such is not the case on $P_\kappa\lambda$ [M] - this result had shed doubt on the traditional definition of $P_\kappa\lambda$ normality some time ago.

§1 Normality and Selectivity

A regressive function f on a set $A \subseteq P_\kappa\lambda$ is one satisfying $f(x) \in x$ for each non-empty $x \in A$, and an ideal I on $P_\kappa\lambda$ is (point) normal if every regressive function on a set A in I^+ is constant on some set in I^+. These are the usual definitions for "regressive" and "normal" found in the literature.

Theorem 1.1 [D. Carr]: $NS_{\kappa\lambda} \upharpoonright B$ is (point) normal for any stationary subset, B, of $P_\kappa\lambda$, and every normal ideal on $P_\kappa\lambda$ extends $NS_{\kappa\lambda}$ (see [C1]).

Thus, the traditional notion of normality on $P_\kappa\lambda$ does yield the analogue to Neumer's theorem on κ. For A and B subsets of $P_\kappa\lambda$ we will define $f : A \to B$ to be set-regressive if $f(x) \subsetneq x$ for each element of A which properly contains some element of B. For $B \subseteq P_\kappa\lambda$ and $y \in P_\kappa\lambda$, $B{\restriction}y$ will denote $B \cap \{x \in P_\kappa\lambda \mid x \subsetneq y\}$. If I is any ideal on $P_\kappa\lambda$ we will define a set $B \subseteq P_\kappa\lambda$ to be a witness index for I if

(1) $B \in I^*$,

and (2) For every set A in I^+, every set-regressive $f : A \to B$ is constant on some set in I^+,

and (3) $\{y \in B \mid \bigcup(B{\restriction}y) = y\} \in I^*$.

An ideal I for which a witness index exists is said to be set-normal.

Proposition 1.2 If I is a set-normal ideal, then I is (point) normal.

proof: Use property (3) above to replace a (point) regressive function f with a set-regressive function g satisfying $f(x) \in g(x)$, then apply κ-additivity of I.

The definition of set-normality goes back to Jech who, in [J2], noted that a (point) normal measure on $P_\kappa\lambda$ is set-normal if and only if it has the partition property. (He could afford to leave condition (3) off.)

An ideal I on $P_\kappa\lambda$ is selective if for every I-small function f with domain $P_\kappa\lambda$ (where f if I-small means that $f^{-1}[f(x)] \in I$ for each $x \in P_\kappa\lambda$) there is a set $Y \in I^*$ on which f is comparably 1:1, which means that $f(x) \neq f(y)$ for every $x,y \in Y$ which are comparable in the $\subsetneq$ ordering. The equivalent in terms of partitions is that for any partition P of $P_\kappa\lambda$ into sets in I there exists a set A in I^* which intersects each piece of P in an anti-chain. Were we to demand that A meet each piece in a singleton, the property would be impossible to satisfy when $\kappa < \lambda$, as witnessed by the partition $P_\kappa\lambda = \dot{\bigcup}_{\delta<\kappa} \kappa^{-1}(\delta)$, where $\kappa(x) = \bigcup(x \cap \kappa)$.

Proposition 1.3 If I is a set-normal ideal, then I is selective.

proof: The proof is just as in the κ case. If I is set-normal with witness index B and f is any function with domain $P_\kappa\lambda$, define k as follows:

$$k(y) = \begin{cases} \text{any } x \subsetneq y \text{ with } x \in B \text{ and } f(x) = f(y), & \text{if such an } x \text{ exists} \\ \text{undefined, otherwise.} \end{cases}$$

Let $A = \{y \in P_\kappa\lambda \mid k(y) \text{ is defined}\}$. If $A \in I$ then f is comparably one-to-one on $B - A \in I^*$. If $A \in I^+$, choose $D \in I^+$ with k constant on D. Then f is constant on D, and so f is not I-small.

Proposition 1.4 When $\lambda > \kappa$, $NS_{\kappa\lambda}$ is not selective, hence not every (point) normal ideal is selective.

proof: Baumgartner and Velleman, independently, have shown that when $\lambda > \kappa$, the κ function (introduced above) can not be comparably one-to-one on a cub (closed and unbounded subset of $P_\kappa\lambda$ - see [Z1]). This function can clearly not be constant on an unbounded subset of $P_\kappa\lambda$.

We'd like a theorem for set-normality analogous to theorem 1.1, and coding sets appear to provide the key. In particular, they show that set-normal ideals frequently exist. A subset A of $P_\kappa\lambda$ is a coding set if it comes equipped with a one-to-one function $c : A \to \lambda$ such that for $x,y \in A$, $x \subsetneq y$ if and only if $c(x) \in y$. A stationary coding set, or "SC", is a coding set that is stationary.

Velleman and Stanley [V] and Shelah [S1 and S2] have shown that SC's consistently exist for a broad variety of κ's and λ's under various hypotheses. Furthermore, Shelah has shown that SC's exist for $P_\kappa\kappa^+$ whenever κ is a successor cardinal $\geq \aleph_3$. Note that 1.3 in conjunction with 1.5 and 1.4 implies that for $\lambda > \kappa$, an SC can never be cub.

Theorem 1.5 $NS_{\kappa\lambda} \upharpoonright B$ is set-normal for any SC,B, and under the hypothesis that $P_\kappa\lambda$ has an unbounded subset of size λ, every set normal ideal extends $NS_{\kappa\lambda} \upharpoonright B$ for some SC, B.

Corollary 1.6 If $P_\kappa\lambda$ has an unbounded set of size λ, then any set-normal restrictions of $NS_{\kappa\lambda}$ is equal to $NS_{\kappa\lambda} \upharpoonright B$ for some SC,B.

The extra hypothesis is fairly mild; if $\lambda = \kappa^+, \kappa^{++}, \ldots \kappa^{\overbrace{++\cdots+}^{n}} \ldots$, for $n \in \omega$, then $P_\kappa\lambda$ has an unbounded subset of size λ. Some interesting definitions arise naturally in the proof of 1.5. We will call an ideal I on $P_\kappa\lambda$...

... outstripping if given any $f : P_\kappa\lambda \to P_\kappa\lambda$ there is a set $A \in I^*$ such that for each $x,y \in A$, if $x \subsetneq y$ then $f(x) \subsetneq y$ (we'll say A outstrips f),

... lean if there exists a set $A \in I^*$ with $|A| = \lambda$,

... and κ-footed if there exists a set $A \in I^*$ such that $|A \upharpoonright x| < \kappa$ for each $x \in P_\kappa\lambda$.

The last property is closely related to one clause in the definition of simplified morass (see [V2]).

Proof of 1.5 If B is a SC then $NS_{\kappa\lambda} \upharpoonright B$ is (point) normal. To see that it is set-normal, let $A \in (NS_{\kappa\lambda} \upharpoonright B)^+$ and let $f : A \to B$ be any set-regressive function. Then $c \circ f$ is a (point) regressive function on $B \cap A$, so if $(c \circ f)(x) = c(y_0)$

for each x in $D \in (NS_{\kappa\lambda} \upharpoonright B)^+$, f is constant on D with value y_0.

Now assume that I is set-normal with witness index B and that Y is a cofinal subset of $P_\kappa\lambda$ with $|Y| = \lambda$. B must witness I's κ-footedness, for if not there would be some $x \in P_\kappa\lambda$ and $i : \kappa \xrightarrow{1:1} B \upharpoonright x$. Then $i \circ \kappa$ would be set-regressive on $B \cap \hat{x}$ (where $\hat{x}$, the cone above x, is $\{y \in P_\kappa\lambda \mid x \subseteq y\}$), but $i \circ \kappa$ can never be constant on a cofinal set.

It follows that I is lean, since $B = \bigcup_{x \in Y} B\upharpoonright x$ is the union of λ-many sets, each of cardinality less than κ. Also, I must be outstripping; let $f : P_\kappa\lambda \to P_\kappa\lambda$ be arbitrary and define k by

$$k(y) = \begin{cases} \text{any } x \subsetneq y \text{ with } x \in B \text{ and } f(x) \text{ not a} \\ \quad \text{proper subset of } y, \text{ if such an } x \text{ exists} \\ \text{undefined, otherwise.} \end{cases}$$

Then $A = \{y \mid k(y) \text{ is undefined}\}$ must be in I^*, for if $P_\kappa\lambda - A$ were in I^+, we could apply set-normality to get an $X \in I^+$ on which k is constant, which is impossible. Now $A \cap B$ outstrips f.

To finish, let $c : B \to \lambda$ be any $1:1$ function and choose $A \subseteq B$ with $A \in I^*$ and such that A outstrips the function $h(x) = x \cup \{c(x)\}$. If we set $E = A \cap C$ where $C = \{y \in P_\kappa\lambda \mid (\forall \delta \in y)(c^{-1}(\delta) \subsetneq y)\}$, then as C is a closed and unbounded subset of $P_\kappa\lambda$ and as I is (point) normal, $E \in I^*$ must be stationary. E is a coding set by construction and I extends $NS_{\kappa\lambda} \upharpoonright E$, so we are done.

This proof breaks down into a series of lemmas. Rather than state them we express the relationships with a diagram. N.B. In Figure 1, below, a single arc joining two arrows indicates that the *conjunction* of the two arrow sources implies the target.

Some Properties of $P_\kappa\lambda$ Ideals (Figure 1)

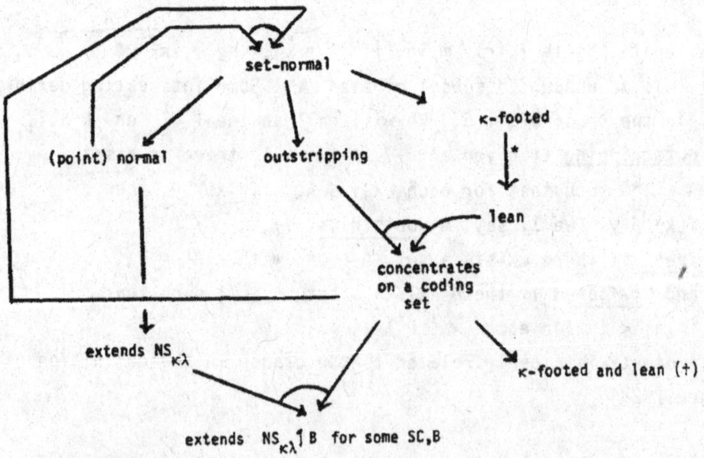

* if $P_\kappa\lambda$ has an unbounded set of cardinality λ.

(†) This is immediate since the coding set B must have $|B| < \lambda$ and $|B|x| < \kappa$ for each $x \in P_\kappa\lambda$

§2 The Q-Point Puzzle

Recall that an ideal, I, on κ is a q-point if every I_κ-small function f with domain κ is 1 : 1 on some set in I*. Equivalently, given any partition P of κ into sets in I_κ, there is a set A in I* intersecting each piece of P in a singleton. If we specialize the definition of outstripping to κ in the obvious way, by requiring for each $f:\kappa \to \kappa$ the existence of some set $A \in I^*$ such that for each $\alpha,\beta \in A$ with $\alpha < \beta, f(\alpha) < \beta$, then we will show in Theorem 2.1 that an ideal I on κ is outstripping if and only if it is a q-point. It is natural, then, to define an ideal I on $P_\kappa\lambda$ to be a q-point if every $I_{\kappa\lambda}$-small function f with domain $P_\kappa\lambda$ is comparably 1 : 1 on some set in I*. Equivalently, given any partition of $P_\kappa\lambda$ into sets each of which is in $I_{\kappa\lambda}$, there exists a set A in I* intersecting each piece of P in an anti-chain. On $P_\kappa\lambda$, outstripping *appears* to be a stronger property than q-point. Is it actually stronger? This is the q-point puzzle, and it is open. Attempts to solve it produced much of the rest of this paper as spin-off.

Theorem 2.1 On κ, q-pointedness is equivalent to outstripping. On $P_\kappa\lambda$, an outstripping ideal is necessarily a q-point.

Theorem 2.2 If I is an outstripping ideal on $P_\kappa\lambda$, then I concentrates on a well-founded subset of $P_\kappa\lambda$.

Theorem 2.3 If I extends $NS_{\kappa\lambda} \restriction B$ for some SC, B, then I is outstripping. (This is the analogue of the theorem on κ which says that any extension of NS_κ is a q-point).

Corollary 2.4 (Immediate from 2.3 and figure 1). For an ideal I that is a lean extension of $NS_{\kappa\lambda}$, I is outstripping if and only if there exists some coding set A in I*.

The reader who attacks the q-point puzzle might wish to first think about the following, presumably easier, question: Can "q-point" replace outstripping in 2.2 or 2.4? In this connection, a tempting conjecture is that an ideal which is both a q-point and concentrates on a coding set must be outstripping, but I have been unable to show this.

Proof of 2.1 Let I be a q-point on κ and $f:\kappa \to \kappa$ be arbitrary. Let $\eta_0, \eta_1, \ldots, \eta_\delta, \ldots$ enumerate a closed and unbounded set of closure points for f $(\alpha < \eta_\delta \to f(\alpha) < \eta_\delta)$. Let $g(\alpha)$ = the least δ with $\alpha < \eta_\delta$. Then g is I_κ-small, and if A is any set on which g is 1 : 1, A clearly outstrips f.

If I is an outstripping ideal on $P_\kappa\lambda$ and f is an $I_{\kappa\lambda}$-small function defined on $P_\kappa\lambda$, define $h : P_\kappa\lambda \to P_\kappa\lambda$ by $h(x)$ = any nubound for $f^{-1}[f(x)]$. Choose $A \in I^*$ which outstrips h. Then f is comparably 1:1 on A. The proof in the κ case is the obvious specialization.

Proof of 2.2 First note that it is always possible to build a well-founded cofinal subset Y of $P_\kappa\lambda$. Enumerate $P_\kappa\lambda$ by its cardinality. Then walk through the enumeration, throwing an element into Y only if it is not a subset of any element already in Y. Now let I be an outstripping ideal on $P_\kappa\lambda$ and define $f : P_\kappa\lambda \to P_\kappa\lambda$ by $f(x)$ = any element y of Y with $x \subsetneqq y$. If A is a set in I^* which outstrips f then A is well-founded, since a descending chain in A would perforce interlace with a descending chain in Y.

Proof of 2.3 It is easy to see that any extension of an outstripping ideal is outstripping (and any extension of a q-point is a q-point). We already know $NS_{\kappa\lambda} \upharpoonright B$ is outstripping when B is a SC, so we are done. Alternately, assume I extends $NS_{\kappa\lambda} \upharpoonright B$ for some SC, B, and let $f : P_\kappa\lambda \to P_\kappa\lambda$. Choose $\overline{f}$ to be any function satisfying $\overline{f} : P_\kappa\lambda \to B$ and $\overline{f}(x) \supseteq f(x)$ for each $x \in P_\kappa\lambda$. Define $k : \lambda \to \lambda$ by $k(\delta) = c(\overline{f}(c^{-1}(\delta)))$, where c is B's coding function and $\delta \in c''B$. Let $C \subseteq P_\kappa\lambda$ be the cub of all elements of $P_\kappa\lambda$ closed under k. Then $C \cap B$ outstrips f and is in I^*.

§3 Quasi-Normality and P^*-points

Recall that an ideal I on κ is a p-point if for every I-small function f with domain κ there exists a set A in I^* on which f is I_κ-small, a q-point if for every I_κ-small function f with domain κ there exists a set A in I^* on which f is one-to-one, and I is selective if for every I-small function f with domain κ there exists a set A in I^* on which f is one-to-one. Equivalently, a p-point ideal allows, for any partition of κ into sets in I, for the existence of a set A in I^* intersecting each piece in a set in I_κ, a q-point ideal allows, for any partition of κ into sets in I_κ, for the existence of a set A in I^* intersecting each piece in a singleton, and a selective ideal allows, for any partition of κ into sets in I, for the existence of a set A in I^* intersecting each piece in a singleton. Thus it is immediate from the definitions that an ideal on κ is selective if and only if it is both a p-point and a q-point. If we now define an ideal I on $P_\kappa\lambda$ to be a p-point if for every I-small function f with domain $P_\kappa\lambda$ there exists a set A in I^* on which f is $I_{\kappa\lambda}$-small, it is equally immediate that an ideal on $P_\kappa\lambda$ is selective if and only if it is both a p-point and a q-point.

These relationships prompt several natural questions. Outstripping appears to be a strengthening of q-pointedness in the $P_\kappa\lambda$ context; are there analogous strengthenings of p-pointedness and of selectivity? Were they already known in the κ context? Will they shed light on the q-point puzzle?

There are indeed such analogues; in [W], Weglorz defined quasi-normality for an ideal on κ and proved it equivalent to selectivity. This turns out to be the κ-analogue of the "strong" version of selectivity. I have found no direct reference in the literature to "p*-point", the κ-analogue of the "strong" version of p-point. Its introduction is probably worthwhile, because it suggests a reconsideration (still in the κ context) of Weglorz's proof that selectivity implies quasi-normality, yielding a new proof on κ which seems to me to be more transparent than the original.

In Figure 2, below, a double arc linking arrows indicates that the conjunction of the two arrow sources is equivalent to the target. In the κ-context, the new

<u>Properties of Ideals on κ</u> (Figure 2)

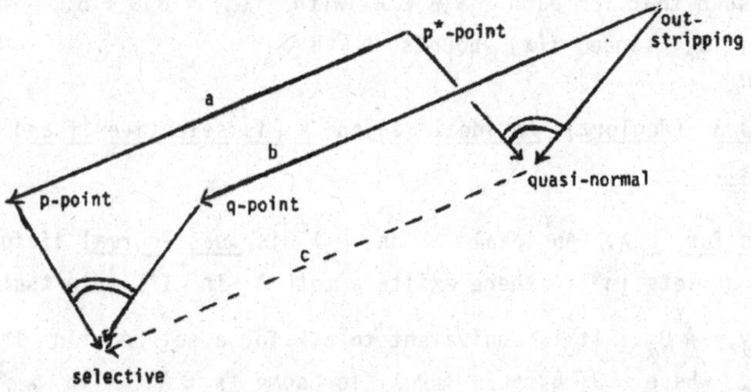

proof of c and its converse consists of the observation (immediate from the definitions) that p^* and outstripping fit together to form quasi-normality (in much the same way that p-point and q-point fit together to form selectivity), together with proofs of a,b and their converses. In the $P_\kappa\lambda$ context, a and b (and hence c) are proved just as on κ, while all three converses are open questions. Thus the attempt to shed light on the q-point puzzle has yielded additional puzzles.

<u>Definitions and Proofs for κ</u>: An ideal I on κ is <u>quasi-normal</u> if for every sequence $\langle B_\alpha\rangle_{\alpha<\kappa}$ of sets in I there exists a set A in I^* such that for every $\alpha,\beta \in A$ with $\alpha < \beta$, $\beta \notin B_\alpha$. (It is equivalent to ask for a set A' in I^* for which $\nabla_{\alpha\in A'} B_\alpha \in I$, where $\nabla_{\alpha\in A'} B_\alpha = \{\beta < \kappa \mid$ for some $\alpha \in A'$ with $\alpha < \beta$, $\beta \in B_\alpha\}$; this is Weglorz's original formulation). An ideal I on κ is a <u>p*-point</u> if for every sequence $\langle B_\alpha\rangle_{\alpha<\kappa}$ of (not necessarily disjoint) sets in I, there exists a set A in I^* and a progressive function $f:\kappa \to \kappa$ such that for every $\alpha,\beta \in A$ with $f(\alpha) < \beta, \beta \notin B_\alpha$, where f is <u>progressive</u> if $f(\alpha) \geq \alpha$ for each α. It is immediate, then, that ...

Observation 3.1 An ideal, I, on κ is quasi-normal if and only if it is outstripping and a p*-point.

Theorem 3.2 An ideal I on κ is a p-point if and only if it is a p*-point.

proof: Assume I is a p-point, and let $\langle B_\alpha \rangle_{\alpha<\kappa}$ be any sequence of subsets of κ each of which is in I. Disjointify the B by setting $B'_\alpha = B_\alpha - \bigcup_{\delta<\alpha} B_\delta$. Viewing the B'_α as a partition of κ into sets in I and applying the formulation of p-pointedness in terms of partitions gives us a set A in I* such that the intersection of A with B'_α is bounded for each α. Let $g(\alpha)$ be any bound for $A \cap B'_\alpha$, and $f(\alpha) = \alpha \cup (\bigcup_{\delta\leq\alpha} g(\delta))$. Then it is easy to see that for $\alpha,\beta \in A$ with $f(\alpha) < \beta, \beta \notin B_\alpha$

In the easy direction, if I is a p*-point and P is a partition of κ into sets each of which is in I, define B_α = the set which α is an element of, and apply p*-pointedness to $\langle B_\alpha \rangle_{\alpha<\kappa}$, to get a set A in I* and a progressive $f:\kappa \to \kappa$ such that for each $\alpha,\beta \in A$ with $f(\alpha) < \beta, \beta \notin B_\alpha$. Then $A \cap B_\alpha$ is in I_κ for any α, since $f(\alpha)$ bounds $A \cap B_\alpha$.

Corollary 3.3 (Weglorz) An ideal I on κ is selective if and only if it is quasi-normal.

Definitions for $P_\kappa\lambda$: An ideal I on $P_\kappa\lambda$ is quasi-normal if for every sequence $\langle B_x \rangle_{x\in P_\kappa\lambda}$ of sets in I there exists a set A in I* such that for every $x,y \in A$ with $x \subsetneq y, y \notin B_x$. It is equivalent to ask for a set A' in I* for which $\nabla_{x\in A'} B_x \in I$, where $\nabla_{x\in A'} B_x = \{y \in P_\kappa\lambda \mid$ for some $x \in A'$ with $x \subsetneq y, y \in B_x\}$.

It is immediate that on $P_\kappa\lambda$ an ideal I is outstripping if and only if for every sequence $\langle B_x \rangle_{x\in P_\kappa\lambda}$ of sets *in* $I_{\kappa\lambda}$ there exists a set A' in I* with $\nabla_{x\in A'} B_x \in I$. In one direction just replace the sequence $\langle B_x \rangle_{x\in P_\kappa\lambda}$ by the function $f(x)$ = any nubound for B_x; in the other replace $f:P_\kappa\lambda \to P_\kappa\lambda$ by the sequence $\langle \text{Co-cone}(f(x)) \rangle_{x\in P_\kappa\lambda}$, where Co-cone(z) is defined to be $\{y \in P_\kappa\lambda \mid$ it is not the case that $z \subsetneq y\})$. This observation lies behind the claim that quasi-normal is to selective as outstripping is to q-point.

An ideal I on $P_\kappa\lambda$ is a p*-point if for every sequence $\langle B_x \rangle_{x\in P_\kappa\lambda}$ of sets in I, there exists a set A in I* and a progressive function $f: P_\kappa\lambda \to P_\kappa\lambda$ such that for each $x,y \in A$ with $f(x) \subsetneq y$, $y \notin B_x$. A function $f:P_\kappa\lambda \to P_\kappa\lambda$ is progressive if $f(x) \supseteq x$ for each x. Incidentally, the requirement that a function be progressive can be dropped in several places in this paper; its presence simplified the exposition.

Observation 3.4 An ideal I on $P_\kappa\lambda$ is quasi-normal if and only if it is both

<u>outstripping and a p*-point</u>.

<u>Theorem 3.5</u> <u>If an ideal I on $P_\kappa\lambda$ is a p*-point, then it is a p-point</u>. <u>(The converse is open - the "p-point puzzle"</u>).

The proof of 3.5 is just as in the proof of this implication in the κ-context.

<u>Corollary 3.6</u> <u>If an ideal I on $P_\kappa\lambda$ is quasi-normal, then it is selective</u>. (The converse is open).

The proof for κ that a p-point is a p*-point entailed disjointifying a non-disjoint collection of sets via set difference, applying p-pointedness to the resulting partition, adding back in the parts of sets that had been subtracted off, and finally observing that this "rejointification" leaves the intersections still in I_κ. The obstacles to the carrying out of this argument for $P_\kappa\lambda$ seem to be the first and last steps.

Fortunately, there is a second way to disjointify a non-disjoint collection: instead of handling overlap between two sets by subtracting one from another, we can merge the two into a larger set, and continue to remerge any two sets that overlap until the sets have no more overlap. In effect, this is what happens in the proof on κ that a q-point is outstripping. The original sets, which do overlap, are the intervals $[\alpha, f(\alpha)]$ while the merged sets are, roughly, the intervals marked off by f's closure points. The first type of disjointification could never have been used in this proof, since there is no hope that rejointification would leave the intersections as singletons.

In the next section this second type of disjointification is applied to both the p-point and q-point puzzles on $P_\kappa\lambda$. However, in the course of merging overlapping sets, they might get too big - i.e., no long lie in $I_{\kappa\lambda}$ (in the case of q-point) or in I (in the case of p-point). So we posit additional properties, the "scandinavian properties" designed to control the growth of merging sets.

§4 <u>The Scandinavian Properties Address the P-point and Q-point Puzzles</u>

Lets re-examine the proof that, in the κ context, a q-point is outstripping. Given a function, f, to be outstripped, first closure points for f are found, then these are used to decompose κ into disjoint blocks, and then q-pointedness is applied to this partition. Both of the first two steps are difficult for $P_\kappa\lambda$. Dan Velleman observed that we wish to put x and y in the same block if $x \subseteq y$ and it is not the case that $f(x) \subsetneq y$, and that we want to keep the blocks as small as possible, since they must each be not unbounded sets for q-pointedness to apply.

For any two place relation, R, on a set, define <u>$(R)^\sim$</u> to be the smallest equivalence relation containing R, and given any function $f : P_\kappa\lambda \to P_\kappa\lambda$ define the relation R_f by xR_fy if both $x \subseteq y$ and it is not the case that $f(x) \subsetneq y$. Dan's suggestion was that we decompose $P_\kappa\lambda$ into the equivalence classes of $(R_f)^\sim$ rather than into some analogue of closure point intervals. In fact, it is straightforward to check that, on κ, any closure point interval is a union of such equivalence classes.

The following example, then, probably explains why there is no straightforward lifting of the proof for κ (that a q-point is outstripping) to the $P_\kappa\lambda$ context. Assume λ has cofinality $\geq \kappa$, and for any $x \in P_\kappa\lambda$ set $\underline{top(x)} = \sup \{\beta+1 | \beta \in x\}$ and let $t : P_\kappa\lambda \to P_\kappa\lambda$ by $t(x) = x \cup \{top(x)\}$. Then $(R_t)^\sim$ turns out to be $P_\kappa\lambda \times P_\kappa\lambda$, so there is but one equivalence class, which is all of $P_\kappa\lambda$. (On κ, nothing like this can happen, since (see 4.2) every "small" relation R produces $(R)^\sim$ equivalence classes which are "small" - i.e., in I_κ.) To see why $(R_t)^\sim$ blows up, note that if x and w are any two elements of $P_\kappa\lambda$, and we set $u = (w \cup x) - \{top(x)\}$ and $v = w - \{top(x)\}$, then xR_tu, vR_tu and vR_tw, so $x(R_t)^\sim w$. Note that $(R_t)^\sim$ relates elements between which there exists a finite chain of elements linked by R_t and reverse R_t. Fortunately, we can attempt to hinder the blowing up of equivalence classes by first restricting R_t to some set in I^*, thus tossing out some of the elements that appear in the middle of the R_t chains.

If R is any two place relation and A any set, define $\underline{R \restriction A}$ to be $R \cap (A \times A)$. An ideal, I, on $P_\kappa\lambda$ will be said to be __danish__ if, given any $f : P_\kappa\lambda \to P_\kappa\lambda$ there exists a set A in I^* such that each equivalence class of $(R_f \restriction A)^\sim$ is in $I_{\kappa\lambda}$. The definition of a danish ideal in the κ context is the obvious specialization.

__Theorem 4.1 An ideal on $P_\kappa\lambda$ is outstripping if and only if it is both danish and a q-point.__

proof: If I is outstripping and $f : P_\kappa\lambda \to P_\kappa\lambda$ let A be a set in I^* which outstrips f. Then the equivalence classes of $(R_f \restriction A)^\sim$ are easily seen to be singletons, hence in $I_{\kappa\lambda}$. Thus, an outstripping ideal is danish. We already knew it was a q-point.

Now suppose I is a danish q-point, and let $f : P_\kappa\lambda \to P_\kappa\lambda$. Choose a set A in I^* such that each equivalence class of $(R_f \restriction A)^\sim$ is in $I_{\kappa\lambda}$. Apply q-pointedness to the resulting partition to get a set A' in I^* intersecting each equivalence class in an anti-chain. This A' outstrips f.

__Theorem 4.2 Every ideal on κ is danish. On ω, I_ω is not danish, but every q-point and every dual to an ultrafilter is. On $P_\kappa\lambda$, $I_{\kappa\lambda}$ is not danish.__

__proof sketches__: Note that on κ, the set A we restrict R to can be all of κ, and the result follows by a simple cardinality argument. It is easy to see that $f(n) = n+1$ shows I_ω to be not danish, while the $t(x)$ discussed earlier shows $I_{\kappa\lambda}$ not danish. The proofs that both q-points and ultrafilters on ω are danish were pointed out to me by Andreas Blass, and are presented in the next section.

It turns out that there is a second property that makes up the difference (if there is one) between p-point and p*-point. Like danishness, it asserts that equivalence classes can be prevented from blowing up. For R any two place relation on $P_\kappa\lambda$ and $f : P_\kappa\lambda \to P_\kappa\lambda$ any function, define $R\restriction f$ by $x(R\restriction f)y$ if xRy and $f(x) \subseteq y$. Let us call a two place relation, R, on $P_\kappa\lambda$ I-small (where I is an ideal on $P_\kappa\lambda$) if for each $x \in P_\kappa\lambda$, $xR = \{y \in P_\kappa\lambda \mid xRy\}$ is a set in I. Now an ideal, I, on $P_\kappa\lambda$ is said to be finish if for every I-small relation R on P there exists a progressive $f : P_\kappa\lambda \to P_\kappa\lambda$ and a set A in I^* such that each equivalence class of $((R\restriction f)\restriction A)^\sim$ is in I. Observe that the following is equivalent to danishness of the ideal I : for each $I_{\kappa\lambda}$-small relation R on $P_\kappa\lambda$, there is a set A in I^* such that each equivalence class of $((R\restriction id)\restriction A)^\sim$ is in $I_{\kappa\lambda}$, where we define $id : P_\kappa\lambda \to P_\kappa\lambda$ by $id(x) = x$ for each x.

<u>Theorem 4.3 An ideal I, on $P_\kappa\lambda$ is a p*-point if and only if it is both finish and a p-point.</u>

<u>proof</u>: If I is a p*-point and R is an I-small relation, then applying p*-pointedness to $\langle xR\rangle_{x\in P_\kappa\lambda}$ yields a progressive $f : P_\kappa\lambda \to P_\kappa\lambda$ and a set A in I^* such that for each $x,y \in A$ with $f(x) \subsetneq y$, $y \notin xR$. The same f and R make the equivalence classes of $((R\restriction f)\restriction A)^\sim$ into singletons, so I is finish.

Now, if I is a finish p-point and $\langle B_x\rangle_{x\in P_\kappa\lambda}$ is a sequence of sets in I, let R be the relation defined by xRy if $y \in B_x$. Apply I's finishness to get a progressive $f : P_\kappa\lambda \to P_\kappa\lambda$ and a set A in I^* such that the equivalence classes of $((R\restriction f)\restriction A)^\sim$ are each in I. Next, apply I's p-pointedness to the partition determined by these equivalence classes, to get a set $A' \subseteq A$, with A' in I^*, which intersects each equivalence class in a set in $I_{\kappa\lambda}$. This induces a function $h : P_\kappa\lambda \to P_\kappa\lambda$ defined by $h(x)$ = any nubound for the intersection of A' with x's equivalence class. If we now define g by $g(x) = f(x) \cup h(x)$ then for any $x,y \in A'$ with $g(x) \subsetneq y$, $y \notin B_x$, and g is progressive because f is. Hence I is a p*-point.

If we now define an ideal I to be scandinavian if it is both danish and finish, the following corollary is immediate:

<u>Corollary 4.4 An ideal on $P_\kappa\lambda$ is quasi-normal if and only if it is both selective and scandinavian.</u>

The results of section 3 and 4 are summarized in Figure 3, next page. At this point, the q-point and p-point puzzles can be rephrased (and remain open):

<u>Is every q-point on $P_\kappa\lambda$ danish?</u>

<u>Is every p-point on $P_\kappa\lambda$ finish?</u>

Is the Q-point Puzzle Resolved? (Figure 3)

As before, a double arc linking arrows indicates that the conjunction of the two arrow sources is equivalent to the target.

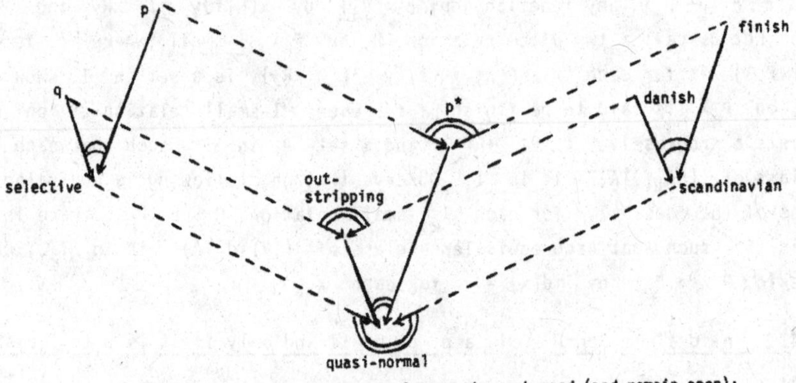

Thus, the q-point and p-point puzzles can be rephrased (and remain open):
Is every q-point on $P_\kappa\lambda$ danish? Is every p-point on $P_\kappa\lambda$ finish?

§5 Galvin's Negative Results - Discussion

An alternate approach to the resolution of the q-point puzzle is suggested by the following proof, due to A. Blass, that on ω both q-points and duals to ultrafilters are danish. Let $f:\omega\to\omega$ and define integers $n_0 < n_1 < \ldots < n_j < \ldots$ by $n_0 = 0$ and $n_{j+1} = \sup(f''\{0,1,\ldots,n_j\})$ for $j \geq 0$. Decompose ω into blocks by $B_0 = \{0\}$ and $B_{j+1} = [(n_j)+1, n_{j+1}]$. Now if I is a dual to an ultrafilter let $E = B_0 \cup B_2 \cup \ldots \cup B_{2k} \cup \ldots$ and $O = B_1 \cup B_3 \cup \ldots \cup B_{2k+1} \cup \ldots$, and let A be whichever of E or O is in I^*. Then $(R_f \upharpoonright A)^\sim$ satisfies that the equivalence class of any integer is a subset of the block that the integer belonged to, since we've deleted alternate blocks too large for f to stretch across. If I is a q-point, consider two partitions of ω, one into sets of the form $B_{2j} \cup B_{2j+1}$ and into B_0 and sets of the form $B_{2j+1} \cup B_{2j+2}$. Apply q-pointedness separately to each partition, getting sets A_1, A_2 each in I^* and each intersecting the pieces of their respective partitions in singletons. Now if we let $A = A_1 \cap A_2$, then A intersects each B_j in at most one point and intersects no two successive blocks. This means A outstrips f and, in particular, $(R_f \upharpoonright A)^\sim$'s equivalence classes are singletons.

Thus ω (and κ) have the property that the range of a function on a bounded subset is bounded, and this property is enough to get outstripping from q-point. While $P_\kappa\lambda$ clearly cannot have the property that the range of a function on a set in $I_{\kappa\lambda}$ is always in $I_{\kappa\lambda}$, it seems reasonable to ask if an ideal I on $P_\kappa\lambda$ can have the following property (*): Given any function $f: P_\kappa\lambda \to P_\kappa\lambda$ there is a set A in I^* such that for each set W in $I_{\kappa\lambda}$ with $W \subseteq A$, $f''W$ is in $I_{\kappa\lambda}$.

It is fairly easy to modify the proof for ω to show that any q-point on $P_{\kappa\lambda}$

with the extra property (*) must be outstripping, but this leaves open the question of whether outstripping ideals need have this property (*) and even whether (*) is possible.

Fred Galvin answered a question I posed at the Toronto meeting whose proceedings this paper appears in. A consequence of his results is that there are reasonable circumstances under which outstripping ideals exist on $P_\kappa\lambda$, yet no ideal on $P_\kappa\lambda$ has property (*) because there exists an $f : P_\kappa\lambda \to P_\kappa\lambda$ such that for any unbounded set A there exists an $X \subseteq A$ with X in $I_{\kappa\lambda}$ and $f''X$ unbounded. Since any A in any I^* must be unbounded, this does rule out (*). Thus (*) cannot replace danish as the "difference" between outstripping and q-point on $P_\kappa\lambda$. Galvin's results are listed later in this section.

The property (*) is related to a question about functions which preserve the $\subsetneq$ order almost everywhere on $P_\kappa\lambda$. Given any $f : P_\kappa\lambda \to P_\kappa\lambda$ and any κ-footed ideal I it is possible (see Theorem 5.10) to find a function h defined on a set A in I^* which dominates f on A and which satisfies that for each x,y if $x \subsetneq y$ then $h(x) \subsetneq h(y)$. We will say that the function h, and the ideal I, have the forwards order-preserving property. For most of the properties we look at, if one can "handle" a function dominating a given function on a set of measure one, one can handle the given function. Thus, if we view ideals which are not κ-footed as being pathological, it is safe to assume that any given function is already forwards order-preserving. Now on a *linearly* ordered set, such as κ, if $\alpha < \beta \to h(\alpha) < h(\beta)$, then $h(\alpha) < h(\beta) \to \alpha < \beta$. Hence, we speak of a function as being order-preserving if it satisfies $\alpha < \beta \leftrightarrow h(\alpha) < h(\beta)$ and we know that any function can be replaced on a set of measure one (which is actually all of κ) by an order-preserving function which dominates it. Need an ideal on $P_\kappa\lambda$ have the backwards order-preserving property - that an arbitrary function from $P_\kappa\lambda$ to $P_\kappa\lambda$ can be replaced, on a set A in I^*, by a function h which dominates it and which satisfies, for each $x,y \in A$, that $h(x) \subsetneq h(y) \to x \subsetneq y$?

It turns out (see Observation 5.11) that a backwards order-preserving ideal must have the property (*). Thus Galvin's results tell us that "reversing the arrow" does not come for free on $P_\kappa\lambda$, and that we cannot assume, without loss of generality, that a given function is backwards order-preserving.

Because Galvin's theorems state that, under a number of often-occurring cardinal arithmetic situations, ideals with property (*) do not exist, my reaction has been to view (*) as less basic than the properties discussed earlier in this paper.

Results of Galvin and W. Fleissner

Let $S_\kappa(\lambda)$ assert that for every function $f : P_\kappa\lambda \to P_\kappa\lambda$ there is a cofinal set $A \subseteq P_\kappa\lambda$ such that for every $B \subseteq A$ with B in $I_{\kappa\lambda}$, $f''B$ is in $I_{\kappa\lambda}$. Note that if $S_\kappa(\lambda)$ fails then no ideal on $P_\kappa\lambda$ can have property (*). We'll say a family F of functions from κ to κ is dominating if given any $g : \kappa \to \kappa$ there is some

$f \in F$ satisfying that $g(\delta) \leq f(\delta)$ for each $\delta < \kappa$, and define d_κ to be the minimum possible cardinality of such a family.

Theorem 5.1 [F. Galvin] If $d_\omega = \aleph_1$ then $S_\omega(\aleph_1)$ fails.

proof: In fact, if $d_\omega = \aleph_1$ then there is a function $f : P_\omega(\aleph_1) \to P_\omega(\aleph_1)$ such that for every cofinal $A \subseteq P_\omega(\aleph_1)$ there is a set $B \subseteq A$ such that $\omega \not\subseteq B$ and $f''B$ is cofinal in $P_\omega(\aleph_1)$. To construct f, first use the dominating family of size $\aleph_1$ to construct a sequence of functions, $h_\alpha : \omega \to P_\omega(\aleph_1)$ for $\alpha < \aleph_1$, satisfying that for each $g : \omega \to P_\omega(\aleph_1)$ there exists an α such that $g(n) \subseteq h_\alpha(n)$ for each $n \in \omega$. Next define $f : P_\omega(\aleph_1) \to P_\omega(\aleph_1)$ by

$$f(x) = \bigcup\{h_\alpha(0) \mid \alpha \in x\} \cup \bigcup\{h_\alpha(n+1) \mid \alpha \in x \text{ and } n \in x \cap \omega\}.$$

Let A be any subset of $P_\omega(\aleph_1)$ satisfying $\bigcup A = \aleph_1$. For each $n \in \omega$ let $B_n = \{x \in A \mid n \notin x\}$. Then for some $n \in \omega$, $f''B_n$ must be cofinal in $P_\omega(\aleph_1)$, for if not let $g : \omega \to P_\omega(\aleph_1)$ by $g(n) =$ any nubound for $f''B_n$. Choose an $\alpha \in \aleph_1$ with $g(n) \subseteq h_\alpha(n)$ for each $n \in \omega$. Then for each $n \in \omega$ and $x \in B_n$, $h_\alpha(n) \not\subseteq f(x)$, i.e. for each $n \in \omega$ and each $x \in A$, if $h_\alpha(n) \subseteq f(x)$ then $x \notin B_n$, i.e. for each $n \in \omega$ and each $x \in A$, if $h_\alpha(n) \subseteq f(x)$ then $n \in x$. Now choose any $x \in A$ with $\alpha \in x$. As $h_\alpha(0) \subseteq f(x)$, $0 \in x$, and for each $n \in \omega$, if $n \in x$ then $h_\alpha(n+1) \subseteq f(x)$, so $n+1 \in x$. Hence $\omega \subseteq x$, contradicting $x \in P_\omega(\aleph_1)$.

Theorem 5.2 If $d_\kappa = \kappa^+$ then $S_\kappa(\kappa^+)$ fails (κ regular, as always).

proof: Modify Galvin's argument by defining $f : P_\kappa\lambda \to P_\kappa\lambda$ by $f(x) = \bigcup\{h_\alpha(0) \mid \alpha \in x\} \cup \bigcup\{h_\alpha(\delta) \mid \delta < \kappa \text{ and } \delta \subseteq x \text{ and } \alpha \in x\}$. Also, for $\delta < \kappa$ set $B_\delta = \{x \in A \mid \delta + 1 \not\subseteq x\}$. With g defined as before and h_α dominating g notice that for each $\delta < \kappa$, if $h_\alpha(\delta) \subseteq f(x)$ then $\delta + 1 \subseteq x$, thus obtaining the same contradiction as above, with κ replacing ω.

Theorem 5.3 [F. Galvin] For regular κ, if $S_\kappa(\lambda^+)$ then $S_\kappa(\lambda)$. (proof omitted)

Noting that if $2^\kappa = \kappa^+$ then $d_\kappa = \kappa^+$, and applying 5.3 in the contrapositive yields, for example, that ...

Corollary 5.4 If $V = L$ then $S_\kappa(\lambda)$ fails for $\lambda = \kappa^+, \kappa^{++}, \ldots, \kappa^{(n)}, \ldots$ for each $n \in \omega$. Hence no $P_\kappa\kappa^{(n)}$ bears an ideal with property (*).

The reader should recall, in this connection, that if $V = L$ then SC's exist on each $P_\kappa\kappa^+$ for κ regular, hence outstripping ideals exist for $P_\kappa\kappa^+$. Thus ...

Corollary 5.5 It is consistent with ZFC that outstripping ideals not have property (*). Hence ZFC $\nvdash$ "An ideal on $P_\kappa\lambda$ is outstripping if and only if it is a q-point and has property (*)".

F. Galvin has also shown ...

Theorem 5.6 If κ is regular and λ has uncountable cofinality $> \kappa$ and if $\{\alpha < \lambda | S_\kappa(|\alpha|)$ fails$\}$ is stationary in λ, then $S_\kappa(\lambda)$ fails. (proof omitted)

The following positive results are also of interest. I do not know whether they can be extended to show $S_\kappa(\lambda)$ consistent for uncountable regular κ, or whether the consistency of $S_\kappa(\lambda)$ implies that (*) can consistently hold of some ideal on $P_\kappa\lambda$.

Theorem 5.7 [F. Galvin] If MA_λ holds, then $S_\omega(\lambda)$.

Let $D(\lambda)$ denote the discrete topological space of cardinality λ. Theorem 5.7 has been strengthened to:

Theorem 5.8 [W. Fleissner] If $D(\omega)^\lambda$ is not the union of λ nowhere dense sets, then $S_\omega(\lambda)$.

Additional results include:

Theorem 5.9 [F. Galvin] If λ is any infinite cardinal, and $\lambda^\omega = \lambda$ or $d_\omega < \lambda < \aleph_\omega$ or $D(\lambda)^\omega$ is the union of λ compact sets, then $S_\omega(\lambda)$ fails.

The Order Preserving Properties

Theorem 5.10 If I is a κ-footed ideal on $P_\kappa\lambda$, then I is forwards order-preserving.

proof: Given $f : P_\kappa\lambda \to P_\kappa\lambda$ and a set A in I^* such that $|A{\upharpoonright}x| < \kappa$ for each $x \in P_\kappa\lambda$, first enumerate A as $\{x_\delta | \delta < |A|\}$. Next define functions $t : P_\kappa\lambda \to \kappa$ and $h : P_\kappa\lambda \to P_\kappa\lambda$ in stages. At the δth stage, first t and then h will be defined on all elements of the set A_δ for which t and h have not already been defined at an earlier stage, where $A_\delta = \{x_\delta\} \cup A{\upharpoonright}x_\delta$. First define $t : A_\delta \to \kappa$ in any way such that the following two conditions are satisfied: (1) for each $z \in A_\delta$ for which t was not defined at an earlier stage, $t(z) \notin \{t(w) | w \in A_\delta$ and $t(w)$ was defined at a stage earlier than $\delta\}$ and $t(z) \notin \{f(w) | w \in A_\delta\}$, and (2) t is a one-to-one function when restricted to those elements of A_δ for which t's value was not defined at an earlier stage. Next, define h on A_δ by

$$h(z) = \bigcup\{f(w) \cup \{t(w)\} | w \in \{z\} \cup A{\upharpoonright}z\},$$

noting that if $h(z)$ were defined at an earlier stage, the new definition will agree. Since h's cumulative definition immediately guarantees that $x \subsetneq y \to h(x) \subseteq h(y)$, we need only check that $x \subsetneq y \to h(x) \neq h(y)$, for which it suffices to show that $x \subsetneq y \to t(y) \notin h(x)$. But if $h(y)$ were first defined at stage α, it is easy to check that $t(y) \notin h(x)$ by breaking into cases depending on whether $t(x)$ were first defined at stage α or at an earlier stage.

Observation 5.11 If I is backwards order-preserving, then I has property (*).

proof: Let $f : P_\kappa\lambda \to P_\kappa\lambda$ be an arbitrary function, and apply backwards order-preserving to obtain a set A in I^* and a function h dominating f on A and satisfying for each $x,y \in A$ that $h(x) \subsetneq h(y) \to x \subsetneq y$. For any $x \in P_\kappa\lambda$ recall that Co-cone(x) equals $\{y \in P_\kappa\lambda \mid$ it is not the case that $x \subsetneq y\}$. Note that a set W is in $I_{\kappa\lambda}$ if and only if there exists an x in $P_\kappa\lambda$ with $W \subseteq$ Co-cone (x). Thus the backwards order-preserving property of h, read in contrapositive form, states that for any $x \in A$, if $y \in A \cap$ Co-cone (x) then $h(y) \in$ Co-cone (h(x)), in other words that $h''(A \cap \text{Co-cone}(x)) \subseteq \text{Co-cone}(h(x))$, so that $h''(A \cap \text{Co-cone}(x))$ is a set in $I_{\kappa\lambda}$. It follows that for any set W in $I_{\kappa\lambda}$ with $W \subseteq A$, $h''W$ is in $I_{\kappa\lambda}$, from which the same holds with h replaced by the original function f. Thus I has property (*).

§6 Solution to a Problem of Menas (Added in proof)

In [M], T.K. Menas asks whether isomorphs of a partition measure need have the partition property. He allows, as isomorphisms for a fine measure μ on $P_\kappa\lambda$ only those functions $k : P_\kappa\lambda \to P_\kappa\lambda$ that satisfy two conditions:

(1) k is 1:1 on some set in μ

(2) $k_*\mu$ is fine

He shows that if μ is a fine partition measure on $P_\kappa\lambda$ and k is such an isomorphism, then $k_*\mu$ has the partition property if and only if (using our terminology) k is backwards order-preserving on some set $A \in \mu$ (i.e. $\forall x,y \in A, k(x) \subseteq k(y) \to x \subseteq y$). Note that if h is 1:1 on A then changing both $\subseteq$'s to $\subsetneq$'s does not change the meaning.

Menas then produces an example of a $\lambda > \kappa$ and an isomorphism k of a normal partition measure μ on $P_\kappa\lambda$ such that $k_*\mu$ does not have the partition property, because k fails to be backwards order-preserving on any set in μ. In the last line of this paper he asks whether this failure of isomorphisms to preserve the partition property can also occur when $\lambda = \kappa^+$.

To see that the answer is yes, we need only assume that $2^\kappa = \kappa^+$, and modify Galvin's function $f(x)$ from the proof of 5.2, so that it meets the two conditions above, while remaining a counterexample to $S_\kappa(\kappa^+)$. The new function $k(x)$ cannot then be backwards order-preserving on any set in $I_{\kappa\lambda^+}$ - this follows from the proof of 5.11. Hence, as Menas points out, $k_*\mu$ could not have the partition property, and the violating partition is given by

$$F(x,y) = \begin{cases} 0 & \text{if } x \subseteq y \text{ and } k^{-1}(x) \subseteq k^{-1}(y) \\ 1 & \text{if } x \subseteq y \text{ and } k^{-1}(x) \not\subseteq k^{-1}(y) \end{cases} \qquad \text{for } x,y \in k''P_\kappa\lambda.$$

To modify f, first note that one can dominate an arbitrary $g : P_\kappa\lambda \to P_\kappa\lambda$ by a 1:1 function $k : P_\kappa\lambda \to P_\kappa\lambda$. Simply enumerate $P_\kappa\lambda$ as $\{x_\delta\}_{\delta<|P_\kappa\lambda|}$ and define k inductively by choosing $k(x_\eta)$ to be any element of $\widehat{g(x_\eta)} - \{k(x_\delta)\}_{\delta<\eta}$, which is easily seen to be non-empty.

If we derive such a k from the function $g(x) = f(x) \cup x$ then the fact that k dominates f insures that it serves as a counterexample to $S_\kappa(\kappa^+)$ while $k(x) \supseteq x$ insures that $k_*(\mu)$ is fine whenever μ is. Thus we have proved:

Theorem 6.1 If κ is κ^+ supercompact and $2^\kappa = \kappa^+$ then every fine measure on $P_\kappa\kappa^+$ is isomorphic, in the sense of Menas, to a measure lacking the partition property. As it is known that normal measures on $P_\kappa\kappa^+$ do have the partition property, it follows that isomorphisms on $P_\kappa\kappa^+$ can fail to preserve the partition property.

References

[BTW] Baumgartner, J. D., Taylor, A. D., and Wagon, S., Structural Properties of Ideals, Dissertationes Mathematicae CXCVII, 1982.

[C1] Carr, Donna M., The Minimal Normal Filter on $P_\kappa\lambda$, Proc. Amer. Math. Soc., 86 (1982), 316-320.

[C2] Carr, Donna M., $P_\kappa\lambda$ Partition Relations, Fund. Math. (to appear).

[C3] Carr, Donna M., A Note on the λ-Shelah Property, Fund. Math. (to appear).

[C4] Carr, Donna M., and Pelletier, Donald H., Towards a Structure Theory for Ideals on $P_\kappa\lambda$, in this volume.

[J1] Jech, Thomas J., Some Combinatorial Problems Concerning Uncountable Cardinals, Ann. Math. Logic 5 (1973), 165-198.

[J2] Jech, Thomas J. Set Theory, Academic Press, 1978.

[M] Menas, T. K., A Combinatorial Property of $P_\kappa\lambda$, J. Sym. Logic, 41 (1976), 225-234.

[S1], [S2] Shelah, S., The Existence of Coding Sets, and More on Stationary Coding, Lecture Notes in Mathematics, no. 1182, Springer Verlag, Berlin, 1986. 188-202 and 224-246.

[V1] Velleman, D., Souslin Trees Constructed from Morasses, Axiomatic Set Theory (Baumgartner, Martin, Shelah, ed.) Contemporary Mathematics, vol. 31, A.M.S., 1984.

[V2] Velleman, D., Simplified Morasses, J. Sym. Logic, 49 (1984) pp. 257-271.

[W] Weglorz, B., Some Properties of Filters, Lecture Notes in Mathematics, no. 619, Springer-Verlag, Berlin , 1977. 311-329.

[Z1] Zwicker, W. S., $P_\kappa\lambda$ Combinatorics I, Axiomatic Set Theory (Baumgartner, Martin, Shelah, ed.) Contemporary Mathematics, vol. 31, A.M.S., 1984, 243-259.

[Z2] Zwicker, W. S. Lecture Notes on the Structural Properties of Ideals on $P_\kappa\lambda$, handwritten notes, July 1984. All results in these notes are included in the current paper.

Schedule

MONDAY, AUGUST 10th

9:00 Registration, Coffee and Doughnuts at Curtis Lecture Hall I

9:30 to 10:30 Jim Baumgartner - Partition Relations on the uncountable

10:55 to 11:40 Arnold Miller - Two results on Analytic sets

11:45 to 12:30 Shai Ben David - Successors of Singular Cardinals may have no Aronszajn trees

LUNCH

2:00 to 3:00 Andreas Blass - Some Applications of Super-perfect set forcing and its relative

3:20 to 3:50 Dennis Burke - On Generalized Metric Spaces

4:00 to 4:30 Y. Kimchi - Consistency relations between measurability order of cardinals and partition properties

DINNER

7:30 Wine and Cheese at the Faculty Club

TUESDAY, AUGUST 11th

9:10 Coffee and Doughnuts

9:30 to 10:30 ...Jim Baumgartner concludes

10:55 to 11:40 Petr Simon - Thin-tall superatomic Boolean algebras

11:45 to 12:30 Neil Hindman - Ultrafilters and Ramsey Theory : an update

LUNCH

2:00 to 3:00 Andreas Blass continues...

3:15 to 3:30 Jiang Shou Li - The Strict p-space Problem

3:35 to 3:50 Ingrid Lindstrom - A Construction of non-well-founded sets within Martin-Lof type theory

4:00 to 4:30 Chaz Schlindwein - Special Non-Special Trees

WEDNESDAY, AUGUST 12th

9:10 Coffee and Doughnuts

9:30 to 10:30 Neil Hindman continues...

10:55 to 11:40 Sabina Koppelberg - κ-cofinalities of Boolean algebras 11:45 to 12:45 ...Andreas Blass concludes

LUNCH

2:15 to 3:00 Stevo Todorcevic - Ramsey Problems for the Uncountable

3:15 to 3:30 Steven Cushing - A Set-Theoretic Argument in the Semantics of Natural Language

3:35 to 3:50 Claude Laflamme - Rate of Convergence of Series

4:00 to 4:30 Carlos diPrisco - Normal Closure for Filters

DINNER

7:30 Problem Session at the Faculty Club

THURSDAY, AUGUST 13th

9:10 Coffee and Doughnuts

9:30 to 10:30 Menachem Magidor - Reflection Properties and Applications of Compactness

10:55 to 12:30 W. Just (with A. Krawczyk) - Triviality conditions for meet-preserving functions on certain Boolean algebras

LUNCH

2:00 to 3:00 ...Neil Hindman concludes

3:15 to 3:30 Qi Feng - On Reflection of Spaces of Uncountable Sets

3:40 to 4:10 Pierre Matet - On Jensen's Diamond

7:00 Thai Banquet

FRIDAY, AUGUST 14th

9:10 Coffee and Doughnuts

9:30 to 10:30 Menachem Magidor continues...

10:55 to 11:25 J. Pawlikowski - Parametrized Ellentuck Theorem

11:35 to 12:20 Alan Mekler - On Abelian groups

LUNCH

2:00 to 3:00 Stevo Todorcevic continues...

3:15 to 3:30 Alain Gaudefrey - A Language Criterion for some Representation Theorems

3:35 to 3:50 Steve Purisch - Hereditarily Collectionwise Hausdorff and Monotone Normality

4:00 to 4:30 Efim Khalinsky - Curves, Surfaces, Boundaries and the Jordan Curve Theorem in Finite Product Spaces

MONDAY, AUGUST 17th

9:10 Coffee and Doughnuts

9:30 to 10:30 ...Menachem Magidor concludes

10:50 to 11:35 Zoltan Balogh - The Moore-Mrowka Hypothesis - Coloring Axioms and Related Matters

11:45 to 12:30 Peter Nyikos - Some classes of sequential spaces

LUNCH

2:00 to 3:00 A. V. Arhangel'skii - Cardinal Invariants in Function spaces, general spaces and Topological Groups

3:15 to 3:30 Frantisek Franek - Infinite Steiner Triple Systems

3:35 to 3:50 A. Blaszczyk - On Some Constructions using Inverse Limits

4:00 to 4:30 Donna Carr - Toward a Structure Theory for Ideals on $P_\kappa(\lambda)$

TUESDAY, AUGUST 18th

9:10 Coffee and Doughnuts

9:30 to 10:30 Mary Ellen Rudin - Metrizability of Manifolds

10:50 to 11:35 Bill Weiss - Topological Partition Relations

11:45 to 12:30 J. P. Levinski - Chang Games, Large Cardinals and the Core Model

LUNCH

2:00 to 3:00 A. V. Arhangel'skii continues...

3:15 to 3:50 ...Stevo Todorcevic concludes

3:55 to 4:40 Jacek Cichon - Two Cardinal Properties of Ideals

WEDNESDAY, AUGUST 19th

9:10 Coffee and Doughnuts

9:30 to 10:30 Jan van Mill - The Works of Eric K. van Douwen

10:50 to 11:35 William Mitchell - Speculations on Core Model for Extenders

11:45 to 12:30 Tomek Bartoczynski - On Measurability of Filters on the Natural Numbers

LUNCH

2:00 to 3:00 Hugh Woodin - Determinacy and Large Cardinals

3:15 to 3:50 Max Burke - Some Applications of Set Theory to Measure Theory

4:00 to 4:30 Alexander Sostak - Fuzzy Topological Spaces

DINNER

7:30 Problem Session at the Faculty Club

THURSDAY, AUGUST 20th

9:10 Coffee and Doughnuts

9:30 to 10:30 Mary Ellen Rudin continues...

10:50 to 11:35 ...A. V. Arhangel'skii concludes

11:45 to 12:30 Jan van Mill

LUNCH

2:00 to 3:00 Hugh Woodin - Saturated Ideals

3:15 to 4:15 Boban Velickovic - Forcing Axioms and Stationary Sets

4:25 to 4:55 Scott Williams - Set-Theoretical Problems in Topological Dynamics

7:00 Barbeque at the Island Yacht Club

FRIDAY, AUGUST 21st

9:10 Coffee and Doughnuts

9:30 to 10:30 ...Mary Ellen Rudin concludes

10:50 to 11:20 J. Burzyk - F-spaces

11:30 to 12:20 Murray Bell - Spaces closely associated to Ideals of Partial Functions

LUNCH

2:00 to 2:30 William Boos - Quantum Theory in Boolean Extensions by Measure Algebras

2:35 to 3:05 William Zwicker - Ideals: p-points,q-points and selectives

3:10 to 3:25 S. Thomeier - Remarks on the Number of Mother Structures and an interesting Operation on the Power Set

List of Participants

- Arhangel'skii, A.V.; Moscow State University Mech.-Math. Faculty, Moscow, USSR
- Aull, Charles E.; Dept. of Mathematics, Virginia Tech., Blacksburg, Va., 24061 , USA (tel:703-961-5409)
- Balogh, Zoltan; Dept. of Mathematics, Kossuth University, Debrecen H-4010, Hungary (tel:913-864-4028(o) 841-0757(h))
- Baloglou, George; Dept. of Mathematics, Univ. of Kansas, Lawrence, KS 66045, USA
- Barbanel, Julius; Dept. of Mathematics, Union College, Schenectady, NY 12308, USA (tel:518-370-6526)
- Bartoszynski, Tomek; Dept. of Mathematics, University of California, Berkeley , Calif. 94720, USA (e-mail:barto@cartan.berkeley.edu)
- Baumgartner, James E.; Dept. of Mathematics, Dartmouth College, Hanover, NH 03755, USA (tel:603-646-3559; e-mail:jeb@dartmouth)
- Beaudoin, Robert E.; Division of Mathematics, Dept. of FAT, Auburn Univ., Auburn Alabama 36849, USA
- Bell, Murray; Dept. of Mathematics, Univ. of Manitoba, Winnipeg, Manitoba R3T 2N2, Canada
- Bilaniuk, Stephan; Dept. of Mathematics, Dartmouth College, Hanover, NH 03755, USA (tel:603-646-2565; e-mail:stefan@dartmouth)
- Blass, Andreas; Dept. of Mathematics, Univ. of Michigan, Ann Arbor, Michigan, USA (tel:313-763-1183; e-mail:BLASS@ub.cc.umich.edu)
- Blaszczyk, Aleksander; Silesian Univ., Inst. of Math., ul. Bankowa 14,40-007, Katowice, Poland
- Boos, Bill; Dept. of Mathematics, Univ. of Iowa, Iowa City, IA 52240, USA
- Burke, Dennis; Dept. of Mathematics, Miami University, Oxford, OH 45056, USA (tel:513-529-3508)
- Burke, Max; Dept. of Mathematics, Univ. of Toronto, Toronto, Ontario M5S 1A1, Canada (tel:416-978-4794)

- Burzyk, J.; Inst. of Mathematics, Polish Academy of Science, Katowice, Poland
- Cichon, Jacek; Dept. of Mathematics, Univ. of Wroclaw, 50384 Wroclaw, Poland
- Cushing, Steven; Dept. of Mathematics, Stonehill College, North Easton, MA 02357, USA
- diPrisco, Carlos A.; Dept. of Mathematics IVIC, Apartado 21827, Caracas 1020-A, Venezuela (02)746376
- Dordal, Peter; Dept. of Mathematics, Univ. of Arizona, Tucson, Arizona, USA
- Dow, Alan; Dept. of Mathematics, York University North York, Ontario M3J 1P3 Canada (tel:416-736-5250; e-mail: dow@yorkvml)
- Eklof, Paul; Dept. of Mathematics, University of California-Irvine, Irvine, CA 92626, USA (e-mail:pceklof@ucicp6(bitnet); tel:814-238-6652(h))
- Feng, Qi; Dept. of Mathematics, Pennsylvania State University, University Park, PA 16802, USA
- Fleissner, Bill; Dept. of Mathematics, University of Kansas, Lawrence, KS 66045, USA
- Franek, Frantisek; Dept. of Computer Science, McMaster University, Hamilton, Ontario, Canada (tel:416-525-9140/3233; e-mail:franya@macs on usnet)
- Frankiewicz, Ryszard; Dept. of Mathematics, Polish Academy of Science, Warsaw, Poland
- Gaudefrey, Alain; Dept. of Physics, York University, North York, Ontario M3J 1P3 Canada
- Goffart, J.S.T.; Dept. of Mathematics, University of Wisconsin, Madison, Wisconsin 53706, USA
- Gorelic, Isaac; Dept. of Mathematics, University of Toronto, Toronto, Ontario M5S 1A1 Canada (tel:416-925-1358(h))
- Grant, Douglass; Dept. of Mathematics, University College of Cape Breton, Sydney, Nova Scotia, Canada
- Hechler, Steve, Dept. of Mathematics, Queens College, Flushing, NY 11367, USA (tel:718-520-7014(o) 516-462-5509(h))
- Hindman, Neil; Dept. of Mathematics, Howard University, Washington, D.C. 20059, USA (tel:202-636-7989(o) 301-593-1755(h))

- Jackson, Steve; Dept. of Mathematics, California Institute of Technology, Pasadena, CA 91125, USA
- Jensen, Ronald; Dept. of Mathematics, All Souls College, Oxford, England
- Just, Winnefred, Dept. of Mathematics, University of Toronto, Toronto, Ontario M5S 1A1, Canada (tel:416-978-5162)
- Kato, Akio; Dept. of Mathematics, National Defense Academy, Yokosuka, Japan
- Kimchi, Yechiel; Dept. of Mathematics, Ohio State University, Columbus OH 43210, USA
- Koppelberg, Sabine; Institut der FU Berlin, Arnimallee 3, 1000 Berlin 33, West Germany
- Laflamme, Claude; Dept. of Mathematics, University of Toronto, Toronto, Ontario M5S 1A1, Canada (tel:416-978-3462(o) 416-783-2727(h))
- Landver, Avner; Dept. of Mathematics, University of Wisconsin, Madison, WI 53706, USA (tel:608-263-7939)
- Larson, Jean; Dept. of Mathematics, University of Florida, Gainesville, Florida 32611, USA (tel:904-392-6172)
- Leary, Chris; Dept. of Mathematics, Oberlin College, Oberlin, OH 44074, USA (tel:216-775-8380; e-mail:ccl@oberlin.edu)
- Levinski, Jean-Pierre; Dept. of Mathematics, Dartmouth College, Hanover, NH 03755, USA (tel:603-646-2293)
- Lim, Chor Hoon; Dept. of Mathematics, University of Wisconsin, Madison, WI 53706, USA
- Lindstrom, Ingrid; Dept. of Mathematics, Uppsala University, Trunbergsv. 3, S-75238, Uppsala, Sweden (tel:018-404792(h), 018-183191(o))
- Lubarsky, Bob; Dept. of Mathematics, Cornell University, Ithaca, NY 14853, USA (tel:607-255-4738(o),607-277-5215(h))
- Magidor, Menachem; Dept. of Computer Science, Hebrew University, Jerusalem, Israel (tel:972-2-584123; e-mail:menachem@humus)
- Matet, Pierre; Freie Universitat Berlin, Institut fur Mathematics, Arnimallee 3, 1000 Berlin 33, West Germany (tel:30-831-2530)
- Mekler, Alan; Dept. of Mathematics, Simon Fraser University, Burnaby, British Columbia V5A 1S6, Canada (e-mail:useramam@sfu.bitnet)

- van Mill, Jan; Vrije Universiteit, de Boelelaan 1081, 1081 HV Amsterdam, Netherlands
- Miller, Arnie; Dept. of Mathematics, University of Wisconsin, Madison, WI 53706, USA (tel:608-262-2925(o) 233-0876(h))
- Mitchell, Bill; Pennsylvania State University, State College, PA 16802, USA (tel:814-238-7916)
- Moran, Gadi; University of Haifa, Haifa 31999 Israel, 972-4-254193, RSMA309@haifauvm.
- Nyikos, Peter; University of South Carolina, Columbia, South Carolina, USA (tel:803-777-5134)
- Pawlikowski, Janusz; Univ. of Wroclaw, Poland
- Pelletier, Don H.; York University, North York, Ontario M3J 1P3, Canada (tel:416-736-5250)(e-mail:DHPELL@Yorkvm1)
- Purisch, Steven; Box 5054, Tennessee Technological University, Cookeville, TN 38505, USA (tel:615-372-3441)
- Qiao, Y.Q.; University of Toronto, Toronto Ontario M5S 1A1 Canada (tel:416-978-3201)
- Roitman, Judy; University of Kansas, Lawrence, KS 66045, USA (tel:913-842-7010)
- Rudin, Mary Ellen; University of Wisconsin, Madison, WI 53706, USA
- Scheepers, Marion; University of Kansas, Lawrence, KS 66045, USA (tel: 913-843-9011(b),913-864-4315(h))
- Schlindwein, Chaz; University of Kansas, Lawrence, KS 66045, USA (tel: 913-841-1693)(e-mail:chaz@ukanvax.bitnet)
- Shouli, Jiang; University of Wisconsin, Madison, WI 53706, USA (tel:608-262-3601) Shandong University, Jinan, Shandong, China
- Simon, Petr; Matematicky Ustav Larlovy University, Sokolovska 83, 18600 Praha 8, Czechoslovakia, (tel:422-2316000)
- Smith, Martin; University of Toronto, Toronto, Ontario M5S 1A1, Canada (tel:416-978-2967)
- Stanley, Lee; Lehigh University, Bethlehem, PA 18015, USA (tel:215-758-3723(b) 867-6431(h)

- Steprans, Juris; York University, North York, Ontario M3J 1P3, Canada (tel:416-736-5250)(e-mail: steprans@yorkvml)

- Szeptycki, Paul; University of Toronto, Toronto, Ontario M5S 1A1, Canada (tel:416-978-3201)

- Tall, Frank; University of Toronto, Toronto, Ontario M5S 1A1, Canada (tel:416-978-3318(o) 762-4829(h)

- Thiele, Ernst Jochen; T. U. Berlin, Breisganer Str 30, 1000 Berlin 38, 8017876, West Germany

- Thomeier, S.; Memorial University, St. John's, Newfoundland A1C 5S7, Canada

- Todorcevic, Stevo; University of Colorado, Boulder, Colorado 80309, USA

- Vaughan, Jerry; University of North Carolina at Greensboro, Greensboro, NC 27412, USA (tel:919-334-5891)(e-mail:vaughanj at uncg)

- Velickovic, Boban; California Institute of Technology, Pasadena, CA 91125, USA

- Watson, Stephen; York University, North York, Ontario M3J 1P3 Canada (tel:416-736-5250)(e-mail watson at yorkvml)

- Weiss, Bill; University of Toronto, Toronto, Ontario M5S 1A1, Canada (tel:416-978-3324)

- Williams, Scott; State University of New York at Buffalo, Buffalo, NY 14214, USA (tel:716-831-2144(o)838-3998(h)

- Woodin, W. Hugh; California Institute of Technology, Pasadena, CA 91125, USA

- Zwicker, William S.; Union College, Schenectady, New York, (tel:518-370-6197)